Diversity of Lower Plants

Diversity of Lower Plants

Editors

Rajan Kumar Gupta

M.Sc., Ph.D. F.I.C.E.R.

Associate Professor

Dept. of Botany, Pt. LMS Govt. P.G. Autonomous College, Rishikesh
(Affiliated to HNB Garhwal University, Srinagar, UK, India)

Mukesh Kumar

M.Sc., M. Phil., Ph.D., FBS

Associate Professor

Dept. of Botany, Sahu Jain Post-Graduate College, Najibabad (U.P.)
(Affiliated to MJP Rohilkhand University, Bareilly, U.P., India)

I.K. International Publishing House Pvt. Ltd.

NEW DELHI • BANGALORE

Published by

I.K. International Publishing House Pvt. Ltd.
S-25, Green Park Extension
Uphaar Cinema Market
New Delhi–110 016 (India)
E-mail : info@ikinternational.com

ISBN 978-93-82332-41-1

Published by Krishan Makhijani for I.K. International Publishing House Pvt. Ltd., S-25, Green Park Extension, Uphaar Cinema Market, New Delhi–110 016 and Printed by Rekha Printers Pvt. Ltd., Okhla Industrial Area, Phase II, New Delhi–110 020.

FOREWORD

Biodiversity of biological diversity is the variety of life on the earth, from the simplest bacterial gene to the vast, complex rainforests of the Amazon. Human beings are an integral part of this diversity, as is the food, medicine, clothing and other biological resources that sustain us. Biodiversity is being threatened at an unprecedented rate due to habitat alternation, degradation and fragmentation and human misuse or overexploitation. It is therefore not surprising that biodiversity conservation in recent years has gained considerable importance all over the world. On the other hand, every country, particularly the developing countries, are facing the need of developing their economies, improving peoples' livelihood or providing food security and supply for a rapidly increasing population.

Our food and energy security strongly depend on biodiversity and so does our vulnerability to natural hazards such as fires and flooding. Biodiversity loss has negative effects on our health, material wealth and it largely limits our freedom of choice. As all cultures gain inspiration from or attach spiritual and religious values to ecosystems or their components, e.g., landscapes, trees, hills, rivers or particular species. Biodiversity loss also strongly influences our social relations.

- Biodiversity is essential for global food security and nutrition and also serves as a safety-net to poor households during crisis.
- Increased diversity of genes within species, e.g., as represented by livestock breeds or strains of plats, reduces risk from diseases and increases potential to adapt to changing climates.
- More than 70,000 plant species are used in traditional and modern medicine.
- The value of global ecosystem services is estimated at $16-$64 trillion.

Every decision we take that affects biodiversity, also affects our lives and the lives of other people. Biodiversity is crucial to human well-being, sustainable development and poverty reduction. But people, particularly those in the developed world, have become so far removed from nature that they have forgotten how much they, and others, rely on it.

In the long term, the value of services lost may greatly exceed the short-term economic benefits that are gained from transforming ecosystems. When we modify an ecosystem to improve the service it provides, this generally also results in changes to other ecosystem services. For example, actions to increase food production can lead to reduced water availability in terms of quantity and quality for other users. This can result in the degradation of many services, such as fisheries, water supply, and protection against natural hazards, seriously affecting people's well-being.

With the loss of plant and animal diversity, we're losing the chance to discover new medicines that could end the suffering of millions of people and save national economies billions of dollars each year. Coral reefs support the richest marine biodiversity in the world. But it is not only fellow species that depend on them; coral reefs also provide food, storm protection, jobs, recreation and other sources of income for more than 500 million people worldwide.

Without biodiversity, we would have no food. Since agriculture began some 12,000 years ago, approximately 7000 plant species and several thousand animal species have been used for human consumption. Maintaining healthy biodiversity plays a significant role in climate change mitigation and the world's protected areas—national parks, marine reserves, wilderness areas, and so on are essential in safeguarding this role.

I pen this message with great pleasure that the publication of the "Diversity of Lower Plants", is indeed a glorious endeavour. It is a storehouse of knowledge for all scientists, students and teachers in a single cover which offers innovative study materials in compact and lucid way.

I am happy to know that Dr. Rajan Kumar Gupta, Department of Botany, Govt. P.G. Autonomous College, Rishikesh and Dr. Mukesh Kumar, Department of Botany, Sahu Jain Post-Graduate College, Najibabad are bringing out the volume of "Diversity of Lower Plants" to cater to the long felt need of scientists, research scholars and students. I congratulate the editors for their maiden effort and wish them all success in their endeavour.

The devotion and dedication of the editors has enabled the book to occupy a prominent position in the academic arena. I am sure that it will attain much greater heights in the days to come.

Dr. Gupta and Dr. Kumar deserve my sincere blessings on this occasion. I wish them all success.

Prof. A.B. Bhatt
Department of Botany
H.N.B. Garhwal Central University
Srinagar, Garhwal

PREFACE

If you plan for one year, plant rice
If you plan for ten years, plant trees
But, if you plan for one hundred years, educate the people.

Biodiversity is the natural biological capital of the mother Earth. It refers to variety of life forms, i.e., plants, animals and microorganisms. It has been defined as 'the variability, among the living organisms from all sources including, inter alia, terrestrial, marine and other aquatic ecosystems and the ecological complexes of which they are part, this includes diversity within species, between species and of ecosystems'. In recent years, biodiversity has become a global issue in all important international foray. The 'Convention on Biological Diversity' a specific document on biodiversity, was discussed and adopted in Rio de Janeiro (Brazil) during the United Nations Conference (June 1992) on Environment and Development, also known as the 'Earth Summit'.

The Earth Summit made the term 'biological diversity' a political catch phrase. Since then it has become clear to a wide public that are increasingly being confronted with a dramatic loss of animal and plant species. At the same time, we are becoming aware that more than 905 of our cohabitats on this planet currently remain undiscovered. A great number of them presumably disappear forever without our ever having known them. Biodiversity includes the variability in both, i.e., plant and animal species.

Phytodiversity at global scale is a balance between the rate of speciation and the rate of extinction. At a regional level, it is a balance between the rate of immigration/invasion and that of local/regional extinction. These rates determine the persistence of the communities. It can be assessed at genetic, population, species, and ecosystem and landscape levels, at three spatial scales, viz., local, regional and global. While the assessment at genetic level, molecular techniques are most profitably used, assessment at other levels depends upon field sampling. Depending upon the objectives, the inventory could focus on single species or multiple taxa. For single species inventories, the focus could be on high impact, indicator, umbrella, flagship, threatened or economically important species.

Phytodiversity provides to the humankind enormous benefits in the form of food, medicine and industrial products, and has the potential for providing many more yet unknown benefits. It also provides an array of essential services through natural ecosystems. These include maintenance of the gaseous composition of the atmosphere, shaping the climate and producing the soil. It is essential for the ecosystem function and stability. There are several hypotheses being tested for relationship between biodiversity and ecosystem function.

Phytodiversity is not uniformly distributed on the earth. There are broad patterns, such as increasing variations in plants and animals from poles to equator. It is more concentrated in the tropics because of conditions which permit higher rates of speciation and lower rates of extinction. About 44% of vascular plants and 35% of vertebrates are concentrated as endemic species in 25 hot spots, which account for only 1.4% of the local land area.

The pool of living diversity is dynamic, it increases when new gene variation is produced, a new species is created or a novel ecosystem formed and it decreases when the genetic variation within the species decreases, a species become extinct or an ecosystem complex is lost. Today, as ever, human beings are dependent on their sustenance, health well being and enjoyment of life on fundamental biological systems and processes. Humanity derives all of its food, many medicines and industrial products from wild or domesticated components of biological diversity. Biotic reservoirs are source of recreation, and tourism and underpin the ecosystem which provides us with many services.

Plant diversity as a global resource remains poorly understood, inadequately documented and often wasted, but still remains immense potential for further development of natural products. With the explosive growth of human populations, the life support system on the earth is becoming increasingly threatened as the global change accelerates. Human use of the world's natural resources to provide food, shelter, clothing and fuel has led to major environmental changes including massive deforestation, air and water pollution and global warming. A reduction in diversity, with the consequent loss of available biological knowledge, erodes our ability to enhance the economic well-being and quality of species is faced by the world community to discover, preserve and understand its diversity before our biological heritage is lost forever.

The Indian region represents an immense variety of climatic and altitudinal zones and is floristically rich. The ecological habitats vary from humid tropical Western Ghats to the hot deserts of Rajasthan and ice-capped mountains of the Himalaya to warm coasts of Peninsular India. These habitats favour a wide variety of plants and animals as well. It is estimated that over 45,000 species of plants are accounted for in India which represent 11% of the known plant species of the world. These are distributed in varied groups, viz., Angiosperms-15000 spp., Gymnosperms-64 spp., Pteridophytes-1022 spp., Bryophytes-2584 spp., Lichens-1600 spp., Fungi-23000 spp., Algae-2500 spp., and Bacteria-850 spp. The flowering plants of India comprise about 6% of the world's known flowering plants. About 315 families (of a total of 400 now defined) and 2250 genera of flowering plants are known to occur in India in different ecosystems. 'Every form of life is unique and warrants respect regardless of its worth to human beings'. The favour and intensity of attention is based on the fact that, over time, the value of the variety may be found in the opportunities it provides humanity for adapting to local and global climate change. Pollution is one of the major ecological threats to the biodiversity. The effect of pollution on plant is more noticeable than on the other forms of life. Being rooted, they are unable to evade deleterious environmental changes although they are endowed with considerable degree of tolerance and stress resistance. Sensitivity varies from species to species, age of the plant, and pollution load to which it is exposed. Around 1336 plant species in India are considered vulnerable and endangered. Habit destruction, overexploitation and environmental pollution are the major factors responsible for it.

The loss of biodiversity is a global crisis. There is hardly any region on the earth that is not facing ecological catastrophes. Biological extinction has been a natural phenomenon in geological history. However, man's intervention has speeded up extinction rates all the more. The destruction of the world's tropical forests, which are disappearing at an alarming rate, is one of today's urgent global environmental issues. The loss of plant/animal diversity has immediate and long-term effects on human survival. The plant genetic resources are being depleted at an alarming pace throughout the world. Current rates of extinction demand immediate concerted efforts for conservation of biodiversity.

We seem to be entering into the phase of mass extinction. Today, we are losing two to five species per hour from tropical forests alone. This amounts to a loss of 16 m populations per year or 1800 populations per hour. If the present trend continues, about 25% of the total 2,50,000 higher plant species will be lost in the next few decades and another 25% by the end of the 21st century. There were around 25,000 species of flowering plants some 125 million years ago, by the late Cretaceous this number had swelled to 1,00,000 and today there are some 2,00,000 species of plants on the Earth. Additionally, while the past extinctions occurred each time over a span of million years or less, the present mass extinction may well occur within a short period of about 200 years, and plants (which make the base of the food chain) will be equally or worse affected.

India's phytodiversity is one of the most significant in the world. India is one of the twelve mega-biodiversity centres in the world and also an important centre of origin of agribiodiversity. It is therefore, very important to study, document and conserve the plant wealth of India and also of the world before its extinction.

The diversity of plant life is essential for most of our terrestrial ecosystems. Humans and most other animals are almost totally dependent on plants, directly or indirectly, as a source of energy through their ability to convert the sun's energy through photosynthesis. Worldwide tens of thousands of species of higher plants, and several hundreds of lower plants, are currently used by humans for a wide diversity of purposes as food, fibre, oil, herbs, spices, industrial crops and fodder for domesticated animals. In the tropics alone it has been estimated that 25,000 – 30,000 species are in use and up to 25,000 species have been used in transitional medicines. In addition, many thousands of species are grown as ornamentals in parks, public and private gardens, as street trees and for shade and shelter. Another important role of plant life is the provision of ecosystem services of protection of water sheds, stabilization of slopes, improvement of soils, modernization of climate and the provision of a habit for much of our wild fauna.

While it is generally accepted today that the conservation of all biodiversity should be our goal, especially through the preservation and sustainable use of natural habitats, this is an ideal that is unlikely to be achieved and there are convincing scientific, economic and sociological reasons for giving priority to the conservation of the major centers of plant diversity throughout the world, especially as this will very often also lead to the conservation of much plants, animals and microorganisms.

The strategies to conserve the biodiversity include establishment of protected area network and corridors with emphasis on appropriate levels of management, reduction of anthropogenic pressure on natural populations by cultivating them elsewhere, programmes of augmentation, reintroduction and introduction of target taxa, restoration of degraded habitats, etc. The conservation strategies can be either, *in situ* conservation of genetic resources within their ecosystem and natural habitat or *ex situ* conservation of components of genetic material of biological diversity outside their natural habitat. The choice of conservation strategy depends on the nature of the material to be conserved, i.e., the life cycle and mode of reproduction, size of individual population and ecological status. *Ex situ* techniques include the establishment of botanical and zoological gardens, banks of pollen, seed tissue culture, DNA, etc. Establishment of forest reserves, national parks, protected areas and on farm conservation of valuable plant varieties is being promoted to facilitate their *in situ* conservation.

The present book is the compilation of 26 chapters in the form of research papers and review articles contributed by the eminent scientists and research workers who are acknowledged internationally in their respective fields. The book has been prepared with the intention of providing a sufficient depth of the subject to satisfy the academic needs of the reader up to a higher level, which will be comprehensive and interesting. There are articles on the various groups of plants, i.e., algae, fungi, cyanobacteria, **lichen**, pteridophyta bacteria, mycorrhiza, etc. The book in its totality, contain useful information for the postgraduate students, teachers, scientists and researchers dealing with basic and applied Botany, Microbiology, Mycology, Environmental Biology, forestry and Agricultural Botany. Moreover, it will prove useful for Government Departments dealing with forestry, social forestry and the cultivation of medicinal plants. We do hope that all those interested in the pursuit of plant diversity and its conservation would find the volume useful and stimulating.

– Rajan K. Gupta and Mukesh Kumar

ACKNOWLEDGEMENTS

The editors are thankful to all the academicians and scientists whose contributions have enriched this volume. We also express our deep sense of gratitude to our parents whose blessings have always prompted us to pursue academic activities deeply. It is quite possible that in a work of this nature, some mistakes might have crept in text inadvertently and for these we owe undiluted responsibility.

We are grateful to all authors for their contribution to present book. We are also thankful to Professor G. S. Paliwal and Professor R. D. Gaur (former Heads), Professor S. C. Tiwari (present Head) and Professor A. B. Bhatt, Department of Botany, HNB Garhwal University, Srinagar (Uttarakhand) for their cooperation and valuable suggestions for the preparation of this volume more informative and useful. We thank them from the core of our heart.

The author (RKG) is thankful to University Grants Commission, New Delhi for the financial support and to Uttarakhand Council for Science & Technology for providing financial assistance in the form of research project. RKG wish to place on record his special thanks to his wife Mrs Alka and two little daughters Akriti and Ayushi and his research team for their cooperation in all his academic and scientific endeavors.

The author (MK) is thankful to the UGC, New Delhi for providing financial assistance in the form of a major research project entitled, "Biodiversity of Cyanophyceae of the Sub-Himalayan Belt". The findings of the research work have been incorporated in this book in the form of chapters on distributional pattern of cyanobacteria. His wife Dr Kiran Sharma, PGT Hindi, Kendriya Vidyalaya, BHEL, Haridwar and two sons Ruchir and Nishaant deserve special thanks for their affection and multifarious cooperation extended during the preparation of this academic and scientific pursuit.

Finally, we will always remain debtor to all our well-wishers for their blessings without which this book would not have come to light.

– **Rajan K. Gupta and Mukesh Kumar**

CONTENTS

EDITORS

Dr. Rajan Kumar Gupta obtained his M.Sc. and Ph.D. degrees from Banaras Hindu University and worked on Ecophysiology of Antarctic Cyanobacteria for his Ph.D. degree with Prof. A.K. Kashyap, Centre of Advanced Study in Botany, Banaras Hindu University, Varanasi. For the past twenty years he has been working on various aspects of Antarctic microflora.

Dr. Gupta was deputed by Govt. of India for his participation as Biological Scientist in Antarctica twice. He has participated in XI and XIV Indian Scientific Expeditions to Antarctica during 1991-92 and 1994-95. He has visited several countries like Mauritius, Japan, Nepal, Thailand, South Africa and Belgium for presentation of his work on different aspects of algae. Dr. Gupta has worked on various aspects of cyanobacteria, i.e., morphology, ecology and nitrogen fixation, biotechnological applications and published more than 60 technical papers in various national and overseas journals and more than 30 chapters in various books.

Dr. Gupta has published three Botany Practical Books and one book on Paryavaran Adhyan (Environmental Studies) and three reference (research) books entitled "Glimpses of Cyanobacteria" and "Advances in Applied Phycology" and "Soil Microflora". Three students have been awarded the D.Phil degree and four are working under his supervision for their D.Phil degree of H.N.B. Garhwal University. He has worked on Use of Cyanobacteria as Biofertilizer in Antarctica as well as in foothills of Garhwal Himalaya and worked on a couple of projects on Cyanobacteria of paddy fields of Dehradun district of Himalaya. Presently, Dr. Gupta is working on a project sponsored by Uttarakhand Council for Science and Technology (UCOST), Dehradun on Diversity of Vesicular-Arbuscular Mycorrhiza. UGC has awarded Research Award to Dr. Gupta for doing research in any laboratory of the country. Dr. Gupta is member of number of organizations in India and abroad. He is the Fellow of the Society for Environment & Ecoplanning and International Botanical Society and chaired various sessions in the conferences in India and abroad. He is in the editorial and advisory board of various journals. Presently, Dr. Gupta is teaching Microbiology and Biotechnology in Department of Botany, Govt. P.G. College, Rishikesh (Dehradun), Uttarakhand.

Dr Mukesh Kumar (b. 1963) obtained his M.Sc. and M. Phil. degrees in Botany, and M.Ed. in Education from Meerut University, Meerut (now C.C.S. University, Meerut). He took his Doctorate degree on Polyhouse Technology from H.N.B. Garhwal University, Srinagar (Uttarakhand). Presently, he is Reader at the Department of Botany, Sahu Jain Post-Graduate College, Najibabad (Bijnor) U.P. Dr Kumar is well known for his researches in the fields of Polyhouse Technology and Cyanobacteria. He is a member/ fellow of several national and international research organizations. Working on various aspects of Cyanobacteria, i.e., distributional pattern, population dynamics and dominance of different genera, he has explored the Biodiversity of Cyanophycean Flora of the Sub-Himalayan belt of Garhwal and Kumaon region of Uttarakhand state of India. He has accomplished a couple of research projects in the fields of his specialization, i.e., Polyhouse Technology and Cyanobacterial Diversity sponsored by the University Grants Commission, New Delhi. Presently, he is working on another project on Ganga water pollution sponsored by the UGC.

He has presented his original research findings in several conferences/seminars and workshops. He has published more than 25 technical papers in various journals and chapters in several reputed books. The great success of his previous books entitled "Glimpses of Cyanobacteria" and "Soil Microflora" has inspired him to work more on the phytodiversity, the most important issues of the earth environment.

He is fellow of Indian Botanical Society and life member of several research journals. Four research scholars have already been awarded their Ph.D. degrees of MJP Rohilkhand University, Bareilly, and four others are currently working for their Ph.Ds under his supervision. Presently, he is actively engaged in the studies on Ganga water pollution with particular reference to the quality of soil, water and phytoplankton growth in and around the stream along with the impact of tourism on various parameters.

Diversity of Lower Plants (2014): pp. 1-23
Editors: Rajan Kumar Gupta & Mukesh Kumar
IK International Publishing House Pvt. Ltd.

1

Biodiversity: Status, Threats and Conservation

MUKESH KUMAR*, ANJALI KHARE**, RAJAN KUMAR GUPTA*** AND A.K. GOYAL****

Department of Botany, Sahu Jain P.G. College, Najibabad (Bijnor) U.P.
**Department of Botany, Advance Institute of Science & Technology, Dehradun*
***Department of Botany, Government P.G. College, Rishikesh, Uttarakhand*
****Regional Higher Education Officer, Bareilly (U.P.)*

ABSTRACT

The term 'biological diversity' means the variability among the living organisms from all sources including inter alia, terrestrial, marine and other aquatic ecosystems and the ecological complexes of which they are part. This includes diversity within species, between species and of the ecosystem. Broadly speaking, the term biodiversity includes three different but closely related aspects, i.e., genetic diversity (diversity within species), species diversity (diversity between species or species richness) and ecosystem diversity (habitat diversity).

The earth summit meeting at Rio de Janeiro in 1992 made the term 'biological diversity' a political earth phrase. Since then it became clear to a wide public that we are increasingly being confronted with a dramatic loss of animal and plant species. At the same time, we are becoming aware that more than 90% of our cohabitants on this planet currently remain undiscovered. A great number of them presumably disappear forever without us ever having known them.

Of these 20 million estimated species, only 8.5% (i.e., 1.7 million) have been identified. Amongst these 1.7 million described organisms, producers constitute only 4% (fairly negligible proportion), decomposers 15% and consumers 81%. When comparing these proportions to the biomass generated by the three organism groups, the significance of the groups of the producers become readily apparent, i.e., highest 90% biomass whereas 4% in species number. In our country, out of total 108,276 identified species of microorganisms, plants and animals, producers, consumers and decomposers constitute 29.6%, 58.4% and 22.0%, respectively of diversity. A country like India is a very important Vavilovian centre of diversity and origin with over 167 important cultivated plant species and some domesticated animals. Out of the 18 'hot spots,' identified in the world, India has two, i.e., Eastern Himalayan region and Western Ghats. The two areas contain 5332 endemic species of higher plants (3500 plus 1600, respectively), mammals, reptiles, amphibians and butterflies.

Keywords: Biodiversity, Earth summit, Megadiversity, Hot spot.

*Communicating Author

1. INTRODUCTION

Biodiversity is a balance between the rate of speciation and the rate of extinction. This rate determines the persistence of the communities. The literal meaning of biodiversity (bio = living, diversity = different or unlike) is the occurrence of different types of living beings. The floristic diversity refers to all taxa of plants of a country at a given time. The microbial diversity on the other hand refers to various types of microorganisms of a model area. As such, the number of biological type constitutes the degree of biodiversity. However, the biodiversity of taxa can be appreciated merely on the basis of the number of types. For example, the genus *Syzygium* has maximum diversity of species in India and Malaysia all of the species are closely related with slight morphological variations. Biodiversity provides to the humankind enormous benefits in the form of food, medicine and industrial products, and has the potential for providing many more yet unknown benefits. It also provides an array of essential services through natural ecosystems. These include maintenance of the gaseous composition of the atmosphere, shaping the climate and producing soil.

The term biodiversity is still fairly young. It was coined in the 1980s as a part of increasingly intense scientific debate on the threat to the earth's ecosystems and species. It originated in September 1986 at a symposium in the USA and, during the preparation for the big Earth Summit in 1992 in Rio de Janeiro. It eventually became the catch phrase of both old and new science, which has now realized that the research and conservation of biological diversity have a political dimension. The Earth Summit (United Nations Conference on Environment and Development, UNCED) contributed greatly to making the term biodiversity known to a wider public.

Plant diversity as a global resource remains poorly understood, inadequately documented and often wasted, but still remains immense potential for further development of natural products. With the explosive growth of human populations the life-support system on earth is becoming increasingly threatened as the rate of global change accelerates. Human use of the world's natural resources to provide food, shelter, clothing and fuel has led to major environmental changes including massive deforestation, air and water pollution and global warming. A reduction in diversity, with its consequent loss of available biological knowledge, erodes our ability to enhance the economic well-being and quality of species faced by the world community to discover, preserve and understand its diversity before our biological heritage is lost forever.

Biodiversity is not uniformly distributed on the earth. There are broad patterns, such as increasing biodiversity from poles to equator. It is more concentrated in the tropics because of the conditions which permit higher rates of speciation and lower rates of extinction. About 44% of vascular plants and 35% of vertebrates are concentrated as endemic species in 25 hotspots, which account for only 1.4% of the global land area. There are several theories to explain the non-uniformity of biodiversity.

India has great latitudinal and altitudinal spread, which means a wide range of temperature conditions. It is one of the 12 megadiversity centres of the world with some 17,500 species of flowering plants (Table 1). It has been inhabited by man since almost prehistoric times. One of the earliest known civilizations flourished on Indian soils at a time when large parts of the world were still inhabited by savage man. Remains of Indus Valley civilization prove that rice, wheat, barley and cotton were already there. Many plants, yielding food, fibre and wood were

Table 1: Megadiversity centres of the world

S. No.	Countries	Number of Angiosperms (Flowering plant species)
01	Brazil	55,000
02	Columbia	35,000
03	China	30,000
04	Mexico	25,000
05	South Africa	23,000
06	Soviet Union (former)	22,000
07	Indonesia	20,000
08	Venezuela	20,000
09	U.S.A.	18,000
10	Equador	18,000
11	India	17,500
12	Australia	15,000

known to the ancient Indians, and several others were recognized because of their curative properties in the treatment of diseases. In Indian flora species endemism is estimated at 33% with about 140 endemic genera but no endemic families. Areas rich in endemism are north-east India, the Western Ghats and the north-western and eastern Himalaya. The Gangetic plains are poor in endemics, while the Andaman and Nicobar Islands contribute to some 250 endemic species. India has a rich variety of wetland habitats which comprises 18% of the total area of the country. About half of the world's aquatic plants occur in India.

India has great diversity of its natural ecosystems. The panorama of Indian forests ranges from evergreen tropical rainforests in the Andaman and Nicobar Islands, the Western Ghats and the north-eastern states, dry alpine scrub high in the Himalaya to the north. Between the two extremes, the country has semi-evergreen rainforests, thorn forests, subtropical pine forests in the lower montane zone and temperate montane forests. India contains a great wealth of biological diversity in its forests, its wetlands and in marine areas but as many as 3000-4000 higher plants may be under a degree of threat.

Conservation efforts adopted nowadays require careful planning and involve a wide range of expertise. It is high time that adequate steps are taken to protect and conserve the biological species of natural habitats particularly in biosphere reserves for which proper strategies have to be formulated. The strategies to conserve biodiversity include establishment of protected area network and corridors with emphasis on appropriate levels of management, reduction of anthropogenic pressure on natural populations by cultivating them elsewhere, programmes of augmentation, reintroduction and introduction of target taxa, restoration of degraded habitats, and *ex situ* techniques, such as establishing botanical and zoological gardens, banks of pollen, seed, tissue culture, DNA, etc.

2. WHAT IS BIODIVERSITY?

Biodiversity refers to variability of life forms—the different plants, animals and microorganisms. The floristic diversity refers to all taxa of plants of a country at a given time. As such, the number of biological type constitutes the degree of diversity. The species number in a community or a region plays an important role in prioritizing the diversity centres for the purpose of conservation.

2.1 Definition of Biodiversity

Biodiversity is studied at three levels, i.e. genetic, species and ecosystem. Genetic diversity refers to the variety of genetic information in all the individual living organisms. It occurs between populations of species as well between species. Species diversity constitutes the recording of various types of species at a given place.

Biodiversity can be defined as "the variability among living organisms from all sources, including terrestrial, marine and other aquatic ecosystems, and the ecological complexes of which they are parts; this includes the diversity within species, between species and of ecosystems."

The term 'biological diversity' has been defined variously. According to the Convention on Biodiversity (CBD) in article 2: 'Biological diversity' means the variability among the living organisms from all sources including inter alia, terrestrial, marine and other aquatic ecosystems and the ecological complexes of which they are a part, this includes diversity within species, between species and of ecosystems. Whittaker (1972), used 'alpha' diversity (diversity of number of species within a given vegetation unit), 'beta diversity' (number of species between two or more sites) and 'gamma diversity' (overall number of species within the individual unit) which make up a complete landscape. In other aspects, biodiversity includes three different but closely related categories:

(i) Genetic diversity (diversity within species) refers to the variation of genes within the species. This includes distinct populations of the same species, genetic variation within population and varieties within a species.

(ii) Species diversity (diversity between species or species richness) refers to the variety of species within a region. This diversity can be measured on the basis of the number of a species in a region.

(iii) Ecosystem diversity (diversity of communities or habitat diversity) is difficult to measure since the boundaries of the communities which constitute the various sub-ecosystems are unclear.

In nutshell, biodiversity refers to the variety of world biological resource, living organism, population, community, ecosystem, landscape and biosphere levels. It also refers to all types of variation among living organisms and the ecosystem complexes in which they occur. It is a function not simply the number of ecosystems and animal species in existence at any given time, but also the genetic differences within individual species. This great diversity of the world's plant and animal species has intrinsic value to human existence and commerce.

2.2 Origin and Distribution of Biodiversity in India

The strategic phytogeographical position of the Indian region has contributed to the diverse floristic composition. A great variety of climatic and altitudinal variations coupled with varied ecological habitats have contributed immensely to the rich vegetation wealth, and varied flora and fauna forming a very unique biodiversity.

India covers an area of 32,87,263 square km with a land frontier of 15,200 km and a coastline of 7516 km. The biodiversity is not uniformly distributed on the earth. There are broad patterns, such as increasing biodiversity from poles to equator. It is more concentrated in the tropics because of conditions pertaining higher rates of speciation and lower rates of extinction. Plant diversity of India is very rich, including about 46,000 species of flora. However, the flowering plants are best representatives of plant kingdom. There are 20,000 species of flowering plants in India, which have drawn more attention. Floral diversity among orchids, insectivorous plants, saprophytes, parasites, rhododendrons, etc. also represent the Indian flora.

Biodiversity represents a wide range of variability in plants, and animals as well. The former is called 'Plant Diversity' and the latter 'Animal Diversity'. By virtue of tropical location, various physical features and climatic types, India represents about 46,000 species of flora and 89,000 species of fauna. It ranks amongst the twelve richest mega-biodiversity centres of the world. The biogeographical classification of India has been attempted variously by earlier workers (Hooker 1854, Chaterjee 1940, Puri 1960, Rao 1994). Rodgers and Panwar (1988) have divided India into 10 biogeographical zones (namely, the Himalayan, Trans-Himalayan, north-east India, Indian Desert, the Semi-Arid zone, the Indo-Gangetic plain, the Western Ghats, the Deccan Peninsula, the Coastal zone and Islands) each of which has its own unique but distinct natural vegetation. There is a great difference between the estimated and identified number of species present in the world.

3. PHYSIOLOGICAL CLASSIFICATION OF INDIAN MAINLAND

The second largest nation of Asia, India covers an area of 32,87,263 km^2. Indian region (60°45′ to 37°6′ N and 69°7′ to 97°25′ E) has a land frontier of about 15,200 km and a coast line of 5400 km. In the north, it is bounded by Tibet, Nepal and Bhutan; in the north-west by Pakistan; in the north-east by Myanmar (formerly Burma); and in the east by Bangladesh. The southern Peninsula extends into the Indian Ocean with the Bay of Bengal lying to the south-east and the Arabian Sea to the south-west. The world is located at the boundary of many biogeographical regions and is therefore one of the richest areas of India in terms of habitat and species diversity. For understanding vegetation diversity of India, a basic knowledge of biogeography is necessary. The biogeographical classification of India has been attempted variously by earlier workers, according to Rodgers & Panwar (1988) the Indian region is divided into 10 biogeographical zones, each with its unique but distinct natural vegetation. The salient vagetational features are given below.

3.1 The Himalayas

It occupies the extreme northern margins of India and influences the climate of the entire region. The climate of India is extremely varied, from the cold perpetually snow-clad mountain peaks in the north to some of the hottest places of the earth in the Thar Desert, from the near zero rainfall areas to the world's wettest areas in the north-east India. The climate of India is dominated by the Asiatic monsoon, a great wind system that seasonally revers the direction, resulting in three well marked seasons, particularly in northern India. From June to October the rains come from south-west whereas between December and February the monsoon blows in the north-eastern direction. Variations in temperature of different regions are also considerable. The summer temperature rises well above 38°C. The minimum winter season temperature in South India rarely goes below 20°C and northwards mean winter temperature drops gradually to about 4°C in the foothills of the Himalayan mountains.

The Himalayan mountain system extending approximately 2400 km from east to west supports a remarkable assemblage of vegetation formation. Roughly 8000 species of flowering plants are estimated to occur here, of which about 5% species are endemic. This region includes broadly subtropical, temperate, sub-alpine and alpine forest types. The subtropical and temperate zone in the West Himalayas is predominated by tall conifers, while the broad leaved species dominate the East Himalayas. Oaks are common in West Himalayas and show a distinct altitudinal distribution (Rao 1974). The slopes of East Himalayas are occupied by colourful rhododendrons (85 species), bamboos, primulas and a variety of orchids — about 700 species (Rao 1994, Hajira & Rao 1990).

3.2 Trans Himalayan

This zone is represented by the cold deserts of Ladakh and Lahul-Spiti. The vegetation is primarily dry alpine scrub formation, chiefly of *Juniparous* sp. The other genera that contribute to the sparse vegetation are *Saxifraga, Draba, Ephedra, Kobresia* and *Carex.*

3.3 North-East India

This biogeographic zone is the most significant one and represents a tradition among the Indian, Indo-Malayan, Indo-Chinese biogeographic regions as well as a meeting place of Himalayan Mountain with that of peninsular India. Therefore, this region acts as a biogeographic gateway for plant migration. This zone is the richest in biological diversity at the community level in endemics. Most of the species contributing to the biological diversity of north-east India are either restricted to the region as a whole or even to a smaller locality as in Khasi and Jaintia Hills, a set of twin hills in the state of Meghalaya, probably the richest botanical habitat in the whole Asia.

About 50% of the species of Indian flora occur in this zone which also includes a portion of Arunachal Pradesh. The presence of a large number of primitive flowering plants, viz. *Magnolia, Michelia*, etc. have prompted Takhtajan (1969) to call it the "Cradle of Flowering Plants". There are nine primitive angiosperm families in this area and include such genera as the vessel-less *Terecentron sinense* extending to North Myanmar and south-west China. According to Hutchinson (1969) *Magnolia pterocarpa* of this area is perhaps the most ancient species of living angiosperms. Besides, many primitive flowering plants, several others like orchids, *Citrus, Zingibers*, bamboos, ferns, members of Musaceae and Cucurbitaceae exhibit maximum diversity here.

The vegetation of north-east Asia is broadly classified into tropical (up to 900 m elevation), subtropical (up to 1500 m elevation), temperate (from 1300 – 2500 m) types, *Shorea assamica, Dipterocarpus macrocarpus, Artocarpus chaplasha, Terminalia myriocarpa* are the important species of the tropical zone. Species of *Quercus, Castanopsis, Rosa, Photinia, Eryobotriya, Pyrus* and *Sorbus* are conspicuous in the temperate vegetation (Rao 1974, Rao 1994, Rao & Murti 1990).

3.4 Indian Desert

The Indian Desert which occupies 6.8% of the land surface supports various characteristics of binding species such as *Leptadaenia pyrotechnica, Aerva tomentosa, Citrulas colocynthis and Crotolaria burhia*. Tree species are sparse and are mainly represented by *Acacia, Tecomella undulata* and multipurpose *Prosopsis cineraria*. Bushes of *Ziziphus, Capparis decidua* and *Cailligonum polygonoides* are also common. *Cenchrus* sp., *Lasiurus syndiscus* are grasses, typical of this zone (Bhandari 1978).

3.5 The Semi-Arid Zone

The semi-arid vegetation chiefly consists of thorn-shurub forests of *Atlanthus excelsa, Capparis decidua, Prosopis cineraria, Flacourtia* sp., *Acacia* sp., *Dichrostachys cinerea, Commiphora wightii, Anogeissus latifolia, Balanites* sp., *Diospiros melanoxylon* and *Boswellia serrata.*

3.6 The Indo-Gangetic Plain

The Indo-Gangetic plain stretches from eastern Rajasthan through Uttar Pradesh to Bihar and Bengal. The natural vegetation has been much replaced by cultivated plants. In the Terai area (foothills) the natural vegetation comprises tall grasses of *Themeda, Saccharum* and *Phragmites* along with a few scattered trees.

3.7 The Western Ghats

The Western Ghats geographic zone comes next to Himalaya in floristic richness and diversity. About 4000 species of flowering plants are expected here, of which 1,500 species are endemic (Subramanyam & Nayar 1974, Nayar 1980, Nayar & Danial 1986). Diverse forest types constitute this zone. The moist deciduous forests (at lower elevation) contain highly valued timber species such as *Dalbergia latifolia*, *Terminalia crenulata, Pterocarpus marsupium* and *Tectona grandis*. The wet evergreen forests include species of *Mesua, Callophyllum, Dipterocarpus* and *Hopea.*

At higher reaches, montane flora, chiefly as 'shola' formation is very common. Species of *Syzygium, Machilus, Meliosma, Elaeocarpus* and *Wendlandica* are common. Along streams and in moist places in evergreen forests 'wet bamboo breaks' including *Oxytenanthera* and *Bambusa* sp. form the natural vegetation. Palm, orchids, peppers, Zingiberaceae members are also abundant. The famous Silent Valley is a storehouse of unparallel vegetational wealth being the only remnants of the tropical wet evergreen rainforests in the country and forms one of the unique 'hot spots' of biodiversity of the country.

3.8 The Deccan Peninsula

A major portion of the Deccan Peninsula is covered by tropical thorn forests, and tropical dry and moist deciduous forests. *Tectona grandis, Anogeissus latifolia*, *Boswellia serrata, Butea monosperma* and several thorny species of the semi-arid zone are common here. *Hardwickia binata*, an endemic Caesalpiniaceae tree species, occur scattered in patches on the drier parts of the Peninsula. The tropical moist deciduous forest chiefly includes *Adina cordifolia, Chloroxylon swietenia, Diospyros exsculpata* and in parts of south-eastern Madhya Pradesh, *Shorea robusta.*

3.9 The Coastal Zone

The coastline of India stretches from Gujarat to Cape Comorin in the west and from Sundarbans to Cape Comorin in the east and includes two major vegetation types, i.e. mangrove forests and beach forests. The most characteristic tree species of beach forests are *Casuarina equisetifolia, Anacardium occidentale* (planted), *Malinkara* sp., *Calophyllum inophyllum* and *Harnandia peltata*. The mangrove forests are found along the deltas, estuaries and backwaters, and chiefly comprise genera like *Rhizophora, Avicenna, Sonneratia, Xylocarpus* and *Ceriops*.

3.10 Islands

The Andaman and Nicobar group of islands in the Bay of Bengal with 348 islands depict an interesting biogeography. The Andaman group is the southern continuation of the Arrakkan-Yoma tectonic unit and the Nicobar group, the northern continuation of Sumatran range. These islands support roughly 2200 species of higher plants. There are about 210 endemic species (nearly 10%) of which 75 are trees. The dense vegetation of the island can be broadly classified as below.

3.10.1 Littoral types

Littoral forests are composed of trees like *Manilkara littoralis, Callophyllum inophyllum, Terminalia catappa, Barringtonia asiatica* and mangroves like *Rhizophora mucronata, R. apiculata, Avicemnia marina, Lumnitzera littorea, Nypa fruticans* and a few others.

3.10.2 Inland types

The inland vegetation includes the evergreen and deciduous forests with a number of economically important timber species such as *Dipterocarpus grandis, Pterocarpus dalbergioides, Terminalia bialata, Albizia lebbeck, Ailanthus kurzii* and a few others.

4. TYPES OF BIODIVERSITY

Biodiversity can be studied at genetic, population, species, ecosystems and landscape levels. Depending upon the objectives, the inventory could focus on single species or multiple taxa. For single species inventories, the focus could be on high impact, indicator, umbrella, flagship, threatened, or economically important species. The four later categories could also become the focus of conservation programmes.

All taxa biodiversity inventory (ATBI) results from intensive studies on a particular site cutting across all taxonomic groups, all biota taxonomic inventory (ABTI) from studies on all species within the selected taxa. Rapid biodiversity assessment (RBA) focuses on certain prominent taxa, mostly through visual encounter surveys (VES), and provides an estimate of biological richness of a particular area. The biodiversity can be studied at the following levels.

4.1 Species Diversity

A species can be defined as a group of interbreeding or potentially interbreeding natural populations that are reproductively isolated from other groups. The species diversity refers to the variety of species within a region. So far only about 1.7 million species have been reported and described as against the estimated 10 to 30 million species present on the earth. Insects and microorganisms account for a major proportion (Stork 1993). The biodiversity of a region is generally assessed in terms of species richness, which includes the total number of species in the given geographical area.

The Botanical Survey of India (BSI), the Forest Survey of India (FSI) and the Zoological Survey of India (ZSI) document the plant and animal diversity (collectively known as 'biodiversity') of the country. Of the estimated 45,000 plant species, about 15,000 species of flowering plants have been described. Estimates of other taxa include 5000 of algae, 1600 of lichens, 20,000 of fungi, 2700 of bryophytes and 600 of pteridophytes. There are about 75,000 species of animals including 50,000 insects, 4000 molluscs, 2000 fishes, 140 amphibians, 420 reptiles, 1200 birds and 340 other invertebrates.

A large number of species are endemic to India. About 70% of the world's total flowering plants occur in 12 countries, which are known as megadiversity countries/centres (Mc Neely *et al.,* 1990). Species diversity increases with the maturity of the ecosystem. The lower life forms are replaced by higher life forms of large size. In California annual grasslands, the relationship between species diversity and dominance is related to increased stability by providing alternate pathways of energy flow in a fluctuating environment. The species diversity increases productive efficiency, while dominance makes the system stable. Plants and animal diversity of the country has been placed in Table 2.

Table 2: Plant and animal diversity of the country

S. No.	Group of plants	Number of species	Group of animals	Number of species
1.	Algae	5,000	Insects	50,000
2.	Fungi	20,000	Molluscs	4,000
3.	Lichens	1,600	Fishes	2,000
4.	Bryophytes	2,700	Amphibians	140
5.	Pteridophytes	600	Reptiles	420
6.	Angiosperms	20,000	Birds	1,200
7.	–	–	Mammals and other invertebrates	340

4.2 Genetic Diversity

Diversity of the living world refers to the variations of genes within species and covers distinct populations of the same species or genetic variations within populations. During sexual reproduction, a set of chromosomes (coming from different parents) is passed to the offspring. The genetic variability of both the parents is combined to form new combinations, as a result the new individuals with changed characters are formed adding to the diversity of the living world. Besides, mutation in an organism also brings out changes in the composition of DNA molecule. The changed DNA produces modified enzymes producing modified cells which ultimately turn into modified organisms. Thus, a large number of mutants in a population are produced. The process of speciation separates and establishes genetic variations into distinct species. The original population is known as 'gene pool', which is divided into two or several gene pools. Every gene pool soon acquires a different cellular, tissue, organ characteristics through mutation and natural selection. Such separation in the primary gene pool inhabits interbreeding between individuals and in course of time these newly segregated populations are changed to such an extent that the interbreeding is completely stopped. Out of these certain individuals, a few individuals survive and reproduce to form new individuals while others may disappear without leaving any trace.

There are 12 mega-gene centres of crop plant diversity. India is one of the 12 regions of diversity of crop plants in the world (Table 2). The Indian agro-diversity has been enriched by a continuous stream of introduction of new crops and cultivars since ancient times. The Mughals, Spaniards and the Portuguese introduced the plants like pear, grape, date palm, garlic, opium, maize, potato, tomato, chilies, etc. Arabs bought spices like clove, coriander, cumin, etc. The British introduced coffee, cocoa, cashew nut, cinchona, etc. Thus, enormous diversity has been built up, which consists of the gene pool of indigenous crop plants, their wild relatives as well as introduction. The Indian gene centre has strong linkage and contiguity with Indo-Chinese-Indonesian, Chinese-Japanese and Central Asian regions. Further, the influx of germplasm from the Mediterranean, African and tropical American regions, has built up enormous locally adapted and selected variability (Chowdhary & Murti, 2000).

4.3 Ecosystem Diversity

An ecosystem is a discrete structural-functional life sustaining system. The biotic and abiotic components in a habitat constitute an interesting environmental system in which inorganic constituents are synthesized into organic structure and through energy exchange life processes, life of various forms exists in the system. Such a dynamic environmental system is known as ecosystem. The ecosystem may be:

4.3.1 Aquatic ecosystem

It can be of two types, i.e. freshwater ecosystem and marine ecosystem. The Indian subcontinent has approximately half of the world's aquatic flowering plants (Lavania *et al.*, 1990).

4.3.2 Terrestrial ecosystem

Terrestrial ecosystem may be of several types including forest ecosystem, desert ecosystem and man-made ecosystem. Dry and wet tropical forests, Mangrove wetlands, aquatic habitats, coral reefs and wild lands like African Savannah are very important for biodiversity point of view. Out of these the tropical rainforests are supposed to contain maximum diversity. Ecosystem diversity is generally assessed by the diversity of the component species, relative abundance of various species as well as types of species.

4.4 Human Cultural Diversity

The cultural diversity includes the diversity in language, religious beliefs, land crop management practised, art, music, social structure, crop selection, diet and various other similar aspects of human society. Different species of living organisms on this planet are the outcome of hundreds of millions of years of evolutionary processes and in the course of time human societies have emerged and adapted to the local environment of a particular region, discovering, using and altering the biotic resources leading to further shaping of biodiversity.

5. LOSS OF BIODIVERSITY IN INDIA

Anthropological pressure on the earth has become the cause of depletion of natural resources and shrinkage of vegetation cover. Biodiversity is suffering from the injudicious demands of human beings. Nature can satisfy the needs of the man but not the greed, which is causing the degradation of the natural habitats and therefore the loss of the biological diversity. The ecosystem loss will immediately affect the present and future generations, as the species lost today may have ornamental, food, medicinal or industrial value, presently not known to the mankind.

The earth's biological resources are under severe pressure and fast dwindling. The loss of biodiversity has an adverse effect on the economy and environment of the earth. India has most distinct biogeographic regions of the world including the Great Himalayas, Indo-Gangetic plains and Peninsular India. The rich and diverse vegetation wealth of India is due to its immensely varying climate and geographical conditions with varied habitats. There are regions with cold arctic conditions in the Himalayas in contrast to hot deserts of Rajasthan. There are areas in the west with almost zero rainfall to the highest rainfall areas of the world in north-east India. Hooker (1904) had well stated that, "The Indian flora is more varied than any other country of equal area in the eastern hemisphere, if not on the globe". The Indian flora has about 5000 endemic species (Chaterjee 1962, Nayar 1980).

6. EXTINCTION OF THE SPECIES

We seem to be entering into a phase of mass extinction. Today, we are losing two to five species per hour from tropical forests alone. This amounts to a loss of 16 million populations per year or 1800 populations per hour. If the present trend continues, about 20% of the total 2,50,000 higher plant species will be lost in the next few decades and another 25% by the end of the 21st century. Mass extinction (25-70% of species each time) has occurred in the past geological-time boundaries, such as Frasnium/Fammennian, Ordovician/Silurian, Permian/Triassic, Triassic/Jurassic and Cretaceous/Tertiary. There were a few species at stake then, however, there were around 25,000 species of flowering plants some 150 million years ago, by the late Cretaceous this number had swelled to 1,00,000 and today there are some 2,00,000 species of plants on the earth. Additionally, while the past extinctions occurred each time over a span of about 200 years, and plants (which make up the base of the food chain) will be equally or worse affected (Singh, 2001).

6.1 Causes of Species Extinction

The main cause of species extinction is the destruction and/or alteration of the natural habitats. It is responsible for the extinction of thousands of flora and fauna each year. The tropical rain-forests are among the most species rich habitats on this planet. About 5-10% of tropical forest species will be lost per decade. The forest area harbouring about 50% of the global species will be reduced to a tenth or less by 2020 (Raven, 1987). The forest area has already been reduced to 55% of its original cover and is being further reduced at a rate of more than 1,00,000 sq km per year (Wilson, 1889). The acceleration in the rate of species extinction is due to a continuous loss of tropical forests. If the deforestation continues at the current rate, about 2-8% of the species of the earth will become extinct in the next 25 years. Present loss of the species is highest in the past 65 million years. According to the report of IUCN, if the current trends continue unchecked, 25% of the world's species may be lost by 2050. The following causes to loss of biodiversity have been reported.

6.1.1 Deforestation

Deforestation mainly results from population settlement, shifting cultivation, development project, demands for fuelwood, for domestic as well as industrial purposes.

6.1.2 Man-induced extinction

Injudicious use of the land and greed of man has devastating effect on biodiversity. Overexploitation of timber logging has also affected the wildlife to a great extent.

6.1.3 Domestication of biodiversity

Traditional farmers of the world have developed an incredible variety of crops and livestock. This too, has been eroded over the last few decades, as lakhs of traditional crop strains and hundreds of domesticated livestock breeds are being replaced by dominant cash crops and laboratory-generated hybrid or transgenic crops.

6.1.4 Traditional and cultural diversity loss

Increasing human needs of food, cloth, shelter, rituals, smell and to resist drought and pests were met by traditional diversity. On the other hand, substantially increasing the grain yield and monetary profits, due to modern hybrids has forced the farmers to look elsewhere for their daily needs, especially fodder.

6.1.5 Specimen collection for research and zoological parks

Plants and animals are collected throughout the world for biological laboratories and zoological parks for study and research in science and medicine, and also for recreation. For example, the animals like monkeys and chimpanzees are sacrificed for research, as they have anatomical, physiological and genetical similarities to human beings.

6.1.6 Introduction of exotic species

Due to introduction of exotic species, native species are subjected to competition for food and space, e.g. the introduction of *Parthenium hysterophorus* has resulted in destruction of habitats of several plants.

6.1.7 Control of pests and predators

Pest control measures generally kill predators that are components of balanced ecosystem; sometimes they also indiscriminately poison non-target species.

6.1.8 Pollution

Pollution alters the natural habitats. Some pollutants like toxic wastes, insecticides, pesticides, sulphur and nitrogen oxides, acid rain, ozone depletion and global warming affect plant and animal species adversely.

6.1.9 Political and socio-economic causes

These causes vary from region to region. They can be linked to government and international support for industries, forestry, agriculture and energy programmes.

6.1.10 Promotion of subsidy for immense monoculture of soya beans

The policy of promoting cash crops destroyed all the other best adapted, healthy and economically sound species.

6.1.11 Development of policies and projects

They rarely bother for the needs of biodiversity protection and conservation.

6.1.12 Human population

Man has always created more problems for biodiversity than he has solved. The current rapid rate of species extinction and depletion of biodiversity is clearly related to human population and activities. Nearly half of the world's terrestrial species live in hot spot territories. In 1995, more than 1.1 billion people were living in hot spots. Population growth in these areas has been proceeding at two and a half times the rate of the world population as a whole. This trend indicates the continuing risk of species loss and causes serious concern in about three quarters of these sensitive areas. Human population and the application of economic processes to natural resources are the main causes for substantial losses of biological diversity.

6.1.13 Exploitation of economically important taxa

Ornamental plants, such as orchids are an important commodity in international trade. Today, the diversity of decorative plant species established in cultivation surpasses the variety of plants grown for food around the world. Heavy exploitation of such ornamental (orchids, cacti, cut-flowers, etc.) and medicinal plants combined with habitat destruction has led to the decline of major commercially important species from nature.

6.1.14 Other causes of biodiversity loss

Some ecological factors are also responsible for the extinction of plant and animal species, which can be summarized as follows:

- Distribution range of species–The smaller the range of distribution, the greater the threat of extinction to species.
- Degree of speciation–The chances of extinction are much higher in specialized species.
- Top consumers–The organisms of this category in the food chain are more susceptible.
- Reproductive rate–Larger organisms tend to produce fewer offspring at greater intervals.
- Overpopulation and poverty.
- A large number of cattle herds have destroyed the ground flora of forests.
- Growing demand for land, habitat alteration and destructive activities.
- Energy crisis.
- Construction of dams and lakes.
- Unplanned developmental work.
- Overexploitation of plants having ornamental, medicinal and other importance.

6.2 Rare and Threatened Species

The World Conservation Monitoring Center, Cambridge has alarmed that unless stringent measures are taken, biological diversity of immense economic and ecological value may be lost at several locations in the 'hot spot zone' of India. Botanical Survey of India has published an account of 623 species of vascular plants from India in Red Data Book of Indian Plants.

With continuous explorations, inventories and assignment of various parameters of floral diversity, BSI has identified about 1500 species of rare and threatened plant species. It has also been pointed out that the species are becoming rare/threatened mainly due to anthropogenic factors like habitat destruction and overexploitation. Current rate of extinction demands immediate concerted efforts for conservation of biodiversity. The number of threatened species in India is given in Table 3.

Table 3: Threatened plant species in India

S. No.	Category	Families	Species
01	Possibly/Presumed extinct	22	62
02	Endangered	43	128
03	Vulnerable	50	110
04	Rare	67	247
05	Indeterminate	37	71
06	Insufficiently known	04	05

After: Nayar & Sastry (1988-90)

7. THE CONCEPT OF HOT SPOTS

Vast tracks of forest lands are disappearing day by day leading to humanity to disaster. The loss of biodiversity is a global crisis. There is hardly any region on the earth that is not facing ecological catastrophes. Biological extinction has been a natural phenomenon in geological history. However, man's intervention has speeded up extinction rates all the more. The distribution of the world's tropical forests is one of today's most urgent global environmental issues. The loss of biodiversity has immediate and long-term effects on human survival. The plant genetic resources are being depleted at an alarming pace throughout the world. The areas featuring exceptional concentrations of endemic species and experiencing exceptional loss of habitat are known as 'hot spots' (Ginsberg 1999).

7.1 Definition of Hot Spot

The term hot spot is generally referred as "fragile areas rich in biodiversity with high degree of endemism and higher incidence of rare and threatened species". Preliminary efforts were limited only to vascular plants present in 18 hot spots of the world (Myers 1988, 1990). Now the number of hot spots has been increased to 25, out of these India has two hot spots, i.e. Eastern Himalayas (Indo-Burma) and the Western Ghats. They are estimated to have about 7000 and 2180 endemic species of plants, respectively (Table 4).

Table 4: The 25 hot spots of the world

S. No.	Hot spots	Original extent of primary vegetation (km^2)	Remaining primary vegetation (% of original extent)	Area protected (km^2) (% of hot spot)	Plant species	Endemic plants (% of global plants, 300,000)
1.	Tropical Andes	1,258,000	314,500(25.0)	79,687 (25.3)	45,000	20,000 (6.7%)
2.	Mesoamerica	1,55,000	231,000(20.0)	138,437(59.9)	24,000	5,000(1.7%)
3.	Carribean	263,500	29,640 (11.3)	29,840(100.0)	12,000	7,000(2.3%)
4.	Brazil's Atlantic Forest	1,227,600	91,300(7.5)	33,084(35.9)	20,000	8,000(2.7%)
5.	Choc/Darien/Western Euador	260,600	63,000(24.2)	16,471(26.1)	9,000	2,550(0.8%)
6.	Brazil's Cerrado	1,783,200	356,630(20.0)	22,000(6.2)	10,000	4,400(1.5%)
7.	Central Chile	300,000	90,000(30.0)	9.167(10.2)	3,429	1,605(0.5%)
8.	California Floristic Province	324,400	80,000	31,443(39.3)	4,426	2,125(0.7%)
9.	Medagascar*	594,150	59,038(9.9)	11,548(19.6)	12,000	9,704(3.2%)
10.	Eastern Arc and Coastal Forests of Tanzania/Kenya	30,000	2,000(6.7)	2,000(100.0)	4,000	1,500(0.5%)
11.	Western African Forests	1,265,000	126,500(10.0)	20,234(16.1)	9,000	2,250(0.8%)
12.	Cape Floristic Province	74,000	18,000(24.3)	14,060(78.1)	8,200	5,682(1.9%)
13.	Succulent Karoo	112,000	30,000(26.8)	2,352(7.8)	4,849	1,940(0.6%)
14.	Mediterranean Basin	2,362,000	110,000(4.7)	42,123(38.3)	25,000	13,000(4.3%)
15.	Caucasus	500,000	50,000(10.0)	14,050(28.1)	6,300	1,600(0.5%)
16.	Sundaland	1,600,000	125,000(7.8)	90,000(72.2)	25,000	15,000(5.0%)
17.	Wallacea	347,000	52,020(15.0)	20.415(39.2)	10,000	1,500(0.5%)
18.	Philippines	300,800	9,023(3.0)	3,910943.3)	7,620	5,832(1.9%)
19.	Indo-Burma	2,060,000	100,000(4.9)	100,000(100.0)	13,500	7,000(2.3%)
20.	South-Central China	800,000	64,000(8.0)	16,562(25.9)	12,000	3,500(1.2%)
21.	Western Ghats/ Sri Lanka	182,000	12,450(6.8)	12,450(100.0)	4,780	2,180(0.7%)
22.	SW Australia	309,850	33,336(10.8)	33,336(100.0)	5,469	4,331(1.4%)
23.	New Caledonia	18,600	5,200(28.0)	526.7(10.1)	3,332	2,551(0.9%)
24.	New Zealand	270,500	59,400(22.0)	52,068(87.7)	2,300	1,865(0.6%)
25.	Polynesia/Micronesia	46,000	10,024(21.8)	4,913(49.0)	6,557	3,334(1.1%)
	Total	17,444,300	2,122,891(12.2)	800,767,(37.7)	!	133,149(44%)

*Medagascar includes the nearby islands of Mauritus, Reunion, Seychelles and Comores.
These totals cannot be summed owing to overlapping between hot spots.

7.2 Criterion for Selection of Hot Spots

To qualify as a hot spot, an area must contain at least 0.5% or 1500 of the world's 300,000 plant species as endemics (Price 1977). A second determinant of hot spots status, applied only after an area has met the 'plants' criterion, is the degree of threat to plants through habitat loss. A hot spot should have lost 70% or more of its primary vegetation, this being the form of habitat that usually contains the most species especially endemics. Eleven hot spots have already lost at least 90% and three have lost 95% vegetation. The 70% cutoff is justified on the grounds that most large-scale concentrations of endemic plant species occur within the 25 hot spots as delineated.

India contains a great wealth of biological diversity in its forests, wetlands and marine areas. Indian forest ecosystem is unique in terms of floristic diversity. Forest cover of India is about 633,397 sq. km representing about 12.27% of India's total geographical area (Chowdhery & Murti 2000). But as many as 3,000-4,000 higher plants may be under a degree of threat.

7.3 Boundaries of Hot Spots

The boundaries of 'hot spots' have been determined by biological communities. Each of the areas features a separate biota or community of species that fits together as a biogeographic unit, e.g. Island groups such as New Caledonia, New Zealand, the Philippines, etc. In other areas the definition of hot spot's boundaries derives from recognized divisions such as Wallace's line between Sundaland and Wallacea, or the Kanger-Pattani line between Indo-Burma and Sundaland. In still areas, the definition reflects a best judgement opinion from experts in the field were larger 'hot spots', for example, the Tropical Andes, Mesoamerica, Indo-Burma and Sundaland to be subdivided into areas of smaller size to form many mini-hot spots.

7.4 Hottest Hot Spots

The number of species threatened with extinction and outstripes, available conservation resources and the situation looks set to become rapidly worse (Ehrlich 1994, Myers 1996). This places a premium on identifying priorities. How can we protect the most species by investing minimum amount of money? The main question is to concentrate on areas where there is greatest need and where the payoff from safeguard measures would also be greatest. Conservationists can engage in a systematic response to the challenge of large-scale extinctions ahead.

The hot spots, which are under greatest threat of extinction of species and/or habitats are termed 'hottest hot spot'. There have been identified eight 'hottest hotspots' in the world (Table 5). The leaders are Medagascar, the Phillipines and Sundaland followed by Brazil's Atlantic Forest and the Caribbean, have small areas, which further highlights their importance. Two additional hot spots, the Tropical Andes and the Mediterranean Basin, should be considered hyper-hot candidates for conservation support in light of their exceptional totals of endemic plants, 20,000 and 13,000, respectively.

Table 5: The eight hottest hot spots

S. No.	Hottest hot spots	Endemic plants
1.	Medagascar	9704
2.	Phillipines	5832
3.	Sundaland	15000
4.	Brazil's Atlantic Forest	8000
5.	Caribbean	7000
6.	Indo-Burma	7000
7.	Western Ghats/Sri Lanka	2180
8.	Eastern Arc and Coastal Forests of Tanzania/Kenya	1500

8. BIODIVERSITY CONSERVATION

Every form of life is unique and warrants respect of its worth to human beings. The favour and intensity of attention is based on the fact that, over time, the greatest value of the variety of life may be found in the opportunities it provides to humanity for adapting to local and global climate change. Biodiversity is a dynamic part of the biosphere supported by biological processes and organic evolution. Conservation is deeply linked with utilization and economic activity. Hence, where feasible, cultivation of plant needs to be worked out in a cost-effective mode for local people to take advantage.

The main strategy for conservation of economically important species is the protection of viable habitats in respective ecosystems. Changing habitats due to the activity of various biotic factors disturb our habitats. In addition, specific measures have been taken to prevent poaching and trade in wildlife products. Special steps to check illegal trade in ivory, rhino horns, musk and peacock feathers have also been taken. India is also a signatory to the Conservation on International Trade in Endangered Species of wild fauna and flora (CITES). 'Project Tiger', 'Save the Barasingha' campaign, and more recently a project on the 'Asiatic Elephant' have been initiated to save and conserve these species. A citrus sanctuary has been set up in Meghalaya. These programmes, though focused on single species, have a wider impact as they conserve habitat and a variety of other species.

The Indian Botanical Garden, Howrah (West Bengal) is over 200 years old. Other important botanical gardens include those in Bangalore and Lucknow. The Zoological Parks are looked as centres of education and recreation. They have also played an important role in the conservation of species such as Manipur Thamin Deer (*Cercus eldi*) and the white winged wood duck (*Cairina scutulata*). Notable examples of successful captive breeding are those of Gangetic Gharian (*Gavialis gangeticus*), turtles and the white tiger.

8.1 Conservation, Domestication and Sustainable Utilization of Endangered Plants

Conservation is deeply linked with utilization and economic activity. Hence, where feasible, cultivation of plants needs to be worked out in a cost-effective mode for local people to take advantage. Since many of these plants grow in forest areas, alpine meadows, waste lands, etc., it is necessary to involve different agencies for an integrated approach adopting both *in situ* and *ex situ* measures for their sustainable utilization and hence conservation.

Generation of a knowledge base for sustainable utilization and conservation of plant genetic resources vital and following points are of importance:

- Assessment of distribution, diversity and extent of availability of a species in a particular area. This entails survey, and *in situ* recording of local ecological conditions under which the plant grows, i.e., demarcating of climatic zones, micro-climate parameters such as rainfall, temperature, pH, wind velocity, sunshine hours, soil characteristics, water availability, associated vegetation and microflora.
- Authentication in terms of molecular, biochemical, genetical and chemical variability to establish distinct ecotypes and determine units of conservation.
- Categorization and prioritization of most vulnerable species.
- Study of reproductive biology in terms of seed biology, vegetative propagation and tissue culture.
- Study of the photosynthetic efficiency/respiration and source-sink relationships during developmental phases of the plant. This information is vital in domestication and adaptation of the plants to cultivated environments.
- Evolving package of practices for agronomy, processing and value addition.
- Integrating evolved package of practices in prevailing cropping systems.
- Evolve strategy for *in situ* conservation.
- Imposing restriction or ban on extraction of plant resources from natural habitats.

Initiatives in the shape of biosphere reserves, natural parks, herbal gardens, gene sanctuaries and gene banks are no doubt important but ultimately the fundamental knowledge of plant's interaction within the communities and ecosystems is valuable. Furthermore, in depth understanding of plant physiology, reproductive behaviours, genetic and molecular basis of variability will help us in efficient and sustainable utilization of our plant bioresources, thereby promoting conservation.

9. TYPES OF CONSERVATION

The conservation can be achieved *in situ* (in nature) and *ex situ* (off site) both. The details of which have been described as follows:

9.1 *In situ* Conservation

In this type of conservation the species remain within the nature and thus adapts itself to the changing natural conditions and competes with other species. The concept of *in situ* conservation as existed in various human societies. International Union for Conservation of Nature and Natural Resources (IUCN) in 1980 defined conservation as "The management of human use of the biosphere so that it may field the greatest sustainable benefits to present generation while maintaining its potential to meet the needs and aspirations of the future generation as well."

In 1982, with the help of United Nations Environment Program (UNEP) and the wild-life fund for nature (WWF) had suggested 10 different protected areas for United Nations. In India Wildlife (Protection) Act came into existence in 1972. It includes the constitution and management of national parks, sanctuaries, game reserves and closed areas. This act not only specifies the procedure of setting up of national parks, etc. but also specifies the management parameters. The work and the powers of the officials have also been fixed. Prior to national act, some states had their own legislations like the Hailey National Park Act of U.P., 1036, under which the Corbett National Park was set up. Indian Forest Act 1027 was also having the provisions of wildlife sanctuaries. In India, maximum threat to flora or plant life lies in such areas, which are subjected to various developmental activities. Different types of protected areas are located in all the major habitat types of the country ranging from deserts to hilly regions, evergreens to deciduous forests. National parks and wildlife sanctuaries are two most important types of protected areas. There are about 85 National Parks and 448 Wildlife Sanctuaries in the country.

9.1.1 *National parks*

They are given higher level of protection. No grazing is permitted in them. No private land holding or right is allowed within a national park. No person is allowed to destroy, exploit or remove any wildlife from the national park except when a permission granted by the Chief Wildlife Warden and the State Government is available. The permission can only be granted on the grounds that such destruction, exploitation or removal of wildlife from national park is essential for the betterment of wildlife there (Section 35 (6) of the Act).

9.1.2 *Sanctuaries*

Sanctuaries are given a lesser level of protcction. Certain type of activities may be permitted in sanctuaries not only for the better protection of wildlife but also for any other good and sufficient reason. No person is allowed to hunt any wild animal in the sanctuaries or remove any wild animal from there provided that the Chief Wildlife Warden is satisfied that such activity is for any good or sufficient reason. The permission is granted on the previous approval by State Government (Section 29 (1)) of the Act.

9.2 *Ex situ* Conservation

The Botanical Gardens and Zoological Parks play an important role in the World Conservation Strategy as *ex situ* centres for conserving endangered plants, preservation of seeds and propagules in seed banks. A network of botanical gardens growing and multiplying the rare, vulnerable and endangered plant species, exchanging them and information related to various aspects with other gardens has proved that the Botanic Gardens play an important role for preserving the floristic and garden resources.

Indian Botanical Garden, Howrah is the oldest botanical garden in India. It was established by Col. Robert Kyad in 1787 at the bank of river Hoogly. It spreads over 110 hectare of land. It is famous for 240 years old Banyan tree (*Ficus benghalensis*) which spreads over 3.5 acres of land with nearly 1900 prop-roots to support its canopy. National Botanical Research Institute,

Lucknow and Tropical Botanical Garden & Research Institute, Trivandrum are actively engaged in conservation and multiplication of rare, threatened and endangered species. Table 6 shows some important Botanical Gardens/Parks situated in different regions of our country.

Besides Botanical Gardens some institutions like NBPGR (National Bureau of Plants Genetic Resources, New Delhi), NBAGAR (National Bureau of Animal Genetic Resources, Karnal) and NBFGR (National Bureau of Fish Genetic Resources, Lucknow) are also actively involved in the conservation of germplasms through gene banks of different economically important plants and domesticated animals.

10. ECONOMIC IMPORTANCE

India is inhabited by man since almost prehistoric times. One of the earliest known civilizations flourished in Indian soil at a time when large parts of the world were still inhabited by savage man. Remains of Indus Valley Civilization prove that rice, wheat, barley and cotton were already there. Many plants yielding food, fibre and wood were known to the ancient Indians, and several others were recognized because of their curative properties in the treatment of diseases.

Today, as ever, human beings are dependent for their sustenance, health, well-being and enjoyment of life on fundamental biological systems and processes. Humanity derives all of its food, many medicines and industrial products from wild or domesticated components of biological diversity. Biotic reservoirs are source of recreation. Tourism and underpin the ecosystem, which provide us with many services. This is an undisputed fact that the plants affect our lives. We eat grains, fruits, vegetables and are clothed by fibres from stem, and leaves. Plants generate the oxygen we breathe, and trees provide us with timber, paper and welcome shade on hot days. Many plants make medicines, brilliant dyes, industrial chemicals and useful oils. The colours and fragrance of flowers and foliage satisfy our aesthetic senses, while plant derived spices such as black pepper, nutmeg, ginger, cloves and cinnamom have enhanced our enjoyment of food. Tea and coffee are world's most popular beverages.

It is difficult to name an area of our existence that is not affected by plants. Plants feed us, house us, clothe us, generate oxygen and cure our illnesses. There are about 17,500 species of higher flowering plants, most of them are utilized by human beings for different purposes in life support.

Besides, some plants are utilized as timber in the construction of human shelter and also in industries. Weed plants are utilized for composting as well as for fuel purposes. Microorganisms are utilized in different areas related to our lives. They are used commercially in dairy, alcohol and medicine industries.

However, plants are not entirely beneficial to human beings, they also damage crops, cause severe allergic reactions, and produce some of the most potent poisons known. Not surprisingly, we have dreamt to recognize plants that help us and those that harm us. Plants have been integral part of all civilizations.

Table 6: List of the botanical gardens/parks in India

S. No.	Names of the botanical gardens/parks in India	States
1.	Botanic Gardens, Waghai-Dangs	Gujarat
2.	Government Botanic Gardens, Trivandrum	Kerala
3.	Botanic Garden, Orchid Research and Development Centre, Tippi, Bhalukpong	Arunachal Pradesh
4.	Horticultural Experiment and Training Centre Saharanpur	Uttar Pradesh
5.	The Botanic Garden, Pondicherry	Pondicherry
6.	Government Garden Ootacamund	Tamil Nadu
7.	National Botanical Research Institute, Lucknow	Uttar Pradesh
8.	The Bryant Park, Kodaikanal	Tamil Nadu
9.	State Horticultural Farm, Kallar	Tamil Nadu
10.	Botanical Gardens, Forest Research Institute & College, Dehradun	Uttarakhand
11.	Lloyd Botanic Garden, Darjeeling	West Bengal
12.	Pomological Station, Coonoor	Tamil Nadu
13.	Sim's Park, Coonoor	Tamil Nadu
14.	Narendra Narayan Park, Cooch Behar	West Bengal
15.	Government Gardens, Chaubattia	Uttarakhand
16.	Government Gardens, Eden Gardens, Kolkata	West Bengal
17.	State Botanical Gardens, Barang	Orissa
18.	Lalbagh Gardens, Bengaluru	Karnataka
19.	Ram Bagh, Amritsar	Punjab
20.	Government Garden (Motilal Nehru Park), Allahabad	Uttar Pradesh
21.	Experimental Garden, Botanical Survey of India: Arunachal Field Station, Itanagar	Arunachal Pradesh
22.	Experimental Garden, Botanical Survey of India, Andman & Nicobar Circle, Port Blair	Andman & Nicobar Island
23.	Experimental Garden, Botanical Survey of India, Central Circle, Allahabad	Uttar Pradesh
24.	National Orchidarium and Experimental Garden, Botanical Survey of India, Southern Circle, Yercaud	Tamil Nadu
25.	Experimental Garden, Botanical Survey of India, Northern Circle, Dehradun	Uttarakhand
26.	Experimental Garden, Botanical Survey of India, Western Circle, Pune	Maharashtra
27.	Barapani Experimental Garden, Botanical Survey of India, Eastern Circle, Shillong	Meghalaya
28.	National Orchidarium and Botanic Garden, 'Woodlands' Botanical Survey of India, Eastern Circle, Shillong	Meghalaya
29.	Tropical Botanic Garden and Research Institute, Trivandrum	Kerala
30.	Indian Botanic Garden, Howrah	West Bengal
31.	Botanical Garden, Bihar Agricultural College, Sabour	Bihar
32.	Mughal Garden, Pinjore	Haryana
33.	Talkatora Garden, New Delhi	New Delhi
34.	Rashtrapati Bhawan Garden (Mughal Garden), New Delhi	New Delhi
35.	Nehru Park, New Delhi	New Delhi
36.	Lodi Garden, New Delhi	New Delhi
37.	Buddha Jayanti Park, New Delhi	New Delhi
38.	Brindavan Gardens, Mysore	Karnataka
39.	Nishat Bagh, Mysore	Karnataka
40.	Agri-Horticultural Society, Madras	Tamil Nadu
41.	Pampavana, Munirabad	Karnataka
42.	Botanic Garden, Central Drug Research Institute, Lucknow	Uttar Pradesh
43.	Umaid Garden, Jodhpur (Wallingdon Garden), Jodhpur	Rajasthan
44.	Nehru Park, Jodhpur	Rajasthan
45.	Mandore Garden, Jodhpur	Rajasthan
46.	Ganga Niwas Garden, Jodhpur	Rajasthan
47.	Botanic Garden, Central Arid Zone Research Institute, Jodhpur	Rajasthan
48.	Telco Garden and Nursery, Jamshedpur	Bihar
49.	Jubilee Park, Jamshedpur	Bihar
50.	Botanic Garsen, National Bureau of Plant Genetic Resources – (NBGGR), Horticultural Station, Delhi	Delhi
51.	Padmaja Naidu Himalayan Zoological Park, Darjeeling	West Bengal

Table 6: Contd.

S. No.	Names of the botanical gardens/parks in India	States
52.	Botanic Garden, Darjeeling	West Bengal
53.	Victoria Memorial Garden, Kolkata	West Bengal
54.	Tollygunge, Club Garden, Kolkata	West Bengal
55.	Raj Bhawan Garden, Kolkata	West Bengal
56.	National Library Garden, Kolkata	West Bengal
57.	Jheel-Meel Safari Park, Kolkata	West Bengal
58.	Jatindra Mohan Park, Kolkata (Tala Jheel Park)	West Bengal
59.	Bidhan Shoshu Udyan, Kolkata	West Bengal
60.	Assembly House Garden, Calcutta	West Bengal
61.	The Agri-Horticultural Society of India, Kolkata	West Bengal
62.	Veermata Jijabai Bhosle Udyan and Pranisangrahalaya, Mumbai	Maharashtra
63.	Garden Complex, Bhel, Bhopal	Madhya Pradesh
64.	Jawahar Kunj Garden, Barrackpore	West Bengal
65.	Cubbon Park, Bengaluru	Karnataka
66.	The Retreat Botanic Garden, Shahibagh, Ahmedabad	Gujarat
67.	Botanic Garden, Agra College, Agra	Uttar Pradesh
68.	Andhra University Experimental Garden, Waltair	Andhra Pradesh
69.	Botanic Garden, School of Studies in Botany, Vikram University, Ujjain	Madhya Pradesh
70.	Kerala University Botanic Garden, Trivandrum	Kerala
71.	University Botanical Garden, South Gujarat University, Surat	Gujarat
72.	Dhauladhar, Botanical Garden and Arboretum, Palampur	Himachal Pradesh
73.	Botanic Garden and Arboretum, Department of Forestry, Himachal Pradesh Krishi Vishwavidyalaya, Solan	Himachal Pradesh
74.	Gardens at Uttarayan Complex: Shanti Niketan, Vishwa-Bharti University, Vishwa-Bharti	West Bengal
75.	University Garden Shanti Niketan Vishwa-Bharti University, Shanti Niketan (Bolpur)	West Bengal
76.	Botanic Garden, University of Sagar, Sagar	Madhya Pradesh
77.	Experimental Gardens, Department of Biosciences, Saurashtra University, Rajkot	Gujarat
78.	Botanic Garden, Department of Biosciences Ravi Shankar University, Raipur	Madhya Pradesh
79.	Botanic Garden, University of Poona, Poona	Maharashtra
80.	Botanic Garden, Punjabi University, Patiala	Punjab
81.	Botanic Gardens, Madurai (Madurai Kamraj University Botanic Gardens)	Tamil Nadu
82.	Botanic Garden (Motibagh), Gujarat Agricultural University, Junagarh	Gujarat
83.	Botanic Garden, Delhi University, Delhi	Delhi
84.	Botanic Garden Bidhan Chandra Krishi Vidyalaya (BCKV) North Bengal Campus, Cooch Behar	West Bengal
85.	University Botanic Gardens, Tamil Nadu Agricultural University (THAU), Coimbatore	Tamil Nadu
86.	Punjab University Botanical Garden, Chandigarh	Punjab/Haryana
87.	Calicut University Botanical Garden, Calicut	Gujarat
88.	Experimental Botanic Garden, Department of Botany, University of Calcutta, Kolkata	West Bengal
89.	Burdwan University Botanic Garden (Gopalbag), Burdwan	West Bengal
90.	Magadh University Botanical Garden, Bodhgaya	Bihar
91.	Utkal University Botanic Garden, Bhubaneshwar	Orrisa
92.	Botanic Garden, Postgraduate Department of Botany, Bhagalpur University, Bhagalpur	Bihar
93.	Magadh Sayajirao University, Botanical Garden, Baroda	Gujarat
94.	Botanical Garden, Gandhi Krishi Vigyan Kendra, Bengaluru	Karnataka
95.	Botanic Garden, Marathwara University, Aurangabad	Maharashtra
96.	Botanic Garden, Guru Nanak Dev University, Amritsar	Punjab
97.	Roxburgh Botanic Garden, University of Allahabad, Allahabad	Uttar Pradesh
98.	Kulbhaskar Ashram Post-Graduate College Botanic Garden, Allahabad	Uttar Pradesh
99.	Aligarh Muslim University, Aligarh	Uttar Pradesh
100.	Gujarat University Garden, Ahmedabad	Gujarat

REFERENCES

Bhandari, M.M. (1978). *Flora of the Indian Desert.* Scientific Publishers, Jodhpur.

Chaterjee, D. (1962). Floristic patterns in Indian vegetation, pp. 32-42, *Proc. Summer School of Botany,* Darjeeling and New Delhi.

Chowdhery, H.J. and Murti, S.K. (2000). Plant Diversity and Conservation in India—An Overview. Bishen Singh Mahendra Pal Singh, Dehradun.

Ehrlich, P.R. (1994). Energy use and biodiversity loss, *Phil Trans, R. Soc. Lond* **344**: 99-104.

Ginsberg, I. (1999). Global conservation priorities. *Cons Biol.* **13**: 5.

Groombridge, B. (1992). Global Biodiversity: Status of the Earth's Living Resources, Report compiled by the World Conservation Monitoring Center, Chapman & Hall, London.

Hajra, P.K. and Rao, R.R. (1990) Distribution of vegetation types in northwest Himalaya with brief remarks on phytogeography and floral resource conservation, *Proc. Indian Acad Sci.* **100:** 263-277.

Hooker, J.D. (1854). *Himalayan Journals*, 2 Vols, London.

Hutchinson, J. (1969). *Evolution and Phylogeny of Flowering Plants*, Academic Press, London and New York.

Hyland, B.P.M. (1983). A revision of *Syzygium* and allied genera (Myrtaceae) in Australia, *Austr. J. Bot. C Supp. Ser.* **9:** 1-164.

Lavania, G.S., Paliwal, S.C. and Gopal, B. (1990). Aquatic vegetation of the Indian subcontinent, pp 29-76. *In:* B Gopal. (Ed) *Ecology and Management of Aquatic Vegetation in the Indian Subcontinent.* Kluwer Academic Publishers, Dordrecht.

McNeely, J.A., Miller, K.R., Reid, W.V., Mittermier, R.A. and Werner, T.B. (1990). Conserving the World Biological Diversity, IUCN Gland, Switzerland.

Myers, N. (1988). Threatened biotas: 'Hot Spots' in tropical forests, *The Environmentalist* **8:** 187-208.

Myers, N. (1990). The biodiversity challenge: Expanded hot-spots analysis, *The Environmentalist* **10:** 143-256.

Myers, N. (1996). Two key challenges for biodiversity discontinuities, *Biodiv. Cons.* **5:** 1025-1034.

Nair, N.C. and Daniel, P. (1986). The floristic diversity of the Western Ghats and its conservation- A review, *Proc. Indian Acad. Sci. Suppl.* pp. 127-163.

Nayar, M.P. (1980). Endemism and patterns of distribution of endemic genera (angiosperms). *J. Econ. Tax Bot.* **1:** 99-110.

Puri, G.S. (1960). *Indian Forest Ecology*, Vols 1-2, Oxford & IBH Publishing Co., New Delhi.

Rao, A.S. (1974). The vegetation and phytogeography of Assam-Burma, pp. 204-286. *In*: M.S. Mani (Ed.). *Ecology and Biogeography in India,* W Jank Publishers, The Hague.

Rao, R.R. (1994). Biodiversity in India (Floristic Aspects), Bishen Singh Mahendrapal Singh, Dehradun.

Rao, R.R. and Murti, S.K. (1990). North-east India, a major center of plant diversity in India. *Indian J. For.* **13:** 214-220.

Raven, P.H. (1987). The Scope of the plant conservation world wide, pp 19-29, *In*: D. Bramwell, O. Hamann, V. Heywood and H. Synge (Eds.). *Botanic Gardens and the World Conservation Strategy,* Academic Press, London.

Rodgers, W.A. and Panwar, H.S. (1990). A Biogeographical Classification for Conservation Planning in India, Wildlife Institute of India, Dehradun.

Singh, J.S. (2001). Paradigms of Biodiversity, paper published in Golden Jubilee Symposium on Biotechnological Innovations in Conservation and Analysis of Plant Diversity, held at the Department of Botany, Delhi University, Delhi.

Stork, N. (1993). How many species are there? *Biodiv. Conserv.* **2:** 215-232.

Subramanyam, K. and Nayar, M.P. (1974). Vegetation and Phytogeography of the Western Ghats, pp. 178-179 *In*: M.S. Mani (Ed.), *Ecology and Biogeography in India,* W. Junk Publishers, The Hague.

Takhtajan, A. (1969). *Flowering Plants Origin and Disposal*, Tr Jefferya, Edinburgh.

Wilson, E.O. (1989). Threats to Biodiversity, *Sci. Amer.* (Sept.) pp. 106-116.

Diversity of Lower Plants (2014): pp. 24-35
Editors: Rajan Kumar Gupta & Mukesh Kumar
IK International Publishing House Pvt. Ltd.

2

Microbial Diversity for Sustainable Crop Production in Hill Region

D.K. Shahi, B.K. Agarwal and K.P. Singh*

Department of Soil Science & Agricultural Chemistry, Birsa Agricultural University, Ranchi–834 006 (Jharkhand),
**College of Forestry and Hill Agriculture, GBPUA&T, Hill Campus, Ranichauri–249 199, Tehri Garhwal, Uttarakhand*

ABSTRACT

The interactions that occur between soil organisms and their relationships with the soil environment are so complex that they will never be fully understood. Nevertheless, our current knowledge shows unequivocally that the soil biota play an important role in nutrient cycling and in improving soil structure. Some components are detrimental to plants but, from a horticultural perspective, there are other components that provide an active form of defence against these pests and pathogens. Thus, the soil biota play a vital role in sustaining a healthy, productive soil capable of supporting a level of plant growth that is appropriate for a particular soil and climate.

Soil ecosystems remain firmly at the foundations of human life-support systems, but are the least understood among the natural ecosystems, and increasingly among the most degraded. Soil erosion, loss of soil organic matter and nutrient depletion are among the leading contributors to impaired soil health, reduced crop yields and poverty in the developing world. Global warming will further exacerbate this problem. Not surprisingly, therefore, soil health tops the list of priorities of UN Millennium Project's hunger task force. Given the current concerns about the sustainability of production systems and soil biological health, and the exciting developments in microbial diversity. I chose this theme to link this trinity and look at examples of how to manage agricultural production systems in ways that sustainability is maintained whilst not compromising on microbial diversity and soil health.

Keywords: Microbial diversity, PGPR, Green manuring, Azotobacter, Acetobacter.

*Communicating Author

1. MICROBIAL DIVERSITY

The importance of soil microorganisms for sustenance of all other life forms needs no emphasis. Microbes are the basis of the biosphere; a staggering 5×10^{31} cells exist, weighing 50 quadrillion metric tonnes, constituting about 60% of the total biomass. Bacteria produce about half the oxygen on the planet of 41,000 Pg (10^{15} g) of carbon in the globe, soils contain 1550 Pg and biota 550 Pg. While the total C, N and P in the plant biomass globally is 560, 12-20 and 1-2 Pg, respectively, in prokaryotes the C, N and P content is 300-500, 70-120 and 7-12 Pg, respectively. Soil organisms act as primary driving agents of nutrient cycling, regulating the dynamics of soil organic matter, soil carbon sequestration and greenhouse gas emissions; modifying soil structure and water regimes; enhancing the amount of nutrient acquisition by vegetation; conferring stress tolerance, resisting pathogens and improving plant health. Microbes break down most of the 45,000 or so chemical compounds that humans use in daily life. They are the source of most antibiotics and of some other drugs and industrial enzymes. It is difficult to overstate their importance.

More than 90% of the planet's genetic biodiversity is resident in soils. With the advent of a microorganisms is finally being unravelled. The genotypic diversity in all the Protists groups is far more than the combined diversity of plant, animal or fungal kingdom. The total estimated microbial species in the world habitat may be as high as 4 million. A gram of soil can contain as many as 10,000 different species. Current estimates for numbers of prokaryotes, which include the bacteria and archaea, range from 300,000 to 1 million species or more but only about 5000 species have been described. Thus, less then 1% of the microorganisms have actually been isolated and identified.

Table 1: General effect of major soil management practices on the diversity and abundance of soil organisms

Increases biodiversity and population	Decreases biodiversity and population
Balance fertilizer use	Fumigants
Proper irrigation	Insecticides
Improved drainage and aeration	Compaction
Animal manures and composts	Soil erosion
Domestic (clean) sewage sludge	Industrial wastes and heavy metals
Residue return to soil surface	Residue burning or removal
Organic mulches	Plastic mulches
Cover crops or mulch fallows	Bare fallows
Crop rotation	Monocroping
Lime on acid soils	Tillage

Source: Brady, N.C. and Well, R.R. (2004). Organisms & Ecology of the Soil, The Nature and Properties of Soils, Published by Pearson Education, Pte. Ltd., Delhi, India, pp. 449-495.

Soil biology: Important relationship with soil quality

- Organic matter
- Residue decomposition
- Soil structure
- Nutrient cycling
- 1 g of soil has 100,000,000 bacteria

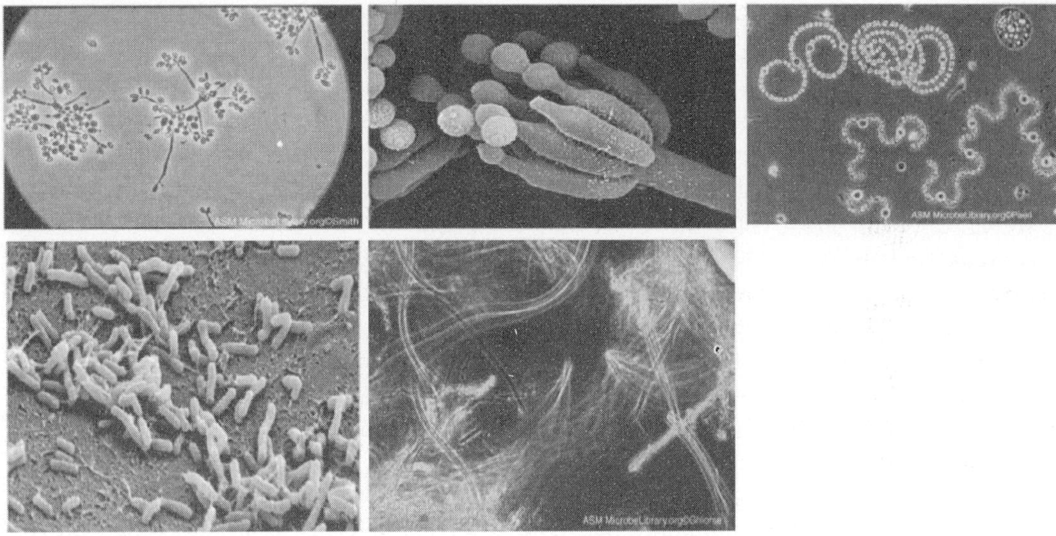

1.1 Assessing Microbial Diversity in Soil Rhizosphere

Traditional methods of studying microorganisms by enrichment culture is of very limited use in exploring diversity since a large majority the organisms cannot be cultured in the available media and will thus remain unknown. Some techniques are now available for getting a better appreciation of the soil microbial community. In community level substrate utilization (CLSU) fingerprints, soil suspensions are inoculated into multi-well plates, e.g. BIOLIG, containing a selection of specific substrates (Garland 1996), but is again limited for diversity studies due to its selectivity for culturable microorganisms and also because of the ability of multiple phylogenetic groups of organisms to use the same substrates. Soil bacterial diversity has also been assessed by analysis of fatty acid methyl ester (FAME) composition of individual isolates as well as total FAME community profile of soils (Schutter and Dick 2000). Recently, however, recombinant DNA techniques have provided methods for characterizing natural microbial communities without the need to cultivate organisms. Total DNA community profile can be fingerprinted by extracting total soil microbial DNA and amplified by polymerase chain reaction (PCR) using bacterial oligonucleotide primers targeting conserved genes on operons (e.g., 16s RNA gene). The amplified products are separated by gel electrophoresis, extracted and sequenced. The sequences are then compared with known sequences of ribosomal RNA to allow for the determination of a 'phylotype'. Additionally, hybridization probes complementary to specific regions of rRNA are synthesized and used to aid in screening other organisms (phylogenetic stains).

1.2 Microbial Diversity Plant Growth Promoting Bacteria

Low soil microbial diversity indicates stressed conditions in soil while high diversity is an indicator of a healthy soil. Dehydrogenase and urease activities were low in alkali soils, which however improved greatly upon the addition of gypsum followed by cropping, growing tree or grasses. Amendment with a readily decomposable carbon source (Sesbania) removed the sodicity stress on microbes and improved the activity dramatically (Rao and Pathak 1996). So, mineralization of added C was similar (~ 38%) throughout the pH range of 8.1 to 10.0 (ESP 2-89). This showed that autochthonous microflora of alkali soils are as effective as those in normal soils in decomposing added organic matter. In saline soils carbon amendment improved the activity in saline soils of low and intermediate salinity upto ECe 26 dS m^{-1} whereas at higher salinity the osmotic and ionic effects were predominant and added organic matter alone was not enough to improve activity. Desalinization immediately improved the activity, although it was not completely restored. A number of bacterial species associated with the plant rhizosphere belonging to genera *Azospirillum*, *Arthrobacter*, *Acinetobacter*, *Bacillus*, *Enterobacter*, *Erwinia*, *Flavobacterium*, *Pseudomonas*, *Rhizobium* and *Serratia* are able to exert a beneficial effect on plant growth.

Table 2: Effect of various organic matter on microbial diversity in soils

Treatments	Bacterial population (x 10^6 g^{-1} soil)	Actinomycetes population (x 10^5 g^{-1} soil)	Fungi population (x 10^5 g^{-1} soil)	N-fixer (x 10^5 g^{-1} soil)	P-solubilizing organisms (x 10^6 g^{-1} soil)	S-oxidiser cfu x 10^5 g^{-1} soil
Control	22	26	75	77	23	05
Berseem straw	67	97	95	150	51	13
Bagasse	59	70	109	109	35	11
Paddy straw	40	65	88	108	37	10
Cassia sonpera	76	84	94	130	46	15
Cestrum diurnum	85	84	102	131	42	24
Ipomia cornea	59	101	109	123	40	17

Source : Biodiversity Strategy Action Plan, W.B.

Maintenance of a diverse and functioning microbial community is important for soil sustainability. Loss of microbial diversity makes biological systems less able to adapt to environmental stresses. In agricultural soils many typical functions such as the decomposition of organic matter seem to follow this, that is, loss of one species is compensated by the presence of others which can take on that function. Further, I hypothesize that the degree to which a particular ecological hypothesis would apply will depend on the degree of soil degradation and propose this as the 'Stress Dependent Biodiversity: Functionality Relationship' (SDBFN). The main tenets of the hypothesis are as follows:

(a) In undisturbed pristine soil ecosystems, with abundant biodiversity, the redundant species hypothesis would operate for most functions.

(b) As the degree of stress on a soil ecosystem increases and biodiversity decreases (due to loss soil organic matter or high or low pH or salinity), while the redundant hypothesis would operate for the robust functions like carbon cycling, the more specialized functions, e.g. ammonification and nitrification would follow the predictable change hypothesis and activity decreases with increase in stress.

(c) Under stress as in drastically disturbed systems like complete removal of surface soil by strip mining and extreme acidification say by acid mine drainage waters or at extremely high salinity coupled by poor leaching, all of which result in considerable loss of soil microbial diversity, both the predictable and idiosyncratic hypothesis would operate for various functions. The restoration of the productivity and soil health of such sites would depend not only on application of specific soil amendments but also inoculation with mixed microbial communities including *Arbuscular mycorrhizae* and plant growth promoting rhizobacteria.

2. NITROGEN-FIXING BACTERIA

Biological nitrogen fixation is estimated to contribute 180×10^6 metric tonnes/year globally, of which eighty per cent comes from symbiotic associations and the rest from free-living or associative systems. The ability to reduce and siphon out such appreciable amounts of nitrogen from the atmospheric reservoir and enrich the soil is confined to bacteria and Archaea. These include; (a) Symbiotic nitrogen-fixing (N_2-fixing) forms, viz. *Rhizobium,* the obligate symbionts in leguminous plants and *Frankia* innon-leguminous trees, and b) Non-symbiotic (free-living, associative or endophytic) N_2-fixing forms such as cyanobacteria, *Azospirillum*, *Azotobacter*, *Acetobacter diazotrophicus*, *Azoarcus*, etc.

2.1 Symbiotic Nitrogen-Fixers

Two groups of nitrogen-fixing bacteria, i.e. rhizobia and *Frankia* have been studied extensively. *Frankia* forms root nodules on more than species of woody plants from 8 different families, however, its symbiotic relationship is not as well understood. Species of *Alnus* and *Casuarina* are globally known to form effective symbiosis with *Frankia.* In India, a technique for isolation of *Frankia* by single spore culture technique was developed, and PCR-RFLP markers were identified for screening actinorhizal symbionts. In the context of rhizobia, considerable change in taxonomic status has come about during the last years. Sahgal and Johri outlined the current status of rhizobial taxonomy and enlisted 36 species distributed among seven genera (*Allorhizobium*, *Azorhizobium*, *Bradyrhizobium*, *Mesorhizobium*, *Methylobacterium*, *Rhizobium* and *Sinorhizobium*) derived, based on the polyphasic taxonomic approach. Although most *Rhizobium* isolates can nodulate more than one host species and also several different bacterial species are often isolated from a single legume, it is only from a few legumes that the symbionts have, so far, been investigated thoroughly. The family Fabaceae (formerly Leguminoseae) is important both ecologically and agriculturally, since it is a major source of biological nitrogen-fixation. Species of *Parasponia* and *Tremma* are the only non-legumes that form an effective symbiosis with *Rhizobium* or *Bradyrhizobium.* A few aquatic legumes bear stem nodules in addition to the normal root nodules. Some important non-symbiotic nitrogen-fixing bacteria include, *Achromobacter*, *Acetobacter*, *Arthrobacter*, *Azospirillum*, *Azotobacter*, *Azomonas*, *Bacillus*, *Beijerinckia*, *Clostridium*, *Derxia*, *Enterobacter*, *Herbaspirillum*, and *Pseudomonas*.

Azotobacter

The family Azotobacteriaceae comprises two genera, namely, *Azomonas* (non-cyst forming) with three species (*A. agilis*, *A. insignis* and *A. macrocytogenes*) and *Azotobacter* (cyst forming) comprising 6 species, namely, *A. chroococcum, A. vinelandii, A. beijerinckii, A. nigricans, A. armeniacus* and *A. paspali. Azotobacter* is generally regarded as a free-living aerobic nitrogen-fixer. *Azotobacter paspali* which was first described by Dobereiner and Pedrosa, has been isolated from the rhizosphere of *Paspalum notatum*, a tetraploid subtropical grass, and is highly host specific. Various crops in India have been inoculated with diazotrophs particularly *Azotobacter* and *Azospirillum*. Application of *Azotobacter* and *Azospirillum* has been reported to improve yields of both annual and perennial grasses. Saikia and Bezbaruah reported increased seed germination of *Cicer arietinum, Phaseolus mungo, Vigna catjung* and *Zea mays*. However, yield improvement is attributed more to the ability of *Azotobacter* to produce plant growth promoting substances such as phytohormone IAA and siderophore azotobactin, rather than to diazotrophic activity.

Azospirillum

Members of the genus *Azospirillum* fix nitrogen under microaerophilic conditions, and are frequently associated with root and rhizosphere of a large number of agriculturally important crops and cereals. Due to their frequent occurrence in the rhizosphere these are known as associative diazotrophs. Despite their N_2-fixing capability (~1–10 kg N/ha), the increase in yield is mainly attributed to improved root development due to the production of growth promoting substances and consequently increased rates of water and mineral uptake. Azospirilla proliferate in the rhizosphere of numerous plant species and the genus *Azospirillum* now contains seven species – *A. brasilense, A. lipoferum, A. amazonense, A. halopraeferens, A. irakense, A. dobereinerae* and *A. largimobile*. *Azospirillum* sp. can contribute towards long-term goal of improving plant-microbe interactions for salinity affected fields and crop productivity. The synthesis and activity of nitrogenases in *A. brasilense* is inhibited by salinity stress.

Acetobacter

Acetobacter diazotropicus (family Acetobacteriaceae), isolated from roots and stems of sugar cane, was first reported as an N_2-fixing bacterium from Brazil, and subsequently from Australia, India, Mexico, Canada and Cuba. Isolation of this bacterium from most tissues of sugar cane, and its absence from the soils of sugar cane fields suggested these to be systemic endophytes. The occurrence of this organism has been reported in sugar-rich plants like *Pennisetum purpureum* and sweet potato and in insects like mealy bugs and leaf hoppers. The colonization of *A. diazotropicus* has also been reported in coffee plants grown through seeds and vegetative propagation. This bacterium successfully colonizes sugar cane varieties in India where the chemical N fertilization is completely avoided for at least two successive years and replaced by organic manures. *Acetobacter* has gained importance as an inoculants for sugar cane.

3. PHOSPHATE SOLUBILIZING MICROORGANISMS

Phosphorus is a major essential macronutrients for biological growth and development. P in soils is immobilized or becomes less soluble either by absorption, chemical precipitation, or both. A survey of Indian soils revealed that 98% of these need phosphorus fertilization either in the form of chemical or biological fertilizer. Although P content in an average soil is 0.05%, only 0.1% of the total P present is available to the plants because of its chemical fixation and low solubility. Application of chemical phosphatic fertilizers is practised though a majority of the soil P reaction products are only sparingly soluble. Under such conditions, microorganisms offer a biological rescue system capable of solubilizing the insoluble inorganic P of soil and make it available to the plants. Phosphate solubilizing microorganisms (PSMs) include largely bacteria and fungi, which can grow in media containing tricalcium, iron and aluminum phosphate, hydroxyapatite, bonemeal, rock phosphate and similar in soluble phosphate compounds as the sole phosphate source. Such microbes not only assimilate P but a large portion of soluble phosphate is released in quantities in excess of their own requirement. The most efficient PSMs belong to genera *Bacillus* and *Pseudomonas* amongst bacteria and *Aspergillus* and *Penicillium* amongst fungi. Among phosphate solubilizing fungi, *Aspergillus niger*, *A. flavus*, *A. nidulans*, *A. awamori*, *A. carbonum*, *A. fumigatus*, *A. terreus* and *A. wentii* have been reported from the rhizosphere of maize, soya bean, chilli, tista soils, acidic lateritic soils and compost. *Paeciliomyces fusisporus*, *Penicillium digitatum*, *P. simplicissimum*, *P. aurantiogriseum*, *Sclerotium rolfisii* and species of *Cephalosporium*, *Alternaria*, *Cylindrocladium*, *Fusarium* and *Rhizoctonia* are other solubilizers of insoluble phosphate. PSM inoculants include species of *Aspergillus*, *Bacillus*, *Escherichia*, *Arthrobacter* and *Pseudomonas* which can add 30–35 kg P_2O_5 ha^{-1}.

4. MYCORRHIZAL ASSOCIATIONS IN HIMALAYAN TREES

Mycorrhizae are beneficial associates of plants that influence plant growth in terms of increased nutrient and water absorption from the soil, protection from pathogens, increased tolerance to soil toxins, and root elongation. Seven types of mycorrhizal associations, namely, ectomycorrhizae, arbuscular mycorrhizae, ecto-endo mycorrhiza, arbutoid mycorrhiza, monotropoid, ericoid and orchidoid mycorrhizae are known (Mukerji *et al.*, 2002). The below ground diversity of arbuscular mycorrhizal fungi is a major factor in the maintenance of plant diversity and ecosystem functioning. Detailed investigations conducted on the arbuscular mycorrhizal (AM) fungi associated with *Taxus baccata* and five species of rhododendrons found in Uttarakhand Himalaya were recently reported (Chaurasia *et al.*, 2005).

5. INCREASING BIOLOGICAL ACTIVITY

Organic conversion begins with a process that encourages increased microbial and arthropod activity within the soil. The elemental composition, structure, and organic matter content of the soil need to be favourable if soil biological activity is to be enhanced. Biological activity begins with the breakdown of soil organic matter. During the decomposition process, the organic molecules in organic matter are broken down into simpler organic molecules that require further decomposition into mineralized nutrients. Organic farmers supply organic matter through incorporation of green manure crops and crop refuse, and the addition of compost.

The use of bio-indicators is becoming an increasingly important way to assess soil health. Pankhurst *et al.* (1997) review the measurement of soil organisms and biotic processes as indicators of soil health. A range of techniques is available to assess soil biological activity. These include measurement of CO_2 respiration; DNA testing to determine the diversity and abundance of microorganisms present; and measurement of the tensile strength of cotton strips buried in the soil. Commercial laboratories offering soil microbial assessment are now becoming more common in Australia.

6. GREEN MANURING

Green manure crops are grown specifically to be cultivated back into the soil to build up soil organic matter and nutrients, and to stimulate biological activity. The type of green manure crop and stage at which it is turned in determine the amount of organic matter or nutrients returned to the soil. A lush, actively growing sward of legumes such as vetch, faba beans or lupins contains large amounts of nitrogen (50-140 kg N-gain/ha) that is released to the soil upon cultivation. The same crop, when allowed to mature, contributes more organic matter but less available nitrogen. If a soil is low in organic matter, then a green manure crop that increases soil organic matter, such as oats, is desirable.

Fig. 1. Green manure crops of oats, faba bean and vetch at organic demonstration site.

Fig. 2. Lucerne under sown eight weeks after maize emergence comes away following maize harvest at organic demonstration site.

7. MAINTENANCE OF SOIL STRUCTURE

The binding substances that hold soil particles together have both mineral and organic origins. Some of the organic 'binding agents' are contributed by soil biota. Fungal hyphae, along with fibrous roots from plants, bind soil particles and small aggregates together into larger units. Polysaccharides produced by microorganisms (e.g., many commonly occurring soil bacteria such as *Bacillus, Pseudomonas, Agrobacterium, Azotobacter* and *Rhizobium*) act as the gums that bind and stabilize aggregates. Plant residues are also broken down by soil biota to create soil aggregates. Mycorrhizal fungi (VAM) contribute significantly to soil aggregate formation and soil stability at both micro and macro levels by enmeshing mineral and organic debris in a network of external hyphae.

8. NUTRIENT CYCLING

Most of the nutrients contained in soil organic matter are in complex organic forms that have to be mineralized to an inorganic form before they can be used by plants. Soil microorganisms play a dominant role in the decomposition of organic materials such as cellulose, hemicellulose, polysaccharides, hydrocarbons, lignins, proteins and amino acids, and are also responsible for nearly all nitrogen and carbon transformations in soil. They are also important in transforming nutrients such as phosphorus, sulphur, iron, potassium, calcium, magnesium, manganese, aluminium and zinc into forms that can be used by plants. Soil microorganisms therefore have a beneficial impact on plant health by releasing nutrients that would otherwise be 'locked away' in dead plant and animal tissue.

The decomposition of organic matter is brought about by a succession of microbial communities. For instance, the sugar fungi (*Mucor* spp.) are the primary colonizers and the cellulose and lignin degraders then follow them. Different nutrients are cycled at different rates by soil biota. Thus, amino acids and various oligosaccharides turn over rapidly and form a fast/active nutrient pool whereas humified soil organic matter forms a slow/passive pool that may take several decades to be degraded. The fast/active pool, which cycles 8-10 times per year, has the greatest impact on plant growth. In agricultural soils, the way these food webs operate depend on the type of farming system. When plant residues are incorporated into soil by cultivation, bacteria start the decomposition process. This immediately results in an increase in protozoa and bacterial feeding nematodes and nutrient cycling proceeds rapidly. On the other hand, when plant residues remain on the surface of the soil (e.g., minimum tillage), a fungal flora develops and populations of fungal feeding organisms increase in the surface layers. In this case mineralization of nutrients proceeds much more slowly.

9. SUPPRESSION OF PLANT PATHOGENS

Soil-borne disease problems are common in soils that have been intensively cropped for decades. Such soils are said to be 'conducive' to disease. They have lost much of their microbial diversity and biological buffering capacity, so many competitors of fungal pathogens and root-feeding nematodes have disappeared. In contrast, a 'disease suppressive soil' has a full complement of beneficial organisms, and the pathogens that cause disease are unable to increase to levels that will cause damage. The organisms involved in disease suppression act in many different ways. Fungi and bacteria are able to displace each other from specific ecological niches in soil by competing for nutrients. A group of bacteria known as the fluorescent pseudomonads (*Pseudomonas* spp.), for example, inhibit the growth of pathogens by limiting their access to soluble ferric iron. Other bacteria (e.g., *Bacillus* spp., and some actinomycetes) produce antibiotics that are detrimental to pathogens. Fungi such as *Trichoderma* and *Gliocladium* are able to parasitise fungal pathogens and are therefore useful biological control agents. Other fungi can parasitise or prey on nematodes. The 'nematode trapping fungi', for example, are able to capture nematodes by producing unique trapping structures in which nematodes become ensnared. Some small arthropods (e.g., mites and collembola) consume fungal spores or feed on parasitic nematodes.

10. BACTERIAL DIVERSITY IN RICE-WHEAT CROPPING SYSTEMS

Rice and wheat, the two most staple food crops of India, are cultivated as wheat-rice cropping sequence worldwide by farmers. However, intensive use of chemical fertilizer has resulted in increased soil salinity leading to deterioration of soil health. Moreover, the cultivation practices of two food grains are completely different; as rice requires waterlogging, which creates micro-aerophilic to anaerobic environment, that may change the rhizosphere microbial community. When wheat is sown in the same field, the microbial community structure has to change to aerobic resulting in alteration in soil biological equilibrium. Wheat-rice ecosystem is therefore of central interest to explore for sustainable agriculture. In view of its global significance in agriculture production and human health, wheat agro ecosystem has been studied extensively from the point of view of bacterial diversity during the last 15–20 years across various regions of the world. In a study involving rhizosphere of *Triticum monococcum* (an ancient wheat cultivar), *T. aestivum* cv Red File (a historical cultivar), and *T. aestivum* cv CDC Teal (a modern cultivar), a continuum in microbial diversity from the ancient races to modern cultivar, was observed.

11. BIODIVERSITY PERSPECTIVE

Nearly half of the 17 ecosystem services within the 16 biomes distinguished as making up the earth's ecosystem are dependent, either directly or indirectly, on life and activities of microorganisms. In terms of the necessity of services, (i) that the biological diversity provides including geochemical cycles and climatic regulation, and (ii) on which is dependent the very existence of human society, it is not too large a cost for nature's bounty. Besides the recognized hot spots like the Western Ghats and north-eastern hill region, India is endowed with other rich biodiversity locales like the Bastar region inhabited by tribals, the mangrove forests of Sunder Bans, wet evergreen rainforest of silent valley of Kerala, Chilka Lake in Orissa, thermal springs in the Central Himalayan region, that are abode to large unexplored microbial diversity. That a major component of biodiversity in the country could be microbial is borne out by the data of the Western Ghat region by Gadgil wherein he has shown that for a possible taxonomic component of a set of 2500 species, 17.6% are microorganisms. The existing culture collections carry a very small component of this diversity but are gradually making hectic efforts to come to terms with the requirements of human resource training, conservation strategies and new leads in exploitation. In particular, the efforts made by the Microbial Type Culture Collection (MTCC) and Gene Bank at the Institute of Microbial Technology, Chandigarh in this direction gives definite hope. In addition, the establishment of a National Bureau of Agriculturally Important Microorganisms (NBAIM) by the Indian Council of Agricultural Research is another positive step in this direction.

12. PESTICIDES

Most of the pesticides in common use in agriculture have been widely tested and there is a plethora of literature which conclusively proves that at recommended doses of application, they have no adverse influence on microbial activity. Hart and Brooks (1996) found in U.K. that 19 years of cumulative annual field application of five pesticides (benomyl, chlorfenvinphos, aldicarb, triadimefon and glyphosate) used continuously either singly or in combination applied at or slightly above the recommended rate had no measurable harmful effects on the soil microbial biomass or its activity. More studies are needed to assess the effects of pesticides on microbial diversity since there are reports of both lack of effects of fungicides and herbicides as well as measurable effects on diversity (Johnsen, *et al.,* 2001).

Table 3: Effect of insecticides on the diversity of micro flora in rice field soil.

Treatments	Bacterial population (x 10^5g–1 soil)	Actinomycetes population (x 10^5g–1 soil)	Fungi population (x 10^4g–1 soil)	N-fixing bacterial population (x 10^4 g–1 soil)	P-solublizing organisms (x 10^5 g–1 soil)
	At 60 days	At 60 days	At 60 days	At 60 days	At 60 days
Control	121	101	13	63	60
HCH	174	144	13	69	95
Phorate	153	132	14	68	86
Carbofuran	129	106	10	60	70
Fenvalerate	116	137	11	61	00
Initial value	102	75	8	40	44

Source : Biodiversity Strategy Action Plan, W.B.

13. PRESENT SCENARIO OF MICROBIAL DIVERSITY IN THE COUNTRY

Various ecosystems that have been surveyed for microbial denizens include, marine coastal areas, endorhizosphere, eriophid mites, fishes, forest plants, freshwater, soil, termite mounds, tree bark, sediments, rumen of domesticated animals, grasslands, food and associated habitats, extreme ecosystems, agricultural crops and anthropogenic systems such as effluents and hospital wastes. Unfortunately, even this little information is available with scientists often working solitarily with almost no interaction. However, considering the varied agro climatic regions in the country, much greater thrust is required to search for the existing and novel/new forms.

The problem is that most exploratory investigations are carried out with the help of graduate and postgraduate students, which are a mobile lot. Therefore, any gene pool recovered during their course of study is generally short-lived. This leads to considerable loss of precious time, effort and money. Although there are several university and research centres which maintain germplasm of various microorganisms, they require financial and manpower support to become functional depositories and service centres.

The Department of Biotechnology has supported a Rhizobial Collection Centre at the Indian Agricultural Research Institute, New Delhi, a Marine Cyanobacterial Collection at the Bharathidasan University, Tiruchirappalli and a Mycorrhizal Collection Centre at the Energy and Resources Institute, New Delhi for some time now. Another major problem has been the availability of trained manpower, both academic and technical, to undertake explorations, characterization and conservation.

14. EMERGING ISSUES

It has been hypothesized that in a future scenario of global warming, the litter produced under elevated CO_2, having a high C:N ratio would be difficult to decompose, but experiments by van Ginkel and Gorissen (1998) showed that microorganisms were found to adopt to changing soil C input under elevated CO_2 and there was no effect on their turnover and behaviour. There were no differences between microbial communities from field plots that contained harvested transgenic canola plants and control plants indicating that the changes in the microbial community structure associated with genetically modified plants were temporary and did not persist into the next season (Dunfield and Germida 2001).

REFERENCES

Chaurasia, B., Pandey, A. and Palni, L.M.S. (2005a). Occurrence of Arbuscular Mycorrhizae in the rhizosphere of Himalayan Yew (*Taxus baccata* L. subsp. *wallichiana* (Zucc.) Pilger)- A Case Study. In: *Basic Research and Applications of Mycorrhizae* (Eds. G.K. Podila and A.K. Varma), IK International Pvt. Ltd., New Delhi, pp. 26-35.

Chaurasia, B., Pandey, A. and Palni, L.M.S. (2005b). Distribution, Colonization and Diversity of Arbuscular Mycorrhizal fungi in Rhododendrons of Central Himalayan region of India. *Forest Ecology and Management* **207(3):** 315-324.

Chaurasia, B., Pandey, A., Palni, L.M.S., Trivedi, P., Kumar, B. and Colvin, N. (2005). Diffusible and volatile compounds produced by an antagonistic *Bacillus subtilis* strain cause structural deformities in pathogenic fungi *in vitro. Microbiological Research* **160(1):** 75-81.

Doran, J.W., Coleman, D.C., Bezdicek, D.F. and Stewart, B.A. (Eds.). (1994). *Defining Soil Quality for a Sustainable Environment. Soil Sci. Soc. Am. Inc.* Wisconsin.

Doran, J.W. and Zeiss, M.R. (2000). Soil health and sustainability: managing the biotic component of soil quality. *Applied Soil Ecology* **15:** 3-11.

Jensen, S., Ovreas, L., Daae, F.L. and Torsvik, V. (1998). Diversity in methane enrichments from agricultural soil revealed by DGGE separation of PCR amplified 16S rDNA fragments. *FEMS Microbiology Ecology.* **26(1):** 17-26.

Pankhurst, C.E., Doube, B.M. and Gupta, V.V.S.R. (Eds.). (1997). *Biological Indicators of Soil Health.* CAB International, Wallingford.

Pankhurst, C.E., Doube, B.M., Gupta, V.V.S.R. and Grace, P.R. (Eds.). (1994). *Soil Biota: Management in Sustainable Farming Systems.* CSIRO, Australia.

Diversity of Lower Plants (2014): pp. 36-53
Editors: Rajan Kumar Gupta & Mukesh Kumar
IK International Publishing House Pvt. Ltd.

3

Cyanobacterial Biodiversity: A Profile

KAUSHAL KISHORE CHOUDHARY*, RAJAN KUMAR GUPTA** AND MUKESH KUMAR***

**Department of Botany, B.R.A. Bihar University, Muzaffarpur–842 001, Bihar*
***Department of Botany, Pt. L.M.S. Govt. P.G. College, Rishikesh–249 201 (U.K.)*
****Department of Botany, Sahu Jain P.G. College, Najibabad–246 763 (U.P.)*

ABSTRACT

Cyanobacteria originated in 'proterozoic era' and occupied all possible biomes on the earth during course of evolution. They possibly evolved as unicells and developed into complex filamentous and branched forms through colonial and parenchymatous forms. They acquired cell differentiation ability and developed two specialized cells – 'heterocyst' and 'akinetes' along with vegetative cells. Heterocyst enabled them to convert atmospheric nitrogen into nitrogenous compounds whereas akinete help them to endure environmental stresses. Although, freshwater and terrestrial habitats are the most productive environment for the proliferation of cyanobacteria, their occurrence in Antarctic and Arctic, thermal springs, brines and marine ecosystem, and alkaline and acidic conditions is well established. The potential of cyanobacteria to establish symbiotic associations with different classes of plants are the additional evolutionary adaptation. Diversifying different habitats, cyanobacteria produce and secrete a large number of biotechnologically important primary and secondary metabolites of human interest. Development in cyanobacterial potential and their great adaptability to diverse habitats have necessitated the documentation of cyanobacterial biodiversity. The adaptability of cyanobacteria is based on their enormous diversity in species and strains and their ability to synthesize structurally and functionally diverse natural products which they have passed on by horizontal gene transfer to other bacterial species during evolution. In spite of enormous biodiversity among cyanobacteria only a small proportion of cyanobacterial species and their natural products have been characterized to date.

This article presents the morphological and habitat biodiversity of cyanobacteria as well as their symbiotic associations with different classes of plants.

Keywords: Biodiversity, Cyanobacteria, Evolution, Morphology, Thermal springs, Symbiotic associations.

***Communicating Author

1. INTRODUCTION

Biodiversity is of immense value to human beings, as it provides food, medicine and industrial raw materials. Biodiversity regulates the ecosystem functioning (Loreau *et al.*, 2001) and determines the distribution and dominance of flora and fauna in a given time and space. Nevertheless only this success of biotechnological programme is also dependent upon the richness of biodiversity. Biodiversity covers an array of organisms particularly phyto- and zoodiversity. Phytodiversity is the most important component of the biosphere that supports life. They are important constituents of nature that regulate the nutrient cycling and builds up the platform for survival of heterotrophic populations. Cyanobacteria, being primary producers, occupy most vital position in the ecosystem as they regulate the ecosystem functioning at greater extent. Discoveries in recent years have disclosed the importance of cyanobacteria in context of primary production and global nitrogen cycling. Cyanobacteria have been believed to be the first oxygen-evolving photoautotroph that modified an anaerobic to aerobic atmosphere. Recent development in cyanobacteria has revealed that these tiny organisms are helpful in combating the challenges arising due to deterioration in air, water and soil ecosystem. The current article has been aimed to describe diversity of cyanobacteria in different habitats together with small emphasis on cellular structure and morphological organization.

2. CYANOBACTERIA

Cyanobacteria are photosynthetic oxygen-evolving and geographically widespread prokaryotes that share the metabolic property of the eukaryotes and cellular structure of prokaryotes. Cyanobacteria colonized the earth's surface in extreme conditions and build up the platform for the proliferation of oxygenic plants and higher forms. Cyanobacteria show plasticity in their genomic and morphological organization. They play a significant role in primary productivity in marine and freshwater ecosystems. With the establishment of agronomic potential of cyanobacteria, distribution (Singh *et al.*, 2001), diversity (Choudhary, 1999) and their role as biofertilizer (Mishra *et al.*, 2005), bioremediating agents (Micheletti *et al.*, 2008) and for bioactive compounds (Choudhary and Bimal, 2008; Mihali *et al.*, 2008) was extensively studied.

2.1 Evolution of Cyanobacteria

Cyanobacteria are oxygenic photosynthetic Gram-negative bacteria formerly known as blue-green algae as they were long thought to belong to the plant kingdom. A feature shared by cyanobacteria and plants is their ability to perform photosynthesis. According to the endosymbiont theory, cyanobacteria are in fact the origin of chloroplasts, which harbour the plant's photosynthetic machinery. As evidenced by fossils records, cyanobacteria are among the earliest life forms on earth and are involved in the formation of the earth's atmosphere. Up to now the production of oxygen by cyanobacterial photosynthesis contributes to keeping the balance of our atmosphere.

Cyanobacteria have been supposed to dominate the earth long before the origin of most of the present-day life forms. Cyanobacteria originated about 3.5 billion years ago in "proterozoic era" known as the 'Age of cyanobacteria' (Schopf, 2000). During their long evolution, these organisms colonized all possible illuminated habitats. It is proposed that the first appearing cyanobacteria were unicellular and freshwater forms (Sanchez-Baracaldo, 2005). Cyanobacteria have also been supposed to be the first organism to carry out oxygenic photosynthesis thereby releasing oxygen into the atmosphere. This event subsequently changed the anaerobic into aerobic environments and prepared the platform for the present-day oxygen dependent life forms. Moreover, a strong evolutionary relationship has been proposed between cyanobacteria and higher plants (Giovannoni *et al.*, 1988; Martin *et al.*, 2002). It is assumed that the cyanobacteria have given rise to photosynthetic organelles in higher plants through symbiosis (Margulis, 1970).

Cyanobacteria are recognized for their ability to convert atmospheric elementary nitrogen into soluble and available forms of ammonia. Cyanobacteria are a fascinating group of organisms that are independent for carbon and nitrogen. The recent advances in cyanobacterial metabolisms have revealed that the nitrogen-fixing ability is not only confined to the heterocystous forms rather this property is also attributed to unicellular and non-heterocystous forms of cyanobacteria (Bergman *et al.*, 1997; Taniuchi *et al.*, 2008). Although cyanobacteria were studied extensively for their nitrogen-fixation in terrestrial and freshwater environments, marine cyanobacteria were recognized only after the establishment of nitrogen-fixing ability of non-heterocystous cyanobacterium, *Trichodesmium* sp. (Karl *et al.*, 2002; Zehr *et al.*, 2001). The recent approaches have also considered the significance of benthic cyanobacteria in primary productivity and nitrogen fixation in microbial mats (Yannarel *et al.*, 2007: Ley *et al.*, 2006). Cyanobacteria originating in freshwater ecosystem have now occupied all kinds of possible biomes.

2.2 Structure, Organization and Cyanobacterial Morphology

Cyanobacteria were originally considered algae because of their microscopic morphology, pigmentation and oxygen evolving photosynthesis in which PS II and PS I are connected in series. Cyanobacterial genome size ranges between 1.6×10^9 and 8.6×10^9 Da which is comparable to bacteria (1.0 to 3.6×10^9 Da) (Herdman *et al.*, 1979). Cyanobacterial morphology ranges from simple unicellular to multicellular, coccoid to profusely branched filaments and colourless to intensely pigmented. Unicellular forms of cyanobacteria show diversity in their arrangement and provide different shapes to resulting colonies or aggregation of cells. Unicellular forms of cyanobacteria are represented by spherical, ovoid and cylindrical cells. The cells, in advanced forms, aggregate together by means of gelatinous substances secreted during growth and give rise to colonial forms. The colony may further be encased in a gelatinous matrix giving rise to more ordered forms of cyanobacteria. The primitive forms of unicellular cyanobacteria are represented by single cells whereas in colonial forms the number of cells may be indefinite and unarranged or definite and arranged manner. The arrangement of cells in colonial forms varies from species to species and is determined by the cell division. Sometimes cells may adhere together to give rise to mats in the fields.

The filamentous forms are represented by members of *Nostocales* and *Stigonematales*. The filaments of cyanobacteria may be either unbranched, falsely branched or truly branched. Filamentous forms may further be divided into heterocystous and non-heterocystous. The unbranched filamentous forms (including falsely branched forms) are represented by *Nostocales* and branched forms by *Stigonematales*. Heterocystous forms are represented by *Nostocaceae* and *Scytonemataceae* (falsely branched) (*Nostocales*) and *Stigonematales*. Filamentous forms are produced by repeated cell division in a single plane (*Nostocales*) or more than one plain (*Stigonematales*) at right angles to the main axis of the filament. Repeated division of cells gives rise to chain of cells that is called a trichome. The trichome may be straight, curved or coiled. Cell shape and size differ among the filamentous cyanobacteria. The cellular composition in filamentous forms may be homogeneous or heterogeneous. The homogeneous forms consist of similar type of cells whereas heterogeneous forms show variability in their cellular composition. The members of *Oscillatoriaceae* show homogeneous forms of filaments whereas members of *Nostocaceae*, *Scytonemataceae* and *Stigonematales* are characterized by heterocysts and akinetes. Heterocysts composed of thick wall and hyaline protoplast provide ability to fix atmospheric nitrogen to ammonium forms whereas akinetes composed of thick walled and reserve materials are developed during unfavourable conditions that help in tolerating the adverse conditions. In the order Stigonematales, the trichomes are multiseriate and orderly branched. The filaments consist of heterocysts and akinetes.

Table 1: Basic morphology and reproduction among the orders of cyanobacteria (Fritsch, 1945)

Order (Family)	Morphology	Reproduction	Plain of division	Genera
Chroococcales	Unicellular or colonial	Binary fission	Single Two/mores	*Gloeothece*, Synechococcus, *Gloeocapsa*, *Microcystis*
Chamaesiphonales	Unicellular or colonial	Budding or multiple fission		*Chamaesiphon*, *Dermocarpa*, *Chroococcidiopsis*
Pleurocapsales	Unicellular or colonial	Budding or multiple fission		*Hyella*, *Pleurocapsa*, *Xenococcus*
Nostocales (Oscillatoriaceae)	Filamentous, non-heterocystous	Trichome fragmentation, hormogonia	Single	*Lyngbya*, *Oscillatoria*, *Phormidium*
(Nostocaceae)	Filamentous, heterocystous	Trichome fragmentation, hormogonia, akinete	Single	*Anabaena*, *Nostoc*, *Cylindrospermum*
(Scytonemataceae)	Filamentous, heterocystous, false branching	Trichome fragmentation, hormogonia, akinete		*Calothrix*,*Rivularia*, *Scytonema*
Stigonematales	Branched, filamentous, heterocystous	Trichome fragmentation, hormogonia, akinete	Two or more	*Hapalosiphon*, *Stigonema*, *Westiellopsis*

The multiplication in cyanobacteria is asexual. In filamentous forms, the multiplication is carried out by trichome fragmentation or by formation of special hormogonia. Hormogonia differ from fragmented filaments and show gliding movements upon their liberation. Hormogonia gradually develop into new trichomes. The basic morphology, reproduction and plain of division occurring in principal groups of cyanobacteria have been described in Table 1.

2.3 Cellular Structure of Cyanobacteria

Cyanobacteria differ from eubacteria in having chloroplasts and from eukaryotes in lacking double layer membrane-bound organelles. Their wall structure is composed of peptidoglycan.

They do not possess a membrane-bound nucleus and lack histone protein associated with nucleic acid. The protein synthesis machinery, i.e. ribosomes are 70S (Allen, 1984) rather than 80S as in eukaryotes. The photosynthetic activity remains localized in thylakoids distributed throughout the periphery of the cell membrane in the cytoplasm. Thylakoids are pigmented structure of the cyanobacteria. The presence of pigments provides them different colours ranging from blue-green to violet-red.

Cyanobacteria are pigmented photosynthetic oxygen evolving organisms capable of carbon fixation with the help of chlorophyll *a* in the presence of light. The pigments are localized in the thylakoids that lie free in the cytoplasm or remain attached to the plasma membrane. The colours of the cells vary from blue-green to violet-red depending on the quantity of the pigments. In addition to chlorophyll *a*, cyanobacteria possess carotenoids and accessory pigments phycobiliproteins (phycocyanin, allophycocyanin and phycoerythrin). The comparative presence of these pigments determines the color of the cells. The phycobiliproteins are embodied in phycobilisomes which are located on the outer surface of the thylakoids (Douglas, 1994). All cyanobacteria contain chlorophyll *a* and phycocyanin. Like eukaryotes, cyanobacteria possess two reaction centres, i.e. photosystem I (PS I) and photosystem II (PS II) in their photosynthetic apparatus. Carotenoids have been supposed to protect the cell from triplet oxygen arisen due to radiation of light. Production of phycobiliproteins (phycocyanin. allophycocyanin and phycoerythrin) is influenced by environmental factors particularly light quality. Chromatic adaptation is largely attributable to a change in the ratio between phycocyanin and phycoerythrin in the phycobilisomes. The presence of pigments along with UV-absorbing pigments makes them sustainable in wide range of habitats.

Cyanobacteria can store essential nutrients and metabolites within their cytoplasm. The electron microscopy study of cyanobacterial cell revealed the presence various prominent cytoplasmic inclusions within their cytoplasm. The important among them are glycogen granules, lipid globules, cyanophycin granules, carboxysomes and polyphosphate bodies (Fay and Baalen, 1987). Reserve foods are accumulated under conditions of excess supply of particular nutrients. The accumulation of glycogen and lipids takes part in absence of supply of nitrogen source.

The nitrogen-fixation among some of the cyanobacteria makes them simplest of living system with respect to nutritional requirements. Certain group of cyanobacteria has been designed with the heterocyst that empowers them to fix atmospheric nitrogen into ammonia. This activity is performed by cyanobacteria with the help of enzyme 'nitrogenase complex' embodied in the heterocyst (Adams and Duggan, 1999). Besides, non-heterocystous forms of cyanobacteria also fix nitrogen under nitrogen-limited conditions (Karl *et al.*, 2002).

Some of the aquatic species of cyanobacteria possess gas vesicles that regulate buoyancy in them. Gas vesicles are cylindrical structures filled with gases. The gas vesicles provide them the ability to maintain them in vertical position in the water column. Gas vesicles assist cyanobacteria in establishing them in vertical column of water against the gravitational, photic, thermal and chemical composition of the water bodies. The gas vesicles increase with decrease in light intensity and slow growth. The other means of movement in cyanobacteria is associated with photomovement by slime secretion.

3. CYANOBACTERIA IN NATURE

Cyanobacteria are geographically widespread in freshwater, marine and terrestrial habitats. Their habitats range from autotrophic to heterotrophic, psychrophilic to thermophilic, alkylophilic to acidophilic, planktonic to barophilic and freshwater to marine (Thajuddin and Subramanian, 2005). Cyanobacteria occur in nature as free-living, endosymbionts and epiphytic forms. They are reported to survive in long periods of darkness and certain cyanobacteria show a distinct ability for heterotrophic nutrition (Fay, 1965). They are dominant forms of many terrestrial and aquatic environments where they play a significant role in maintaining functional processes of ecosystems and nutrient cycling (Whitton, 1992).

The prominent habitats of cyanobacteria are sodic and marine environments of tropical and subtropical regions of the world. During their evolutionary period, they have adopted themselves in freshwater, marine, hot springs and cold regions of Antarctic and Arctic. They inhabit the environments where no other organisms including microbes can survive. They are the most important constituents of marine ecosystems that regulate the nutrient cycling. Most marine forms dominate along the shores as benthic vegetation. Marine forms are characterized by the ability to withstand the salty condition of the ecosystems. Some of the freshwater forms have also been reported to withstand the high concentration of sodium salts. Cyanobacteria also dominate the epilimnion and metalimnion regions of the ponds and estuaries. Others may colonize the rock surface whereas some other forms might form mat on the surface of rocks and soils.

Cyanobacteria have an extraordinary ability to colonize the infertile substrates such as volcanic ash, desert sand and rocks (Garcia-Pichel *et al.*, 2001). Their ability to flourish the extreme conditions of high temperature and cold make them an extraordinary living system of the nature. They have been found to be present in the thermal springs (Castenholtz, 1996), mountain streams (Power *et al.*, 1988), Arctic and Antarctic lakes (Taton *et al.*, 2006) and snow and ice (Laamanen, 1996). Cyanobacteria also persist in symbiotic associations with plants of different groups. Symbiotic relations with algae, fungi, bryophytes, pteridophytes, gymnosperms and angiosperms are of common occurrence in the nature (Rai, 1990). The endosymbiotic association of photosynthetic cyanobacteria has led to the assumption of origin of photosynthetic eukaryotes with being engulfed by a phagotrophic host (Douglas, 1994). The ability to survive at high temperature in hot springs support the early evolution of cyanobacteria and being the first organisms on the earth to release the molecular oxygen for origin of other biological diversity.

3.1 Cyanobacteria in Freshwater Ecosystems

Cyanobacteria belonging to orders *Chroococcales* and family *Oscillatoriaceae* and *Nostocaceae* are commonly found in freshwater as planktonic forms (Thajuddin and Subramanian, 2005). Freshwater cyanobacteria have been reported from sea level to high altitudes. The buoyancy nature of the cyanobacteria makes them compatible for flourishing on water surface. The presence of gas vacuoles helps in floating on water surface and supports the buoyancy nature of the organisms. They are abundantly found in temple tanks, ponds, and lakes. The recent studies on freshwater cyanobacteria have demonstrated that the freshwater cyanobacteria are associated with bloom formation in the water bodies marked by change of colour of the water surface. The

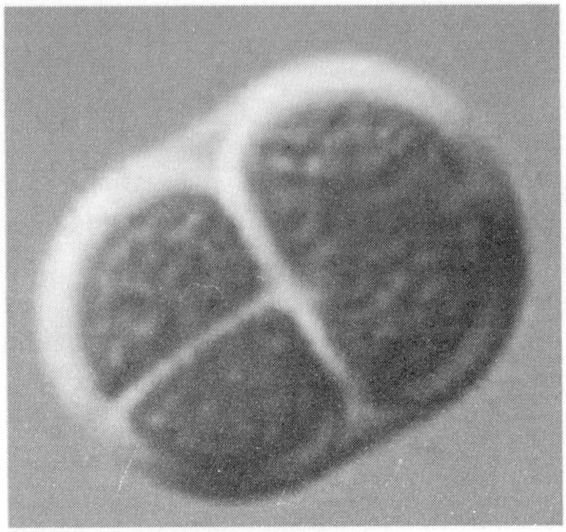

Chroococcus

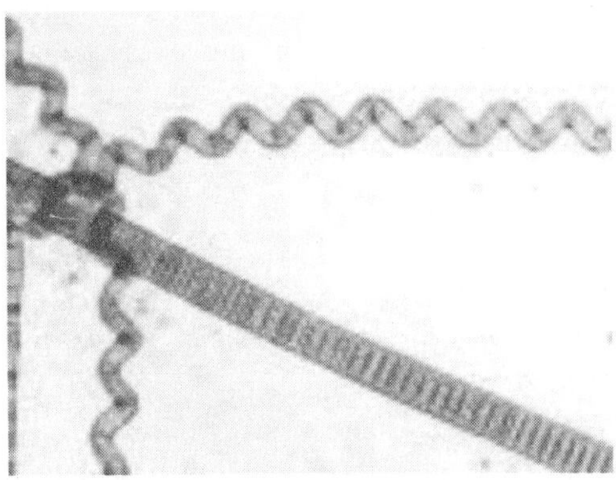

Spirulina with a filament of *Oscillatoria*

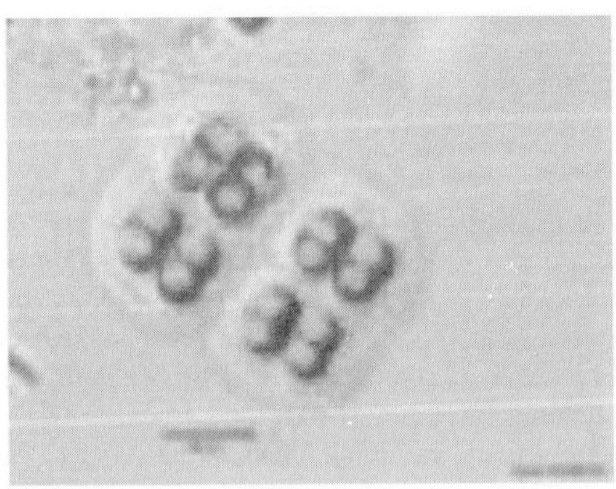

Gloeocapsa

common cyanobacterial forms appearing in the bloom affected water bodies are *Anabaena*, *Anabaenopsis*, *Aphanizomenon*, *Arthrospira*, *Coelospaerium*, *Gloeotrichia*, *Lyngbya*, *Microcystis*, *Nostoc*, *Nodularia*, *Oscillatoria*, *Spirulina gomontina* and *Lyngbya*. *Microcystis* is the common dominant form found in all kinds of bloom affected water bodies producing toxic substances. The source of nitrogen favouring growth of non-heterocystous forms of cyanobacteria in bloom forming water reservoirs are the drainage from the surrounding areas, precipitation bringing in dissolved free ammonia from atmosphere and autochthonous release of nitrates from the dead populations. Nitrates form a preferred non-toxic source of combined nitrogen for most cyanobacteria and are produced as a result of high rate of nitrification under waterlogged conditions. Gas-vacuoles containing cyanobacteria forms dense growth on the water surface of ponds, lakes and water reservoirs and cause serious nuisance because of visual appearance, production of toxins (Carmichael, 1994) and unpleasant odour produced by substances such as geosmin (Juttner, 1987). The appearance of bloom-forming cyanobacteria in stagnant water such as ponds, lakes and water reservoirs cause blockade of drinking water supplies by choking the flow of water in sand filters and cause mortality of fish and animals by creating anoxia or hypoxia inside the water bodies.

3.2 Cyanobacteria in Terrestrial Ecosystems

Although cyanobacteria inhabit a variety of terrestrial environments, soil is the most productive habitats for the cyanobacterial growth particularly moist or waterlogged conditions. Cyanobacteria have been reported from first ten inches of topsoil, from moist soil surfaces and waterlogged soils as epiphytic forms. Among the different habitats, the rice field is the ideal habitat for the growth of cyanobacteria particularly nitrogen-fixing forms. The common nitrogen-fixers reported from the rice fields throughout the world belonged to genus *Anabaena*, *Aulosira*, *Calothrix*, *Cylindrospermum*, *Gloeocapsa*, *Nostoc*, *Rivularia*, *Scytonema* and *Tolypothrix* (Choudhary, 2009; Singh *et al.*, 2001). Besides, certain non-heterocystous and unicellular forms that belong to genus *Lyngbya*, *Microcoleus*, *Porphyrosiphon* and *Schizothrix* are common in rice fields. Cyanobacteria have also been reported from sugar cane and maize fields in India. Cyanobacteria are very important constituents of the soil that forms biological soil crusts. Cyanobacteria help to reduce erosion by helping bind the particles of soil together. Cyanobacteria also contribute in increasing the retention of rainwater and reducing the soil water loss by evaporation. When conditions are wet, they become active and move through the soil, leaving behind sticky filaments which particles of soil cling to. When the filaments become wet, they absorb water and swell up to ten times their original size which helps to store moisture within the upper layer of soil where many plants root systems and other organisms live. Cyanobacteria also play a direct role in aiding plant survival and growth. Levels of nitrogen in arid environments are often low, and limit the growth of plants. Cyanobacteria are important because they help to convert atmospheric nitrogen into a form that can be utilized by plants. In this conversion, nitrogen (N_2) from the air is fixed into an organic form such as nitrate or ammonia. Many cyanobacteria have formed symbiotic relationships with other plants. Cyanobacteria enable plants to photosynthesize and are responsible for the plant's green colour. This relationship between plant and cyanobacteria is believed to have first occurred during the Precambrian era. Cyanobacteria live in the roots of some plants such as legumes as well, and provide nitrogen directly for plants to use. They have also formed symbiotic relationships with fungi and lichens.

3.3 Cyanobacteria in Marine Ecosystems

Cyanobacteria are widespread and abundant in most marine habitats. Their ability to grow in seawater is presumably related to a preference for alkaline conditions and an ability to tolerate high salt concentrations. Thajuddin and Subramanian (1992) have extensively surveyed the marine cyanobacterial biodiversity of shore, deeper sea, stagnant seawater ponds and puddles, backwater and saltpans of India. They reported six new species of marine cyanobacteria, viz. *Ammeatoidea normanii, Dichothrix spiralis, Oscillatoria borneti, Pseudoanchobrosa fluminenensis, Radaisia violacea* and *Siphononema polonicum*, from Indian coasts. Cyanobacteria of marine ecosystems were commonly represented by *Calothrix scopularum, Lyngbya confervoides, L. martansiana, Microcoleus chthonoplastes, M. acustissimus, Oscillatoria nigroviridis, O. salina, O. tenuis, Phormidium valderianum, Spirulina subsalsa* sp. *labynithiformis and Pseudanabaena schemidleii* (Thajuddin and Subramanian, 1992, 1995). A bloom forming cyanobacteria, *Trichodesmium*, have been reported from tropical marine ecosystems (Hoffman, 1999; Thajuddin and Subramanian, 2002). Marine forms of cyanobacteria are a rich source of diverse bioactive compounds. Some of them are important as herbivore deterrents (Hoffman, 1999).

3.4 Cyanobacteria in Brines

The brines are most unusual and highly saline extreme environments that are hostile to most life forms. These environments provide an ideal habitat for organisms that are halophilic (salt loving). Cyanobacteria are one of the dominating forms of hypersaline environments. They are represented by non-heterocystous forms. No heterocystous forms have been reported from hypersaline environments. The dominating forms in hypersaline environments are *Microcoleus* sp. and *Lyngbya* sp. Besides, *Aphanothece halophytica, Synechocystis, Schizothrix* and *Oscillatoria* sp. are common along with little abundance of *Pleurocapsa* and *Johannesbaptistia* sp (Paerl *et al.*, 2003). Andrea *et al.* (2008) have reported the abundance of *Phormidium* and *Geitlerinema* in the Great Salt Plains (GSP) in north-central Oklahoma, USA. *Synechococcus* sp. reported from hypersaline alkaline lakes of Wadi Natrun perform anoxygenic photosynthesis in the presence of sulphide (Imhoff *et al.*, 1978). In Indian saltworks, non-heterocystous and filamentous forms such as *Lyngbya majuscula, Oscillatoria salina* and *Spirulina platensis* is common surface-scum forming species on pond's surface (Rahaman *et al.*, 1993). The hypersaline cyanobacteria

Algae growing in saline extreme environments

Cyanobacteria growing in saline water

produce and secrete the high concentration of oxycarotenoids, especially canthaxanthin (Cohen and Ideses, 1989), and massive amounts of slime (De Phillips *et al.*, 1993) of industrial significance.

3.5 Cyanobacteria in Hot Springs

Cyanobacteria inhabit the environmental condition near-boiling temperatures of hot springs in highly organized microbial mats. Steunou *et al.* (2006) reported single-celled cyanobacterium, *Synechococcus*, from Yellowstone's Octopus spring having water temperature about 160°F (71°C). Three cyanobacterial strains belonging to genus *Chroococcidiopsis* were identified from hot springs of Northern Thailand (Hayashi *et al.*, 1994). The optimum temperature and pH for growth of one of these strains, TS-821, was around 50°C and 8 respectively. The laboratory study showed that the strain was capable of dinitrogen fixation, and is tolerant against nitrite (5 mM), sulphate (0.6 mM) and a high concentration of carbon dioxide in the gas phase (0.4 atm). Sompong *et al.* (2005) identified 19 genera and 36 species from nine thermal spring districts in northern Thailand along the thermal gradient from 30 to 80°C. The strains which represented all springs were *Synechococcus lividus* and *Synechocooccus* sp. (> 40 and < 80°C) and *Phormidium boryanum* (> 30 and < 60°C). Recently, microcystin producing toxic strains of cyanobacteria representing *Oscillatoria limosa* and *Synechococcus lividus* from hot springs in Saudi Arabia was reported (Mohamed, 2008).

3.6 Cyanobacteria in Cold Ecosystems

Although cyanobacteria are known to colonize a remarkably wide thermal range of habitats, they are also found in polar tundra environments, ice shelves, glaciers and ice capped lakes (Fay 1983). In Antarctica, cyanobacteria often dominate the total ecosystem biomass and productivity of lakes, ponds and streams (Vincent and James, 1996; Wynn-Williams 1990). These cold-water environments are characterized by extreme variations in energy supply, from continuous darkness in winter to the short growing season of continuous light during summer. McLean (1918) concluded that there was a predominance of cyanobacteria in the Antarctic benthic and soil ecosystems relative to the Arctic. Cyanobacteria play a significant role in nitrogen and carbon economy of polar desert soils. The soils of Antarctic usually favoured the growth forms of *Nostoc commune* and members of Oscillatoriales along with *Nodularia*, *Calothrix*, *Scytonema* and *Tolypothrix*. The common cyanobacteria inhabiting the Arctic and the Antarctic region belong to unicellular, non-heterocystous filamentous and heterocystous filamentous forms. They are represented by *Anabaena*, *Aphanocapsa*, *Chroococcidiopsis*, *Gloeocapsa*, *Lyngbya*, *Mastigocladus*, *Microchaete*, *Microcoleus*, *Oscillatoria*, *Phormidium*, *Plectonema*, *Pseudanabaena*, *Nodularia*, *Nostoc*, *Rivularia*, *Schizothrix*, *Scytonema*, *Stigonema*, *Synechococcus* and *Tolypothrix* (Vincent, 2000).

3.7 Cyanobacteria in Extreme pH

Cyanobacteria have been reported from both extreme low pH (acidophilic) and high pH (alkaliphilic) condition. Gerasimenko *et al.* (1996) reported a wide diversity of alkaliphilic cyanobacteria from soda lakes of Tuva (16 genera and 34 species). Singh (1995) has purified the urease from the alkaliphilic cyanobacterium *Nostoc calcicola*. *Spirulina* is the common

Boiling hot springs in Yellowstone National Park. The orange-red coloration is caused by thriving colonies of photosynthetic cyanobacteria.

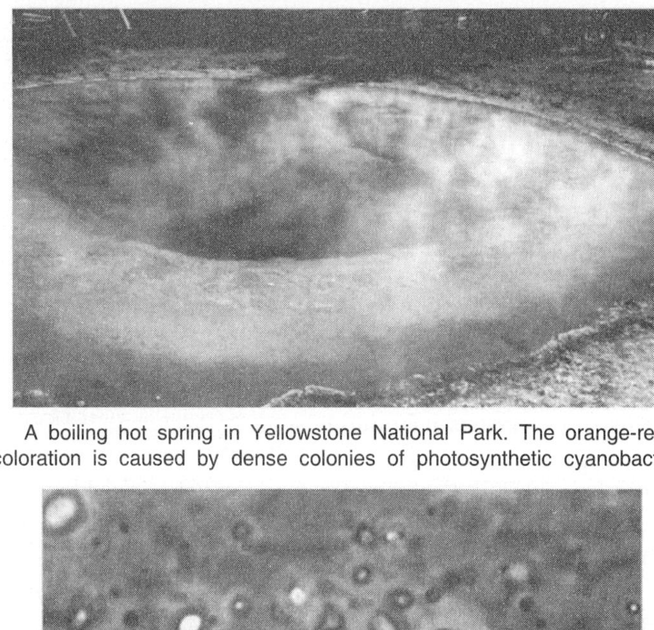

A boiling hot spring in Yellowstone National Park. The orange-red coloration is caused by dense colonies of photosynthetic cyanobacteria.

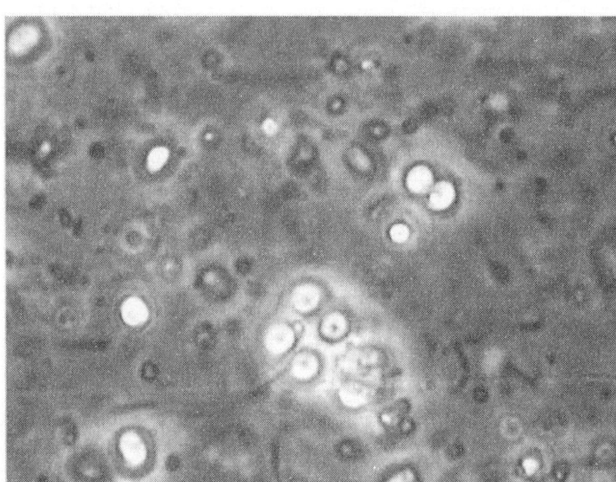

Coccoid cyanobacterium with encapsulated cell packets. *Cyanothece minervae*

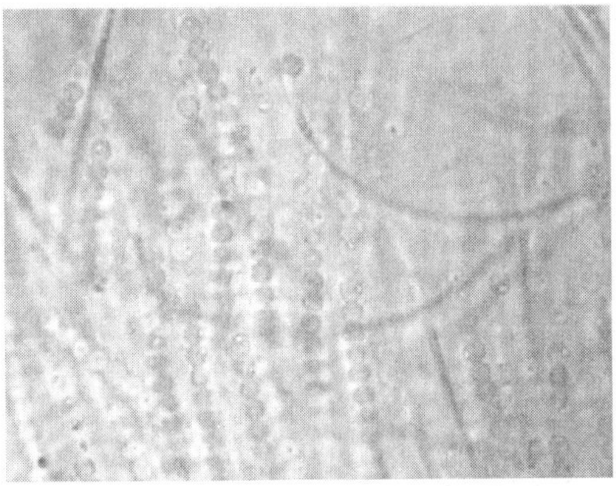

Filament bundled to form strands of an Anabaena-like cyanobacterium

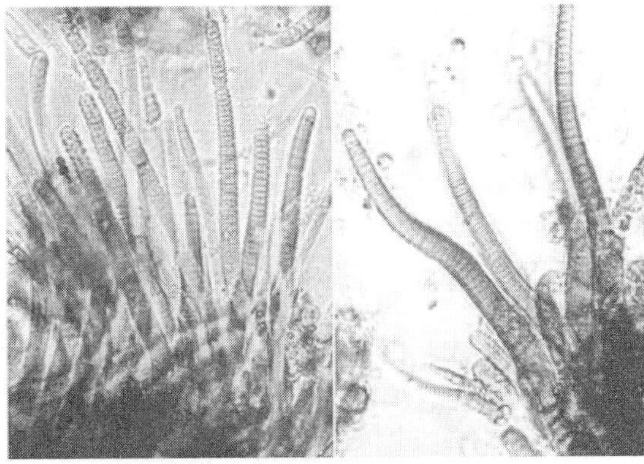

Magnified view of filamentous oscillatorian cyanobacteria resembling Rivularia

Lichens growing in Antarctica

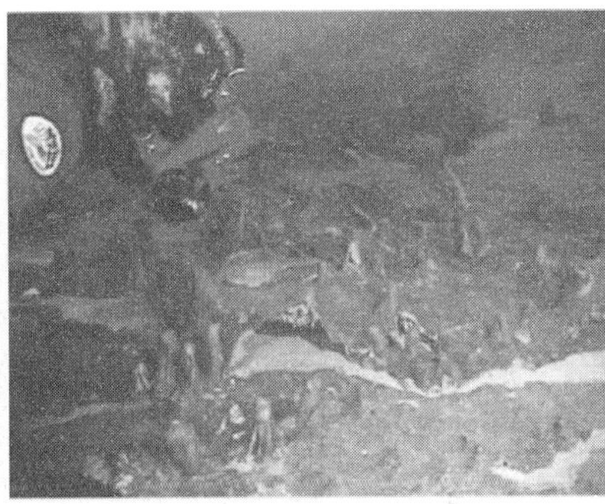

Cyanobacterial mats in cold oceanic beds

example of alkaliphilic cyanobacterium occurring in the soda lake and serving as human food of high nutritional value (Ciferri and Tiboni, 1985). *Synechococcus* sp. is also predominant in alkaline environments. Cyanobacteria are also present in acid lakes (pH 5-4.1) and have also found to dominate at low pH. Steinberg *et al.* (1998) have defined *Chroococcus turgidus*, *Mastigocladus* sp., *Oscillatoria* sp., *Spirulina* sp. (pH 2.9) and *Synechococcus* sp. (pH 4) in acid lakes in Germany. The existence of *Anabaena*, *Westiellopsis*, *Nostoc*, *Oscillatoria*, *Microcystis*, *Calothrix* and *Phormidium* in acidic rice grown soils (pH 4.9-6.2) of Tamil Nadu from India is well established (Gopalaswamy *et al.*, 2007).

3.8 Cyanobacteria in Associations

Cyanobacteria have distinctly been reported in association with different groups of plants and animals. They associate with algae, fungi (lichens), bryophytes, pteridophytes, gymnosperms, angiosperms and animals to form nitrogen-fixing symbioses (Rai *et al.*, 2000). The endosymbiotic association of cyanobacteria with coral reef sponges was reported by Wilkinson

Cyanobacteria growing in extreme pH

(1978). Wilkinson and Fay (1979) detected the nitrogenase activity in two sponges with cyanobacteria but not in sponges without cyanobacteria. Symbiotic associations of blue-green algae with other algae are also prevalent in nature. Some examples are endosymbiotic associations of small filamentous cyanobacterium *Richelia intracellularis* in cells of the marine diatom *Rhizosolenia* spp., *Nostoc symbioticum* in a siphonous green alga, *Geosiphon pyriformis*. Some species such as *Calothrix*, *Cyanodictyon*, *Lyngbya* and *Phormidium* occurs in the mucilage of other algae (Thajuddin and Subramanian, 2005). *Rhopalodia gibba* is a freshwater centric diatom in which unicellular cyanosymbiont, *Cyanothece* appears intracellularly as inclusion bodies (Prechtl *et al.*, 2004).

Cyanobacterial associations with fungi (lichens) particularly with ascomycetes and basidiomycetes are important events of natural evolution. Green algae may also form such associations with fungi (lichens). Thus, lichens are the associations of two organisms, a fungal component (mycobiont) and a green alga or cyanobacterium (phycobiont). The most common cyanobacteria associated with lichens are *Calothrix*, *Dichothrix*, *Nostoc*, *Scytonema* and *Stigonema* (Hitch and Millbank, 1975; Millbank, 1977). Millbank (1977) reported the associations of *Calothrix* sp with *Lichina*; *Dichothrix* with *Placynthium*; *Nostoc* sp. with *Collema*, *Leptogium*, *Lobaria*, *Peltigera*, etc; *Scytonema* with *Dendriscocaulon*, *Massalongia* and *Polychidium*; and *Stigonema* with *Ephebe*.

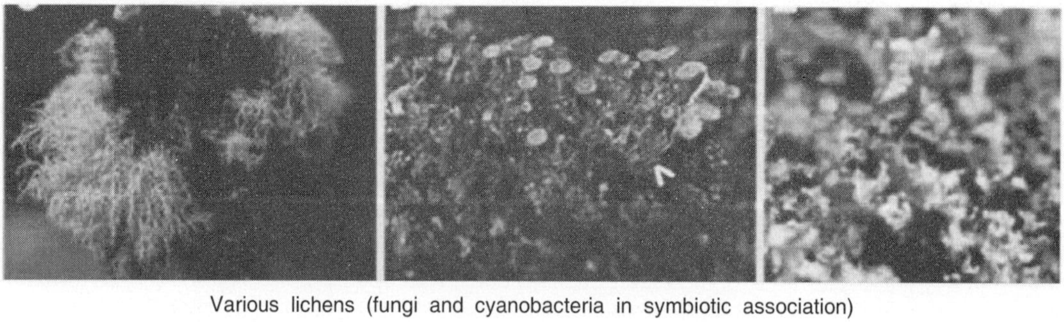

Various lichens (fungi and cyanobacteria in symbiotic association)

Cyanobacteria have been reported to form epiphytic or endophytic associations with some of the small non-vascular plants of bryophytes. Meeks (1990) has reported the association of cyanobacteria with only four liverworts (out of 340 genera of liverworts). Out of four, two genera viz. *Merchantia* and *Porella* forms epiphytic associations whereas *Blasia* and *Cavicularia* forms endophytic associations with cyanobacteria. Furthermore, Meeks (1990) has reported the endophytic associations of cyanobacteria with four genera, viz. *Anthoceros*, *Phaeoceros*, *Notothylus* and *Dendroceros* out of six hornwort genera. The common symbiotically competent cyanobacteria are hormogonia-forming strains of genus *Nostoc*, although *Calothrix* and *Chlorogloeopsis* strains have been shown to reconstitute the symbiosis with *Blasia* and *Phaeoceros* (Adams *et al.*, 2006; West and Adams, 1997). The mosses like *Sphagnum*, *Bryum* sp., *Weisia controversa*, *Funaria hygrometrica*, *Polytrichum juniperinum*, *Ceratodon purpureus*, *Campylopus introflexus* have been also reported to exhibit epiphytic or endophytic associations with cyanobacteria like *Nostoc*, *Anabaena*, *Scytonema*, *Chroococcus*, *Phormidium*, *Oscillatoria* and *Lyngbya* (Madhusoodan and Dominik, 1996; Solheim and Zielke, 2002). Cyanobacterial associations with pteridophytes have been well studied with aquatic ferns *Azolla*. The common cyanobiont present in the leaf cavity of *Azolla* is a filamentous and heterocystoys cyanobacterium, *Anabaena azollae* (Braun-Howland and Nierzwicki-Bauer, 1990). The *Azolla* plant is extensively used as biofertilizer due to nitrogen-fixation and biomass production.

The gymnosperms belonging to family *Cycadaceae* are characterized by the nine genera and 150 species that produce coralloid roots. Coralloid roots are morphologically different from

Cycas forming nodules in roots

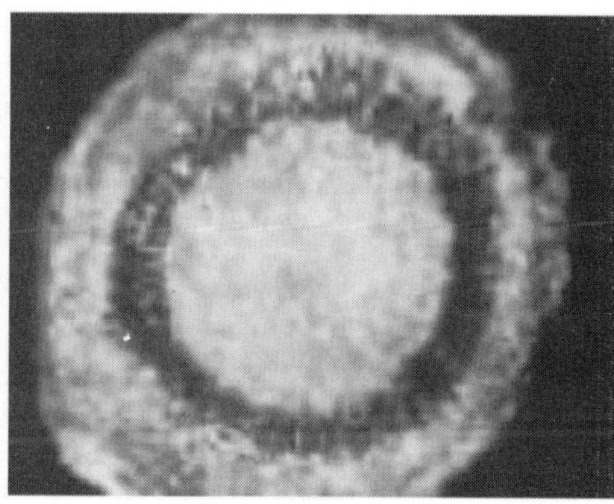

Cross section of root nodule

Gunnera growing in Hawaii

Azolla having Anabaena azollae

normal roots and contain nodule-like structure. Cyanobacteria infect these roots and establish themselves intercellularly or intracellularly in coralloid roots. The common cyanobacteria infecting gymnosperms belong to genus *Anabaena* and *Nostoc*. Cyanobacterial associations were reported with *Ceratozamia*, *Cycas*, *Dioon*, *Macrozamia* and *Zamia* (Lindblad and Bergman, 1990; Ow *et al.*, 1999). Cyanobacterial associations with angiosperms are presently restricted to one monogeneric family, the Gunneraceae. The cyanobacterium that forms microsymbiont with genus *Gunnera* is the *Nostoc punctiforme* (Bergman *et al.*, 1992).

The above described cyanobacterial diversity in different habitats suggested that cyanobacteria has possibly adapted diverse habitats by modifying their metabolic activity. Diversifying different habitats, they produce and secrete a large number of primary and secondary metabolites of human interest. The cyanobacterial potential to produce metabolites in response to habitats have attracted worldwide cyanobacteriologists to exploit these tiny organisms for the present-day biotechnological programme and for basic understanding of metabolic alterations. This article will certainly help in providing an insight into the distribution of cyanobacteria in different ecosystems.

REFERENCES

Adams, D.G. and Duggan, P.S. (1999). Heterocyst and akinete differentiation in cyanobacteria. *New Phytol.* **144:** 3-33.

Adams, D.G., Bergman, B., Nierzwicki-Bauer, S.A., Rai, A.N. and Schubler, A. (2006). Cyanobacterial-plant symbiosis. In: Dworkin, M., Falkow, S., Rosenberg, E., Karl-Heinz, S. and Stackebrandt, E. (Eds.), *Prokaryotes* **1:** 331-363.

Allen, M.M. (1984). Cyanobacterial cell inclusions. *Ann. Rev. Microbiol.* **38:** 1-25.

Andrea, K., Julie, B., Mark, B. and William, H. (2008). Cyanobacterial diversity and halotolerance in a variable hypersaline environments. *Microb. Ecol.* **55:** 453-465.

Bergman, B., Johansson, C. and Söderbäck, E. (1992). The *Nostoc-Gunnera* symbiosis. *New Phytol.* **122:** 379-400.

Bergman, B., Gallon, J.R., Rai, A.N. and Stal, L.J. (1997). N_2 Fixation by non-heterocystous cyanobacteria. *FEMS Microbiol. Rev.* **19:** 139-185.

Braun-Howland, E.B. and Nierzwicki-Bauer, S.A. (1990). Biochemistry, physiology, ultrastructure and molecular biology of the *Azolla-Anabaena* symbiosis. In: A.N. Rai, (Ed). *Handbook of Symbiotic Cyanobacteria*, CRC Press, Boca Raton, FL, pp. 65-118.

Carmichael, W.W. (1994). The toxins of cyanobacteria. *Sci. Am.* **270:** 78-86.

Castenholz, R.W. (1996). Endemism and biodiversity of thermophilic cyanobacteria. *Nova Hedwigia Beih.* **112:** 33-47.

Choudhary, K.K. (1999). Ex-situ conservation of cyanobacterial germplasm of north Bihar, India, Ph.D. thesis, BRA Bihar University, Muzaffarpur, Bihar, India.

Choudhary, K.K. (2009). Ecological and biotechnological relevance of cyanobacteria. In: Gupta, R.K., Kumar, M. and Vyas, D. (Eds.), Soil Microflora, Daya Publishing House, New Delhi, pp. 324-339.

Choudhary, K.K. and Bimal, R. (2008). Phenolics production by cells of the cyanobacterium *Anabaena doliolum* Bharadwaza immobilized in calcium alginate. *Seaweed Res. Util.* **30:** 93-102.

Ciferri, O. and Tiboni, O. (1985). The biochemistry and industrial potentials of *Spirulina*. *Annu. Rev. Microbiol.* **39:** 503-526.

Cohen, Y. and Ideses, R. (1989). Hypersaline cyanobacterial mats: environmental manipulations for the production of specific carotenoids and specific polysachharides. In: Miyachi, S., Karube, Y. and Ishida, I. (Eds.), *Current Topics in Marine Biotechnology, Japanese Society for Marine Biotechnology*, pp. 135-138.

De Philippis, R., Margheri, M.C., Pelosi, E. and Ventura, S. (1993). Exopolysachharide production by a unicellular cyanobacterium isolated from a hypersaline habitat. *J. Appl. Phycol.* **5:** 387-394.

Douglas, S.E. (1994). Chloroplast origins and evolution. In: D.A. Bryant (Ed.). *The Molecular Biology of Cyanobacteria.,* Kluwer Academic Publishers, Dordrecht. 91-118.

Fay, P. (1965). Heterotrophy and nitrogen fixation in *Chlorogloea fritschii*. *J. Gen. Microbiol.* **39:** 11-20.

Fay, P. (1983). The Blue-greens (Cyanophyta—Cyanobacteria). Edward Arnold (Pubs.), Baltimore, MD. pp. 88.

Fay, P. and Van Baalen, C. (1987). *The Cyanobacteria.* Elsevier, Amsterdam, pp. 534.

Fritsch, F.E. (1945). The Structure and Reproduction of the Algae, Cambridge, England: Cambridge University Press.

Garcia-Pichel, F., López-Cortés, A. and Nübel, U. (2001). Phylogenetic and Morphological Diversity of Cyanobacteria in Soil Desert Crusts from the Colorado Plateau. *Appl. Environ. Microbiol.* **67:** 1902-1910.

Gerasimenko, L.M., Dubinin, A.V. and Zavarzin, G.A. (1996). Alkaliphilic cyanobacteria from soda lakes of Tuva and their ecophysiology. *Microbiology* **65:** 736-740.

Giovannoni, S.J., Turner, S., Olsen, G.J., Barns, S., Lane, D.J. and Pace, N.R. (1988). Evolutionary relationships among cyanobacteria and green chloroplasts. *J. Bacteriol.* **170:** 3584-3592.

Gopalaswamy, G., Karthikeyan, C.V., Raghu, R., Udayasuriyan, V. and Apte, S.K. (2007). Identification of acid-stress-tolerant proteins from promising cyanobacterial isolates. *J. Appl. Phycol.* **19:** 631-639.

Hayashi, N.R., Peerapornpisal, Y., Nishihara, H., Ishii, M., Igarashi, Y. and Kodama, T. (1994). Isolation and cultivation of thermophilic cyanobacteria from hot springs of Northern Thailand. *J. Ferment. Bioeng.* **78:** 179-181.

Herdman, M., Janvier, M., Rippka, R. and Stanier, R.Y. (1979). Genome size of cyanobacteria. *J. Gen. Microbiol.* **111:** 73-85.

Hitch, C.J.B. and Millbank, J.W. (1975). Nitrogenase metabolism in lichens. VII; Nitrogenase activity and heterocyst frequency in lichens with blue–green phycobionts. *New Phytol.* **73:** 239-244.

Hoffman, L. (1999). Marine cyanobacteria in tropical regions: diversity and ecology. *Eur. J. Phycol.* **34:** 371-379.

Imhoff, J.F., Hashwa, F. and Truper, H.G. (1978). Isolation of extremely halophilic phototrophic bacteria from the alkaline Wadi Natrun. *Arch. Hydrobiol.* **84:** 381-388.

Juttner, F. (1987). Volatile organic substances. In: Fay, P. and Van Baalen, C. (Eds.). *The Cyanobacteria*, Elsevier, Amsterdam, pp. 453-469.

Karl, D., Michaels, A., Bergman, B., Capone, D., Carpenter, E., Letelifer, R., Lipschultz, F., Paerl, H., Sigman, D. and Stal, L. (2002). Dinitrogen fixation in the world's oceans. *Biogeochemistry* **57/58:** 47-98.

Laamanen, M. (1996). Cyanoprokaryotes in the Baltic Sea ice and winter plankton. *Arch. Hydrobiol. Suppl. 117, Algological Studies* **83:** 423-433.

Ley, R.E., Harris, J.K., Wilcox, J., Spear, J.R., Miller, S.R., Bebout, B.M., Maresca, J.A., Bryant, D.A., Sogin, M.L. and Pace, N.R. (2006). Unexpected Diversity and Complexity of the Guerrero Negro Hypersaline Microbial Mat. *Appl. Environ. Microb.* **72:** 3685-3695.

Lindblad, P. and Bergman, B. (1990). The cycad-cyanobacterial symbiosis. In: Rai, A.N. (Ed.). *Handbook of Symbiotic Cyanobacteria,* CRC Press, Boca Raton, FL, pp. 137-159.

Loreau, M., Naeem, S., Inchausti, P., Bengtsson, J., Grime, J.P., Hector, A., Hooper, D.U., Huston, M.A., Raffaelli, D., Schmid, B., Tilman, D. and Wardle, D.A. (2001). Biodiversity and Ecosystem Functioning: Current Knowledge and Future Challenges. *Science* **294:** 804-808.

McLean, A.L. (1918). Bacteria of ice and snow in Antarctica. *Nature* **102:** 35-39.

Madhusoodan, P.V. and Dominik, T.K. (1996). Epiphytic cyanobacteria on mosses from Western Ghats of Kerala. *J. Econ. Tax Bot.* **20:** 355-359.

Margulis, L. (1970). Origin of eukaryotic cells: evidence and research implications for a theory of the origin and evolution of microbial, plant and animal cells on the Precambrian earth. New Haven, Yale University Press.

Martin, W., Rujan, T., Richly, E., Hansen, A., Cornelsen, S., Lins, S., Liester, D., Stoebe, B., Hasegawa, M. and Penny, D. (2002). Evolutionary analysis of Arabidopsis, cyanobacterial, and chloroplast genomes reveals plastid phylogeny and thousands of cyanobacterial genes in the nucleus. *Proc. Natl. Acad. Sci.* USA. **99:** 12246-51.

Meeks, J.C. (1990). Cyanobacterial-bryophyte associations. In: Rai, A.N. (Ed). CRC Handbook of Cyanobacteria CRC Press, Boca Raton, Florida, pp. 43-63.

Micheletti, E., Colica, G., Viti, C., Tamagnini, P. and De Philippis, R. (2008). Selectivity in the heavy metal removal by exopolysaccharide-producing cyanobacteria. *J. Appl. Microbiol.* **105:** 88-94.

Mihali, T.K., Kellmann, R., Muenchhoff, J., Barrow, K.D. and Neilan, B.A. (2008). Characterization of the gene cluster responsible for cylindrospermopsin biosynthesis. *Appl. Environ. Microbiol.* **74:** 716-722.

Millbank, J.W. (1972). Nitrogen metabolism in lichens IV. The nitrogenase activity of the *Nostoc* phycobiont in *Peltigera canina. New Phytol.* **71:** 1-10.

Mishra, U., Choudhary, K.K., Pabbi, S., Dhar, D.W. and Singh, P.K. (2005). Influence of Blue-Green Algae and Azolla inoculation on specific soil enzymes under paddy cultivation. *Asian Jr. Microbiol. Biotechnol. Env. Sc.* **7:** 9-12.

Mohamed, Z.A. (2008). Toxic cyanobacteria and cyanotoxins in public hot springs in Saudi Arabia. *Toxicon* **51:** 17-27.

Ow, M.C., Gantar, M. and Elhai, J. (1999). Reconstitution of cycad-cyanobacterial association. *Symbiosis* **27:** 125-134.

Paerl, H.W., Steppe, T.F., Buchan, K.C. and Potts, M. (2003). Hypersaline cyanobacterial mats as indicators of elevated tropical hurricane activity and associated climate change. *AMBIO: A J. Human Environment* **32:** 87-90.

Power, M.E., Stewart, A.J. and Matthews, W.J. (1988). Grazer control of algae in an Ozark mountain stream: effects of short-term exclusion. *Ecology* **69:** 1894-1898.

Prechtl, J., Kneip, C., Lockhart, P., Wenderoth, K. and Maier, U.G. (2004). Intracellular spheroid bodies of *Rhopalodia gibba* have nitrogen-fixing apparatus of cyanobacterial origin. *Mol. Biol. Evol.* **21:** 1477-1481.

Rahaman, A.A., Ambikadevi, M. and Sosamma-Esso, (1993). Biological management of Indian solar saltworks. In: Seventh symposium on salt, Vol. I. Elsevier, Amsterdam. pp. 633-643.

Rai, A.N. (1990). Cyanobacteria in symbiosis.In: Rai, A.N. (Ed). *Handbook of Symbiotic Cyanobacteria*, CRC Press, Boca Raton, FL, pp. 1-7.

Rai, A.N., Soderback, E. and Bergman, B. (2000). Tansley Review No. 116, Cyanobacterium plant symbioses. *New Phytol.* **147:** 449-481.

Sanchez-Baracaldo, P., Haynes, P.K. and Blank, C.E. (2005). Morphological and habitat evolution in the cyanobacteria using a compartmentalization approach. *Geobiology* **3:** 145-165.

Schopf, J.W. (2000). The fossil record: tracing the roots of the cyanobacterial lineage. In: Whitton, B.A. and Potts, M. (Eds.). The ecology of cyanobacteria: their diversity in time and space, Boston, Kluwer Academic, pp. 13-35.

Singh, S. (1995). Partial purification and some properties of urease from the alkaliphilic cyanobacterium *Nostoc calcicola. Folia Microbiol. Prague* **40:** 529-533.

Singh, B.V., Choudhary, K.K., Dhar, D.W. and Singh, P.K. (2001). Occurrence of some *Nostocales* from 24 Parganas of West Bengal. *Phykos.* **40:** 83-87.

Solheim, B. and Zielke, M. (2002). Associations between cyanobacteria and mosses. In: Rai, A.N., Bergman, B. and Rasmussen, U. (Eds.). Cyanobacteria in Symbiosis, Kluwer Academic Publishers, Netherlands, pp. 137-152.

Sompong, U., Hawkins, P., Besley, C. and Peerapornpisal, Y. (2005). The distribution of cyanobacteria across physical and chemical gradients in hot springs in northern Thailand. *FEMS Microbiol. Ecol.* **52:** 365-376.

Steinberg, C.E.W., Schafer, H. and Beisker, W. (1998). Do acid-tolerant cyanobacteria exist? *Acta Hydrochem Hydrobiol* **26:** 13-19.

Steunou Anne-Soisig, Bhaya, D., Bateson, M.M., Melendrez, M.C., Ward, D.M., Brecht, E., Peters, J.W., Kühl, M. and Grossman, A.R. (2006). *In situ* analysis of nitrogen fixation and metabolic switching in unicellular thermophilic cyanobacteria inhabiting hot spring microbial mats. *PNAS* **103:** 2398-2403.

Taniuchi, Y., Yoshikawa, S., Maeda, S-I., Omata, T. and Ohki, K. (2008). Diazotrophy under continuous light in a marine unicellular diazotrophic cyanobacterium, Gloeothece sp. 68DGA. *Microbiology* **154:** 1859-1865.

Taton, A., Grubisic, S., Balthasart, P., Hodgson, D.A., Laybourn-Parry, J. and Wilmotte, A. (2006). Biogeographical distribution and ecological ranges of benthic cyanobacteria in East Antarctic lakes. *FEMS Microbiol. Ecol.* **57:** 272-289.

Thajuddin, N. and Subramanian, G. (1992). Survey of cyanobacterial flora of the southern east coast of India. *Bot. Mar.* **35:** 305-311.

Thajuddin, N. and Subramanian, G. (1995). Additions to the new reports of marine cyanobacteria from the east coast of India. *Phykos.* **34:** 33-37.

Thajuddin, N. and Subramanian, G. (2002). The enigmatic bloom forming marine cyanobacterium *Trichodesmium.* In: Sahoo, D. and Pandey, P.C. (Eds.). *Advances in Marine and Antarctic Sciences*, APH Publishing Co, New Delhi, pp. 57-89.

Thajuddin, N. and Subramanian, G. (2005). Cyanobacterial biodiversity and potential applications in biotechnology. *Curr. Sci.* **89:** 47-57.

Vincent, W.F. (2000). Cyanobacterial dominance in the polar regions. In: Whitton, B. and Potts, M. (Eds.). Ecology of the Cyanobacteria: their diversity in space and time. Kluwer Academic Press, Dordrecht, The Netherlands. pp. 321-340.

Vincent, W.F. and James, M..R (1996). Biodiversity in extreme aquatic environments: lakes, ponds and streams of the Ross Sea Sectore, Antarctica. *Biodiv. Conserv.* **55:** 1451-1471.

West, N. and Adams, D.G. (1997). Phenotypic and genotypic comparison of symbiotic and free-living cyanobacteria from a single field site. *Appl. Environ. Microbiol.* **63:** 4479-4484.

Whitton, B.A. (1992). Diversity, ecology and taxonomy of the cyanobacteria. *In*: Mann, N.H. and Carr, N.G. (Eds.). Photosynthetic Prokaryotes, Plenum, New York, 1-51.

Wilkinson, C.R. (1978). Microbial associations in sponges. III. Ultrastructure of the *in situ* associations in coral reef sponges. *Mar. Biol.* **49:** 177-185.

Wilkinson, C.R. and Fay, P. (1979). Nitrogen fixation in coral reef sponges with symbiotic cyanobacteria. *Nature* **279:** 527-529.

Wynnn-Williams, D.D. (1990). Ecological aspects of Antarctic microbiology. *Adv. Microb. Ecol.* **11:** 71-146.

Yannarell, A.C., Steppe, T.F. and Paerl, H.W. (2007). Disturbance and recovery of microbial community structure and function following Hurricane Frances. *Environ. Microbiol.* **9:** 376-383.

Zehr, J.P., Waterbury, J.B., Turner, P.J., Montoya, J.P., Omoregie, E., Steward, G.F., Hansen, A. and Karl, D.M. (2001). Unicellular cyanobacteria fix N_2 in the subtropical North Pacific Ocean. *Nature* **412:** 635-638.

Diversity of Lower Plants (2014): pp. 54-63
Editors: Rajan Kumar Gupta & Mukesh Kumar
IK International Publishing House Pvt. Ltd.

4

Freshwater Cyanobacterial Diversity of Doon Valley

ANJALI KHARE*, MUKESH KUMAR** AND RAJAN KUMAR GUPTA***

**Department of Botany, Advance Institute of Science & Technology, Dehradun*
**Department of Botany, Sahu Jain College, Najibabad–246 763, U.P.*
***Department of Botany, Government P.G. College, Rishikesh, Uttarakhand*

ABSTRACT

India has over 1,15,000 identified and described species of plants and animals. There is a great diversity in plants ranging from algae to angiosperms. Cyanobacteria are a group of prokaryotes that possess a higher plant-type oxygenic photosynthesis. In addition to being a key player in aquatic productivity, several of these organisms are capable of fixing atmospheric nitrogen either as free-living organisms or in symbiosis with many other species including protists animals and plants.

The present chapter explores the freshwater cyanobacterial biodiversity of Doon Valley. All in all, 62 species of 3 orders and 22 genera belonging to 5 families have been recorded from the research sites.

Keywords: Cyanobacteria, Himalaya, Hot springs, Associations

1. INTRODUCTION

Biological diversity may be defined as the sum total of species richness, i.e., the number of species of plants, animals and microorganisms occurring in a given habitat. The Indian region (80°-30°N and 60°-97.5°E) having a geographical area of 329 million hectares is quite rich in biodiversity with a substantial percentage of endemic flora and fauna.

Our country has over 1,15,000 identified and described species of plants and animals. There is a great diversity in plants ranging from algae to angiosperms. The algae are green, autotrophic, photosynthetic, thalloid and chlorophyll bearing simple lower plants which ranges from unicellular forms to aggregation of cells, filaments or parenchymatous. Algae is mainly divided into four

*Communicating Author

groups, i.e., Green algae (Chlorophyceae), Blue-green algae (Cyanophyceae), Brown algae (Phaeophyceae) and Red algae (Rhodophyceae).

The cyanobacteria (the earlier blue-green algae), or the blue-green bacteria, represent a group of photosynthetic, mostly photolysis-mediated oxygen-evolving monerans (prokaryotes). These are the only organisms able to perform oxygenic photosynthesis that can also fix nitrogen. These organisms are amongst the oldest organisms known dating back to the early Precambrian period 3.6×10^9 years ago (Schopf & Waiter 1982) and probably played a crucial role in the evolution of higher plants. Cyanobacteria are a group of prokaryotes that possess a higher plant-type oxygenic photosynthesis. In addition to being a key player in aquatic productivity, several of these organisms are capable of fixing atmospheric nitrogen either as free-living organisms or in symbiosis with many other species including protists, animals and plants (Sinha & Häder 1997). They use the enzyme nitrogenase to reduce atmospheric nitrogen into ammonium ions (NH_4^+), which they make available for aquatic eukaryotic phytoplankton as well as higher plants (Kashyap *et al.*, 1991; Sinha *et al.*, 1996; Sinha & Häder, 1996; Kumar *et al.*, 1996a). The agricultural potential of cyanobacteria has been recognized as a biological fertilizer for wet soils such as in rice paddies (Banerjee & Häder, 1996). Cyanobacteria are cosmopolitan and must possess a high potential of adaptation to diverse environmental factors.

The cyanobacteria possess various **distinguishing characters**, which can be summarized as under.

(a) Cyanobacteria can grow in diverse habitats, but one striking feature in their occurrence and predominance in habitats alternating between photoaerobic and photoanaerobic conditions can be correlated with their preference for low oxygen tension and low redox-potential. These properties stem from their recently discovered dual-capacity of oxygenic photosynthesis and facultative anoxygenic photosynthesis.

(b) The cyanobacteria possess various morphologically distinctive structures, e.g. akinetes and heterocysts.

(c) The main cell wall constituent of cyanobacteria is peptidoglycan.

(d) The cyanobacterial cytoplasm is traversed extensively by flattened vesicular structures called thylakoids or lamellae, the photosynthetic sites.

(e) The principal photosynthetic pigment of all cyanobacteria is chlorophyll 'a'. Besides, they possess β-carotene and other accessory pigments, namely, phycobiliproteins. The phycobiliproteins are phycocyanin (PC), allophycocyanin (AP), allophycocyanin B (APB), and phycoerythrin.

(f) Most filamentous cyanobacteria show a gliding motility at some stage of development; they lack flagella.

Many cyanobacteria are able to survive in hostile environments, such as African soda lakes. Some cyanobacteria adjust their buoyancy by means of **gas vacuoles**, enabling them to adjust their position in the water column, floating near the surface during the day for photosynthesis and sinking deeper at night to harvest nutrients. Found in almost every conceivable habitat, from oceans to freshwater to bare rock to soil, cyanobacteria produce the compounds responsible for "earthy" odours we detect in soil and some bodies of water (such as those being cyanobacterially cleaned at water treatment plants). The greenish slime on the side of your damp flower pot, the wall of your house or the trunk of that big tree is more likely to be

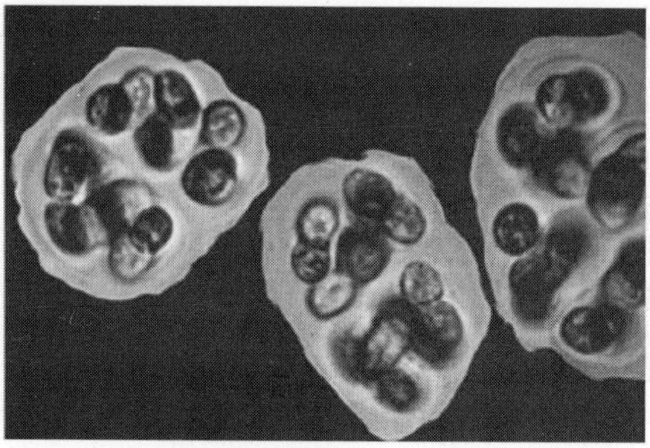

(*Synechococcus* sp.)

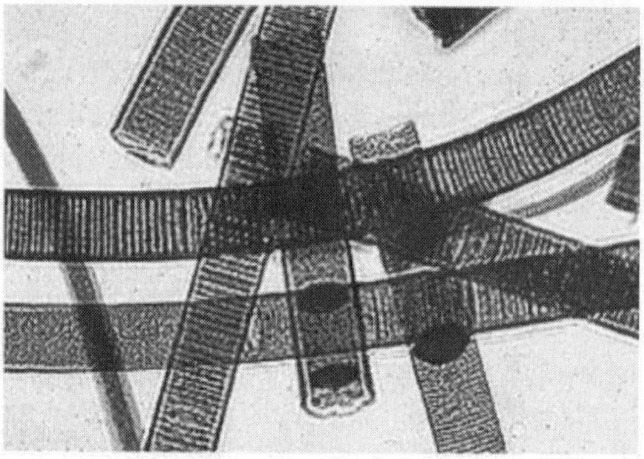

(*Oscillatoria* sp.)

cyanobacteria than anything else. Cyanobacteria have even been found on the fur of *polar bears*, to which they impart a greenish tinge!

Cyanobacteria may be single-celled or colonial. Depending upon the species and environmental conditions, colonies may form filaments, sheets or even hollow balls. Some filamentous colonies show the ability to differentiate into three different cell types. Vegetative cells are the normal, photosynthetic cells formed under favourable growing conditions. Climate-resistant spores may form when environmental conditions become harsh. A third type of cell, a thick-walled heterocyst, contains the enzyme nitrogenase, vital for nitrogen fixation. They come in a variety of forms, both singular and colonial.

Spirulina is big business, with thousands of metric tons grown in outdoor ponds and harvested annually in the USA, Mexico, Thailand, India and China. It is used as a food supplement, high in protein, sold in health food stores as a dried powder or in tablet form. However, many other species of cyanobacteria produce toxins—blue-green pond scum has been linked to the poisoning of cattle, dogs and occasionally people.

Finally, a number of cyanobacteria have formed symbiosis with other organisms such as liverworts, ferns and cycads, however, their best known symbiosis is with various fungi to form numerous lichens.

Cyanobacteria have tremendous potential in environment management; as soil conditioners, biofertilizers; biomonitors of soil fertility; water quality; amelioratory agents, feed for animals, protein supplements and rehabilitation of degraded ecosystems through bioabsorption of metals (Whitton & Potts 2000).

(*Aphanothece* sp.)

(*Oscillatoria* sp.)

(*Merismopedia* sp.)

(*Spirulina* sp.)

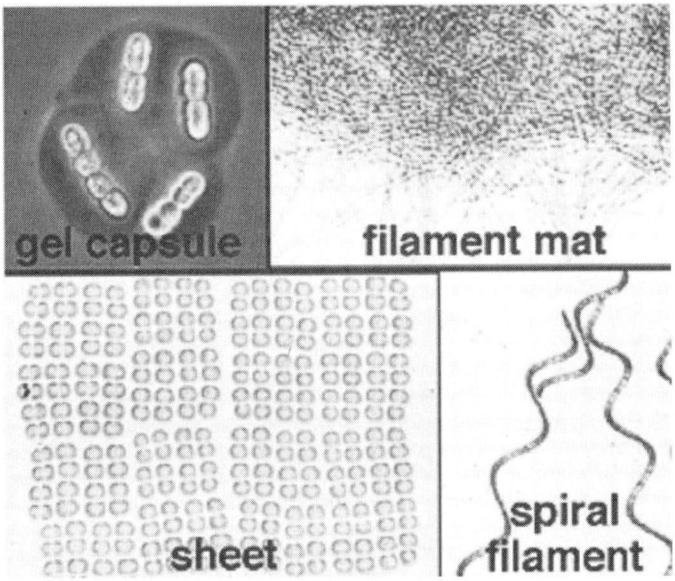

Sheet formation (*Merismopedia* sp.)

Himalayas

World's youngest mountain, the Himalayas stretch across the northern India from Jammu and Kashmir to Arunachal Pradesh for about 2400 km. It has been a perennial source of attraction, curiosity and challenge to human intellect throughout the ages. Amongst several assets, the vegetation of Himalayas provides an everlasting and interesting area of investigation. The diversity, copiousness and uniqueness of the plant components in Himalayas have retained sound and aesthetic environment in itself. However, in the past couple of decades, excessive exploitation of vegetation, unplanned land use, natural disaster and several developmental processes have accelerated in the deterioration of biodiversity and harmonious ecosystem of the Himalayas.

The Himalayas have unique physiographic setting based on specific lithology, strike-trend and joint pattern of rocks. The crescent shape of the mountain range is traced back to the pushing movement of the Assam ranges and the Aravallis at the time when the Indian Peninsula was moving towards the north. In general, the topology appears to be immature and rugged with densely vegetated slopes. There are three well-defined physiographic belts in the Himalayas, i.e., Greater Himalayas, Lesser Himalayas and Shiwalik, lying to each other from south to north and each being a distinct geological unit.

Garhwal Himalayas

Garhwal has its prehistory hidden in the heavy veils of time. Recent archaeological excavations near Srinagar and Mordhwaj date some settlements back to 800 B.C. Geographically, Tons river separates Garhwal from Himachal Pradesh in the west and district boundaries of Kumaon in the east. Garhwal is a former district of British India and later of the Indian state of Uttar Pradesh and now of Uttarakhand. The name Garhwal has been used interchangeably with the name Kumaon Himalaya. Today, Garhwal consists of Uttarkashi, Chamoli, Pauri, and Tehri districts and is spread over some 55,000 sq. kilometres.

Description of State Uttarakhand

Uttarakhand is one of the most beautiful and enchanting regions of northern India. Nature has endowed this land with so much beauty and spiritual bliss that it is also known as Dev Bhoomi, the land of gods. In Sanskrit 'Himalaya' means 'abode of snow', truly characterizing the vast permanent fields above the snow line. In the heart of these majestic mountains lies the state of Uttarakhand with Kumaon region in its east and Garhwal in the west. Geographically, Uttarakhand is situated between 77°34' to 81°02' E longitude and 28°04' to 31°27' N latitude.

Uttarakhand is predominantly a hilly state with 88% of hilly area. The climate of the state varies from subtropical in valleys to temperate on higher slopes. Total area of Uttarakhand is 53483 sq. km, which is 17.3% of the total area of India. The state is very rich in natural resources specially forest and water, as it has many glaciers, dense forests, mountain peaks and a network of mighty rivers, viz., Ganga, Yamuna, Ramganga, Kosi, etc. A total of 64.6% of the area is under forest cover.

Description of Sites

In spite of their vast diversity, the members of Cyanophyceae have remained neglected. While working on the distribution pattern of cyanobacteria in inland marshes of northern Belize Rej Mankova *et al.* (2004) revealed it as a neglected component of biodiversity in the defined area.

District-Wise Details

Dehradun is one of the most beautiful open valleys enclosed by the Shiwalik hills and outer scarp of the Himalaya. This district lies between 29°57′–30°57′ N latitude and 77°40′–78°20′ E longitude. The total area of the district is about 3080 km^2 out of which 479 km^2 is forest cover. The district is bounded by Tehri Garhwal in the north, Shiwalik submountain in the south, Himachal Pradesh in the west and Pauri Garhwal in the east. The altitude of the district varies from 250 m to 2850 m aMSL. Khadamba situated about 20 km north from Chakrata is the highest peak. The name "Dehra" refers to the town which sprang up around a temple presumably built by the religious Guru Ram Rai in 1699 A.D. and Doon (Dun) means a valley. The Doon Valley is surrounded by the Himalaya on the north, the Shiwalik hills on the south, the river Ganges on the east and the river Yamuna on the west. Dehradun has typical weather of a north Indian city: cold winter, rainy monsoon and a balmy spring. The city had a wonderful climate till a few decades ago but a change is noticed in the weather pattern in the last 2 decades, with the growth of city.

Site-Wise Details

The study was made from four sites, namely, Mothronwala, Dakpatthar, Kaansro and Sahastradhara. All the sites are well covered with water rivulets and other water bodies.

2. METHODOLOGY

With a view to study the ecology and taxonomy of the cyanobacterial flora, following methods have been employed. Firstly, identification of algal rich localities was done. The algal samples were collected from different sites of various freshwater habitats, such as ponds, pools, rivers, etc., and then preserved in 4% formalin in plastic tubes which on returning to the laboratory were transferred to the suitable medium for further studies. Microscopic slides were prepared and subjected to identification up to species level with the help of monograph given by Desikachary (1959). Meteorological parameters (min temp, max temp, rainfall and relative humidity have been collected. (Fig. 4)

3. OBSERVATIONS

A considerable variation has been recorded in the meteorological data at Doon Valley during different seasons of the year. The average mean temperature has been recorded 29.5°C as maximum and 13.93°C as minimum with an average rainfall of 9.27 cm. The average relative humidity was recorded 82%. (Table 1)

The various members collected during the present study have been grouped as per the order and the species included in these, following the outline presented in the volume Cyanophyta by Desikachary (1959). All in all, 62 species of 3 orders and 22 genera belonging to 5 families have been recorded from the research sites. A list of identified species is given in the Table 2. camera lucida drawing has been made. (Figs. 2 & 3)

Table 1: Meteorological parameters of the Doon Valley

Parameters	Summer	Winter	Rainy	Mean
Average Atmospheric Temperature (°C)				
Maximum	34.4	15.15	90.3	29.5
Minimum	5.65	27.2	21.0	13.93
Average Rainfall (cm)	4.09	1.015	22.72	9.27
Average Relative Humidity (%)	80.75	71.0	94.25	82

Table 2: Flow charts showing taxonomic enumeration of the taxa recorded from the four sites of Doon Valley

PHYLUM: CYANOPHYTA
CLASS: CYANOPHYCEAE
ORDER: CHROOCOCCALES
FAMILY: CHROOCOCCACEAE
GENUS: *Microcystis*
1. M. ramosa
2. M. holsatica
3. M. pulverea
4. M. viridis
5. M. flos-aquae
GENUS: *Chroococcus*
6. C. minor
7. C. multicoloratus
GENUS: *Aphanocapsa*
8. A. elachista
9. A. elachista var. conferta
10. A. delicatissima
GENUS: *Aphanothece*
11. *A. costagni*
12. *A. conferta*
13. *A. saxicola*
GENUS: *Synechococcus*
14. *S. elongatus*
15. *S. aeruginosus*
GENUS: *Coelosphaerium*
16. *C. kuetzingianum*
GENUS: *Merismopedia*
17. *M. minima*
FAMILY: ENTOPHYSALIDACEAE
GENUS: *Chlorogloea*
18. *C. microcystoides*
ORDER: NOSTOCALES
FAMILY: OSCILLATORIACEAE
GENUS: *Spirulina*
19. *S. subsalsa*
20. *S. gigantea*
21. *S. meneghiniana*
22. *S. laxissima forma major*
23. *S. princeps*
GENUS: *Oscillatoria*
24. *O. animalis*
25. *O. brevis*
26. *O. deflexa*
27. *O. obscura*
28. *O. pseudogeminata*
29. *O. amoena*
30. *O. rubescens*
31. *O. limosa*
32. *O.decolorata*
33. *O. nigra*
34. *O. princeps*
35. *O. subbrevis*
36. O. *gruunowiana var. articulata*

Contd.

Table 2: Contd.

37. *O. curviceps*
38. *O. schultzii*
39. *O. jasorvensis*
40. *O. sancta*
41. *O. proteus*

GENUS: *Lyngbya*

43. *L. cryptovaginata*
44. *L. circumcreta*
45. *L. birgei*
46. *L. scotti*

GENUS: *Schizothrix*

47. *S. muelleri*

GENUS: *Microcoleus*

48. *M. chthonoplastes*
49. *M. lacustris*

GENUS: *Symploca*

50. *S. muralis* var. *minor*
51. *S. muralis* forma Rao

GENUS: *Phormidium*

52. *P. truncicola*

FAMILY: NOSTOCACEAE

SUB-FAMILY: ANABAENAE

GENUS: *Anabaena*

53. *A. spiroides*
54. *A. tanganyikae*

GENUS: *Plectonema*

55. *P. radiosum*
56. *P. hansgirgi*

GENUS: *Cylindrospermum*

57. *C. majus*

GENUS: *Nostoc*

58. *N. carneum*

SUB-FAMILY: AULOSIRAE

GENUS: *Aulosira*

59. *A. prolifica*
60. *A. pseudoramosa*

FAMILY: SCYTONEMATACEAE

GENUS: *Tolypothrix*

61. *T. ceylonica*

ORDER: STIGONEMATALES

FAMILY: STIGONEMATACEAE

GENUS: *Camptylonema*

62. *C. indicum*

4. RESULT AND DISCUSSION

The Doon Valley is rich in rivers, rivulets and stagnant water bodies of different nature. The favourable climatic conditions have resulted in high fertility to these water bodies, inviting luxurious growth of diverse algae, particularly the blue-green algae. (Fig. 1)

It must be emphasized that the increasing population and industrial growth has led to environmental pollution, which is causing fast and enormous changes in this range. In view of this realization, an eco-taxonomic study of the cyanobacterial flora has been undertaken to serve as a base for future analysis. It is well known that nutrient pool is high in Doon Valley, as such the distribution and growth of cyanobacterial flora is luxurious in aquatic ecosystems. During the present study, the cyanobacterial flora of four different sites situated in Doon Valley have been recorded in relation to various parameters.

Doon Valley lies between 29°57′–30°57′ N latitude and 77°40′–78°20′ E longitude. The altitude ranges from 250 m to 2850 m amsL. The average mean temperature has been recorded 29.5°C as maximum and 13.93°C as minimum with an average rainfall of 9.27 cm. The average relative humidity was measured 82%. The overall data shows that the maximum temperature has been observed in the month of June (36.7°C) and minimum in the month of January & February (15.3°C). Maximum rainfall has been observed in the month of August (49.22 cm) and minimum or no rainfall has been observed in the month of November (36.0 cm). In contrast to this, the maximum humidity has been recorded in the months of July, August & February (95%) and the minimum in the month of June (80%) (Table 1).

During the study of the cyanobacterial flora of all six research sites, the maximum number of species has been observed in Sahastradhara and the minimum number has been observed in Dakpathar. This may be due to certain ecological parameters and the chemical constituents of the water present in the water bodies which may or may not promote the growth of cyanobacteria.

Fig. 1. Different algae growing in different habitats

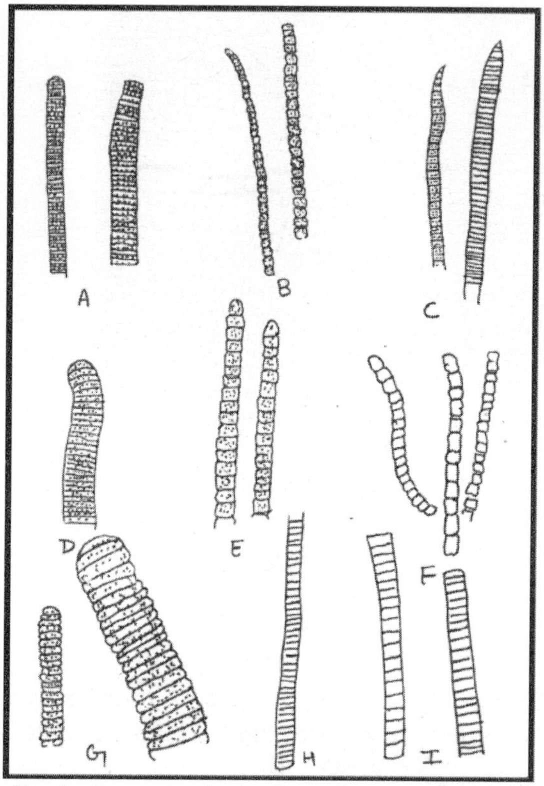

A ***Oscillatoria princeps***
B ***O. schultzii***
C ***O. animalis***
D ***O. curviceps***
E ***O. proteus***
F ***O. pseudogeminata***
G ***O. sancta***
H ***O. grunowiana***
I ***O. jasorvensis***

Fig. 2. Camera-lucida drawings of some cyanobacteria recorded from Doon Valley.

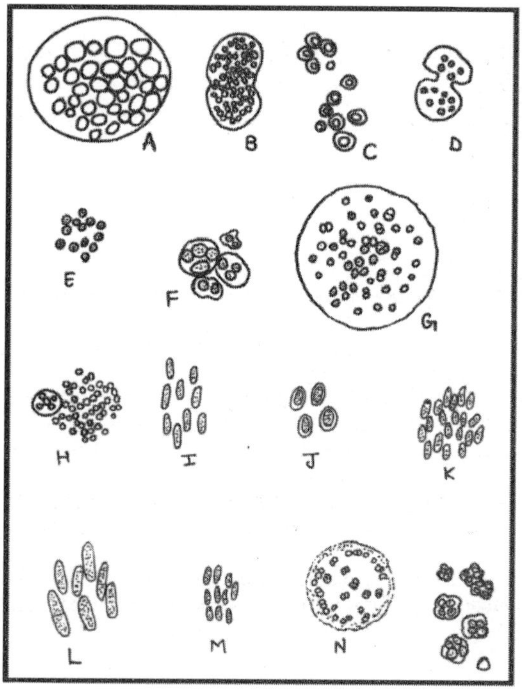

A ***Microcystis viridis***
B ***M. holsatica***
C ***Chroococcus multicolorotus***
D ***M. ramosa***
E ***M. Pulverea***
F ***C. minor***
G ***Aphanocapsa elachista***
H ***A. delicatissima***
I ***Aphanothece castagni***
J ***A. saxicola***
K ***A. conferta***
L ***Synechococcus aervginosus***
M ***S. elongatus***
N ***Coelosphaerium kuetzingianum***
O ***Merismopedia minima***

Fig. 3. Camera-lucida drawings of some cyanobacteria recorded from Doon Valley.

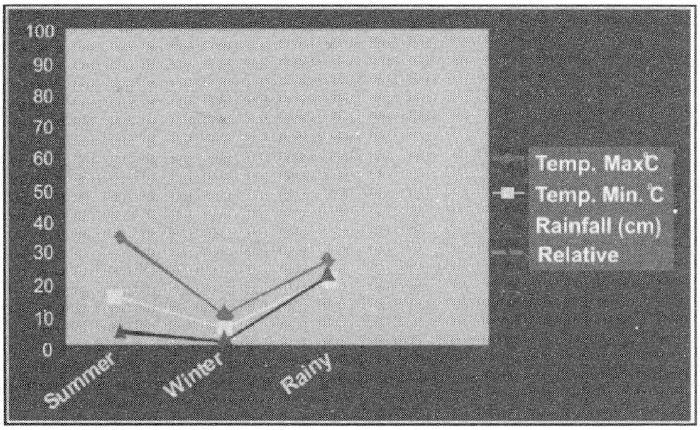

Fig. 4. Meteorological parameters of the Doon Valley.

Numerical analysis of the cyanobacterial flora recorded from 4 research sites in Doon Valley, indicates that 62 BGA species of 22 genera from 5 families belonging to 3 orders growing in various habitats have been reported from this region.

5. CONCLUSION

The nutrient pool is high in Doon Valley, as such the distribution and growth of cyanobacterial flora is luxurious in aquatic ecosystems. The maximum number of species has been observed at Sahastradhara and the minimum at Dakpatthar. This may be due to certain ecological parameters and the chemical constituents of the water present in the water bodies which may or may not promote the growth of cyanobacteria.

All in all, 62 species of 3 orders and 22 genera belonging to 5 families have been recorded from the research sites. A list of identified species is given in Table 2.

It will be, therefore, beneficial to carry out further studies so as to explore the biodiversity of cyanobacteria in this region.

REFERENCES

Banerjee, M. and Häder, D.P. (1996). Effects of UV radiation on the rice field cyanobacterium, *Aulosira fertilissima. Environmental and Experimental Botany* **36:** 281-291.

Kashyap, A.K., Pandey, K.D. and Gupta, R.K. (1991). Nitrogenase activity of the Antarctic cyanobacterium *Nostoc commune*: influence of temperature. *Folia Microbiologica* **36:** 557-560.

Kumar, A., Tyagi, M.B., Srinivas, G., Singh, N., Kumar, H.D., Sinha, R.P. and Häder, D.P. (1996[a]). UVB shielding role of FeCl3 and certain cyanobacterial pigments. *Photochemistry and Photobiology* **64:** 321-325.

Sinha, R.P. and Häder, D.P. (1996). Photobiology and ecophysiology of rice field cyanobacteria. *Photochemistry and Photobiology* **64:** 887-896.

Sinha, R.P., Singh, N., Kumar, A., Kumar, H.D., Häder, M. and Häder, D.P. (1996). Effects of UV irradiation on certain physiological and biochemical processes in cyanobacteria. *J. Photochemistry and Photobiology, B: Biology* **32:** 107-113.

Sinha, R.P. and Häder, D.P. (1997). Impacts of UV-B irradiation on rice-field cyanobacteria. In: *The Effects of Ozone Depletion on Aquatic Ecosystems. Ch. 9*, Häder, D.-P. (Ed.). Environmental Intelligence Unit, Academic Press and RG Landes Company, Austin, 189-198.

Diversity of Lower Plants (2014): pp. 64-77
Editors: Rajan Kumar Gupta & Mukesh Kumar
IK International Publishing House Pvt. Ltd.

5

Physico-Chemical Parameters in Relation to the Distributional Pattern of Cyanobacteria

ANJALI KHARE*, MUKESH KUMAR** AND P.K. SINGH***

**Department of Botany, Advanced Institute of Science & Technology, Dehradun*
***Department of Botany, Sahu Jain P.G. College, Najibabad–246 763 (U.P.)*
****Department of Botany, Hindu College, Moradabad–244 001 (U.P.)*

ABSTRACT

Cyanobacteria are a group of haploid oxygenic photosynthetic blue-green algal prokaryotes. They belong to the largest, most diverse and widely distributed group of prokaryotes. These belong to the most archaic organisms on the earth. Cyanobacteria occur in a variety of habitats and their growth is influenced by several physical and chemical factors of the environment. This chapter discusses the influence of several physical and chemical factors on the growth of cyanobacterial diversity of Himalayas with special reference to Kumaon region of Uttarakhand.

Keywords: Cyanobacteria, Himalaya, Pedological studies, Electrical conductivity

1. CYANOBACTERIA

Cyanobacteria, a group of haploid oxygenic photosynthetic blue-green algal prokaryotes, can be sharply distinguished from other groups of algae since they have low state of cell differentiation, blue-green colour of the cells, production of cyanophycean starch and absence of an organized nucleus in its cells (Singh, 1961). They belong to the largest, most diverse and widely distributed group of prokaryotes.

Cyanobacteria are considered the most archaic organisms on the earth as they dominated the biota in the Proterozoic era. Due to the occurrence of blue-green algae, the Proterozoic era is also known as "Age of Cyanobacteria" (van den Hoek *et al.* 1993). They occur in a great

*Communicating Author

variety of natural habitats, viz. freshwater, marine, epilithic, epipelic, epiphytic, endophytic and thermophilic, etc., but are often abundant in freshwater, seawater and terrestrial environments. (Fogg *et al.,* 1973).

Cyanobacteria are also called blue-green algae because they contain blue and green pigments. They are an ancient and heterogeneous group of prokaryotic photosynthetic organisms. The oxygen-containing atmosphere of earth initially developed due to the oxygenic photosynthesis of cyanobacteria, similar to that of higher plants. Due to the similarities between chloroplasts and cyanobacteria, the latter are believed to be ancestors of higher plant chloroplasts. Chloroplasts probably originated from an endosymbiotic event between a cyanobacterium and another organism some hundreds of millions years ago. An oxidative atmosphere is therefore (in an evolutionary perspective) a relatively new phenomenon, and most enzymatic pathways generally used by the present-day organisms were developed in an earlier anoxygenic environment. Due to its relatively late evolutionary appearance oxygen, and especially its derivatives, are harmful to many cellular components. A number of oxidative stress responses have evolved to avoid such cellular damages.

Cyanobacteria are very important organisms for the health and growth of many plants. They are one of the very few groups of organisms that can convert inert atmospheric nitrogen into an organic form, such as nitrate or ammonia. It is these "fixed" forms of nitrogen, which plants need for their growth, and must obtain from the soil. Fertilizers work the way they do in part because they contain additional fixed nitrogen which plants can then absorb through their roots. Nitrification cannot occur in the presence of oxygen, so nitrogen is fixed in specialized cells called heterocysts. These cells have an especially thickened wall that contains an anaerobic environment. You can see these larger cells among the filaments of *Nostoc*. Many plants, especially legumes, have formed symbiotic relations with nitrifying bacteria, providing specialized tissues in their roots or stems to house the bacteria, in return for organic nitrogen. This has been used to great advantage in the cultivation of rice, where the floating fern *Azolla* is actively distributed among the rice paddies. The fern houses colonies of the cyanobacterium *Anabaena* in its leaves, where it fixes nitrogen. The ferns then provide an inexpensive natural fertilizer and nitrogen source for the rice plants when they die at the end of the season.

The cyanobacterium *Spirulina*, has long been valued as a food source. It is high in protein, and can be cultivated in ponds quite easily. In tropical countries, it is an important part of the diet, and was eaten regularly by the Aztecs; it is also served in several Oriental dishes. In the US, the popularity of *Spirulina* is primarily as a "health food", being sold in stores as a dried powder or in tablet form.

Cyanobacteria appear as the first plant community, as xeroseres and hydroseres, leading to the associations of higher plant associations. The different environmental conditions have triggered the cyanobacterial mass occurrence. Amongst them, the hydrogen ion concentration is an important factor followed by temperature, light, soil type and available nutrients. The mass occurrence of cyanobacteria has been reported to increase the frequency as well as the intensity of biotic community due to eutrophication (Kahru *et al.,* 1994; Finni *et al.,* 2001). The increasing light and temperature conditions stimulate the growth of cyanobacteria. The literature already suggests that the higher pH of water, scarcity of carbon dioxide and high amount of phosphorus is conducive to the production of the biomass of cyanobacteria and algae.

The diversity of cyanobacteria can be seen in the multitude of structural and functional aspects of cell morphology and in variations in metabolic strategies, motility, cell division, developmental biology, etc. These produce extracellular substances like cyanotoxins, which illustrates the diverse nature of their interaction with other organisms (Rizvi & Rizvi 1992). To date, 16S r-RNA has given the most detailed information on the relationships within the cyanobacteria (Rudi *et al.,* 1997).

2. HIMALAYAS

World's youngest mountain, the Himalayas stretch across northern India from Jammu and Kashmir to Arunachal Pradesh for about 2400 km. It has been a perennial source of attraction, curiosity and challenge to human intellect throughout the ages. Amongst several assets, the vegetation of the Himalayas provides an everlasting and interesting area of investigation. The diversity, copiousness and uniqueness of the plant components in the Himalayas have retained sound and aesthetic environment in itself. However, in the past a couple of decades, excessive exploitation of vegetation, unplanned land use, natural disaster and several developmental processes have accelerated in the deterioration of biodiversity and harmonious ecosystem of the Himalayas. In fact, we don't know the extent of deterioration of vegetation or individual species, since no proper botanical records for several localities or regions are still available, one such botanically interesting and little known region in the foothills of Kumaon Himalayas, which sustains unique and rich vegetation in various habitats ranging from the tracts of Tarai-Bhabar to the trans-Himalayas.

2.1 The Sub-Himalayan Belt

The Shiwalik range forms a long chain of narrow and low mountain ranges of about 600-1500 m amsl, running almost parallel to the main Himalayas with a width of 15-50 km. The Shiwalik Himalayas has Tarai and Bhabar region adjacent to it. The former includes long and narrow strips of low-lying plains of about 10-25 km width. The depositions of land materials are mainly derived from the Bhabar land and separate the upper limits of Gangetic plains. This is the wet and swampy belt of seepage where fine sand, silt and clay are deposited by flowing streams. The Bhabar belt includes the foot part of Shiwalik hills, and usually covers 20-45 km in width, in an undulating topography. The belt is distinguishable by huge deposits of boulders under the porous gravelly soil. Several streams originate from the tract, flowing towards Tarai and adjacent plains.

The soils of the foothills of the Himalayas represent wide range of variations in their texture, colour, vegetation and various biological and chemical interactions taking place on them continuously. In general, the soil of Himalayan mountains, on the slopes above 30° represents merely a thin surface horizon. Such a skeletal soil has medium to coarse texture depending upon core materials. The valleys and gentle slopes have considerable depth and have developed from both the colluviums and alluvium. The texture of such a soil is generally coarse. Usually, the south facing slopes are too precipitous and well exposed to denudation; hence a shallow, dry and less fertile soil, whereas the north facing slopes support a deep, moist and fertile soil. In the sub-montane and adjacent plains including the Tarai-Bhabar, four major groups of soil categorized, e.g., Red-loam, Brown forest soil, Podosols and Transitional Podosols & Weisenboden. The Montane zones, representing warm temperate or cool temperate climates, have two major types of soils—the brown forest soil under humid and dense broad-leaved forests and Padozolic soils under sub-humid coniferous forests. The mineral and soil particle compositions vary from place, being more variable in the former type of soil.

In general, the climate of the Himalayan region differs from the adjacent Gangetic plains by a shorter or less severe summer, more or less higher precipitation and colder or more prolonged winter. The climatic factors, i.e. precipitation, temperature, relative humidity and wind, in association with the elevation (valleys to mountain ranges from foothills to the mountain zones), proximity to Greater Himalayas, slope aspects, drainage, and vegetation, etc., cause variations in climates at local or even micro levels. In the foothills and outer Himalayas, it is almost similar to sub-tropical gangetic plains, except the higher annual precipitation (200-250 m per annum) and lower average temperature of 19–21°C, covering an elevation of 300-900 m amsl. The major output of precipitation is in the rainfall, besides occasional occurrence of dew, hailstorms, fog, frost, snowfall, etc.

2.2 Kumaon Himalayas

Lying between the Latitudes 28°44' to 30°49' N and Longitudes 78°45' to 81°1' E, Kumaon is situated at the trijunction of Nepal, Tibet and India. A natural water-divide separates it from Tibet, the Kali River defines its eastern border with Nepal, high transverse mountain spurs separate it from the Chamoli and Pauri districts of Garhwal and the southern limit of Tarai belt demarcates its southern boundary. Thus, it constitutes a distinct geographical entity of great strategic significance and is spread over 21,035 sq. km area. The altitudinal range varies from 204 m to 7,436 m amsl. The Kumaon division comprises four districts—Nainital, Udham Singh Nagar, Almora and Pithoragarh, which are divided into 14 tehsils and 41 blocks. The region exhibits socio-cultural unity amidst physical diversities.

The study area is situated in south-east to north of Kumaon Himalayas (between 20°00' - 30°00' N to 78°80'–81° E), Uttarakhand, India. It comprises a group of low-lying hills with rivers, springs, streams, rivulets, pits and ponds, etc. The elevation of different research sites ranges between 200-500 m amsl. The present work is a maiden attempt in respect to this region and includes the findings on ecological and taxonomic parameters of this important group of algae. The selection of the locality was made with the realization that the Himalayas represent one of youngest mountain chains of the world, ranging from Jammu and Kashmir to Arunachal Pradesh, covering a length about 2400 km. Evidently, such a vast area offers excellent habitats for an extremely large variety of vegetation, which has been consistently providing an unparallel attraction to the entire world, coupled with a challenge to explore them from different points of views. The foothills of Kumaon Himalaya constitute the region, which has still been explored only to a limited extent and where the holy Ganga and Yamuna attain the prime and start flowing in the foothills and subsequently the plains in a slow and serene manner. Thus, the area gets well irrigated and becomes rich in various types of vegetation, especially with respect to the algal flora, besides the angiosperms.

The present investigation had been planned to study the ecology and biodiversity of cyanobacteria of the foothills of Kumaon Himalayas from four research sites, i.e., Kashipur,

Rudrapur, Ramnagar and Haldwani. The first research site, Kashipur is situated at an altitude of 235 m amsl, and 20°13' N latitude and 28°59' E longitude. The second research site, Rudrapur is situated at an altitude of 244 m amsl, and a latitude of 29°03' N and longitude of 79°31' E. The third research site, Ramnagar is situated at an altitude of 330 m amsl, latitude of 29°05' E and longitude of 79°05' E whereas the fourth research site, Haldwani is located at a latitude of 29°13' N and longitude of 79°31' E at 432 m amsl. The area chosen for the analysis is a part of sub-Himalayan belt of Kumaon Division. During this analysis, an overall attempt has been made to survey the distribution pattern of the members of cyanobacteria available in this region encompassing both the attitude and a host of richness.

3. METEOROLOGICAL STUDIES

The climate of various research sites was determined by studying the different climatic or meteorological factors such as temperature, rainfall, relative humidity and intensity of light, etc. variation in climatic conditions has a direct correlation with variability of vegetation types. The following parameters have been studied for the present research work.

3.1 Temperature

The climatic conditions of the foothills of Kumaon Himalayas are subtropical, as such the temperature plays a very important role. It regulates the major biological activities taking place in the body of organisms. The temperature variation is quite evident in the Himalayas so much so that the Himalayas have a general zonation of vegetation from lower to the higher altitudes, as tropical subtropical and alpine. Nevertheless, there are practically no sharp boundaries between these vegetation zones due to difference in topography, soil and geology. Thus, in the Himalayan range the temperature together with altitude and other factors become quite important in respect of diversity and distribution of vegetation.

3.2 Rainfall (Precipitation)

Rainfall is the chief source of soil water available to plants. It occurs as an interchange of water between the earth's surfaces and the atmosphere forming the water cycle. In India, the rainfall is caused by the monsoon. The amount of annual rainfall greatly determines the seasons of the year. It may be important in the regional distribution of the vegetation. Thus, precipitation plays a crucial role in the development and growth of vegetation of a particular area.

3.3 Relative Humidity

Atmospheric moisture in the form of invisible vapour constitutes the humidity and is expressed in terms of "relative humidity" or the amount of moisture in the air as percentage of the amount, which the air can hold at the saturation tolerance at the existing temperature, the high temperature increases the capacity of air to retain moisture and causes lower humidity, whereas, the low temperature causes higher relative humidity by decreasing the folding capacity of air for moisture.

3.4 Intensity of Light

Light is well known for its effects on the basic physiological processes of plants and thus plays an important role in the species composition and development of vegetation. It affects the nature and distribution of vegetation through its effects on photosynthesis, chlorophyll production, number and position of chloroplasts, etc.

4. PEDOLOGICAL STUDIES

Each type of soil has its own physical and chemical characteristics, which affect the vegetation on its leading to a distinct flora. Consequently, it becomes very important to analyze the physical and chemical parameters of the soil.

For pedological studies, the soil samples from different research sites, with or without visual plant community/blue-green algae were collected from the surface as well as from different depths of about 10-15 cm (Singh 1961). The samples were stored in sterile plastic bottles (2.5 × 5.5 cm) at 4°C and transported to the laboratory for physical as well as chemical analysis. The physical and chemical characters of the soil present at different sites were studied in the following manner:

4.1 Physical Parameters

The following physical parameters were taken for study:

(i) Texture

The soil texture was determined by the relative proportion of mineral particles of different sizes present in the soil using the sieve method.

(ii) Temperature

Soil temperature is the result of heat gained by the absorption of solar heat energy. It was determined with the help of "mercury glass thermometer" with pointed steel for digging the soil.

(iii) Determination of pH

The pH values of the soil show much correlation with the soil type and profile horizon. It affects the plant growth, lime requirement and mineral nutrients in the soil. The pH of the soil was determined in soil suspension following the methodology outlined by Piper (1966).

(iv) Electrical Conductivity

The conductivity is the measure of the ability of an aqueous solution to carry an electric current. This ability depends on the presence of the ions, on their total concentration, mobility and valency, and on the temperature of measurement. The electrical conductivity of soil was determined in soil-water (1:5) extract with the help of a Digital Electrical Conductivity Meter (Symtronics Digital Electrical Conductivity Meter Ltd., India).

4.2 Chemical Parameters

The following chemical parameters were taken for the study:

(i) Total Nitrogen (as N)

The total nitrogen has been calculated according to the Kjeldahl method.

(ii) Available Phosphorus

Available phosphorus can be calculated from soil by sodium bicarbonate. The estimation of phosphorus has been done by Olsen method.

(iii) Available Potassium

The available potassium is usually determined in neutral normal ammonium acetate extract of soil. Available potassium changes in potassium acetate in this extract. It is estimated by photometer.

5. GENERAL WATER STUDIES

Water samples from streams, rivers, and ponds were collected under aseptic condition in sterile plastic containers (2.5 × 5.0 cm) and stored at 4°C, then they were transported to the laboratory for further study.

5.1 Physical parameters

The following physical parameters were taken for study:

(a) pH

The pH of the water body directly affects the life cycle of blue-green algae. pH of water samples was determined in the field and also in the laboratory with the help of a digital pH meter equipped with soil state combination electrode [HANNA Instruments (Mauritius) Ltd., Made in Portugal].

(b) Temperature

Water temperature affects the structure of proteins/enzymes taking part in various biological pathways. The temperature was determined by using the mercury glass thermometer.

6. TAXONOMIC STUDIES

The samples with visible growth of blue-green algae have been collected into sterile plastic bottles (2.5 × 5.0 cm), on returning to the laboratory; they were washed thoroughly in water and transferred to airtight glass containers, and preserved in 4% formalin solution. The microslides were prepared from the fixed as well as fresh algal materials for microscopic observations.

Microscopic slides of algal samples were observed under a trinocular research microscope (Olympus Ltd., India). Camera-Lucida drawings have been sketched and the drawings were analyzed on the basis of morphological observations. Consulting the pertinent literature in the field (Anand 1989; Desikachary 1959; Fogg *et al.,* 1973; Geitler 1932 & Rippka *et al.,* 1979), the cyanobacterial specimens were identified at the level of class, order, family, genus and species and morphological variations of different specimens were also analyzed.

7. RESULT AND GENERAL DISCUSSION

The foothills of Kumaon Himalayas along with the border of Uttar Pradesh and Uttarakhand are rich in rivers, rivulets and stagnant water bodies of different nature. The favourable climatic conditions and mineral rich alluvial soil basin of the entire sub-Himalayan belt have resulted in high fertility to these water bodies, inviting luxurious growth of diverse algae, particularly the blue-green algae. It is well known that nutrient pool is high in the foothills of Kumaon Himalayas, as such the distribution and growth of cyanobacterial flora is luxurious in terrestrial and aquatic ecosystems. The present investigation has been undertaken to study the diversity of Cyanobacteria of this particular region. A perusal of literature has rewarded that there is a direct relationship between the algal vegetation and various ecological factors. This has been substantiated under the present work; also, since distributional pattern of cyanobacterial flora is very well correlated with the meteorological, pedological and aquatic features of the region.

Of all four research sites, Kashipur is located at the lowest altitude (235 m amsl), while Haldwani is located at the highest altitude (432 m amsl). Rudrapur and Ramnagar are situated in (244 m amsl, and 330 m amsl altitude), respectively. The soil is loam, clay and silt in Kashipur, sandy loam and silt sand in Rudrapur, silty loam in Ramnagar and loamy sand and clay in Haldwani. Haldwani experiences the lowest annual temperature (27.5°C), as against Kashipur with the highest value (32.0°C). Rudrapur and Ramnagar experience the annual average temperature ranges, in between Haldwani and Kashipur, i.e. 31.3°C and 30.8°C, respectively. Haldwani has the lowest annual average rainfall (14.6 cm). On the other hand, Kashipur experiences highest (15.5 cm). Rudrapur and Ramnagar recorded 14.7 cm and 14.9 cm annual average rainfall, respectively. Kashipur experiences highest annual average relative humidity (78.7%) and Haldwani the lowest (76.2%). Ramnagar has highest annual average light intensity (850 Lux) and Rudrapur lowest (648 Lux). Kashipur and Haldwani have 773 and 722 Lux, respectively. Various physico-chemical parameters for all sites have been recorded. The soil of all the sites is slightly alkaline in nature (pH above 7.5). Kashipur has highest soil temperature (31°C), whereas Haldwani has lowest (26.0°C). The water temperature recorded at Kashipur is highest (29.6°C) and Haldwani lowest (22°C). Rudrapur and Ramnagar have 25.0°C and 24.5°C water temperature, respectively. In respect to electrical conductivity, soil of Pattharchata Nala from Rudrapur has exhibited highest electrical conductivity (1230 micros/mhos/cm) and Ramnagar lowest (190 micro/mhos/cm). The soil of Awas-Vikas Colony (Kashipur) maintain electrical conductivity 850 micro/mhos/cm, that of Nagarpalika (Kashipur) have 360 micro/mhos/cm and Haldwani have 200 micro/mhos/cm, respectively. Awas-Vikas Colony of Kashipur has highest amount of nitrogen (0.25% by mass) and Haldwani has lowest (0.03% by mass). Nagarpalika areas of Kashipur; Ramnagar; Pattharchata Nala and Danpur village of Rudrapur have 0.12;

0.11; 0.05 and 0.04% nitrogen by mass in the soil. The data in respect to the available phosphorus and potassium has also been recorded. The soil of Awas-Vikas Colony of Kashipur has highest phosphorus (0.0055%) and soil of Danpur village of Rudrapur has highest potassium (0.056%). Ramnagar has lowest phosphorus (0.0004%) and Haldwani has lowest potassium (0.0020%). The soil of Awas-Vikas Colony of Kashipur has 0.0194% potassium. The soil of Nagarpalika area of Kashipur has 0.0037% and 0.0120% phosphorus and potassium, respectively. Pattharchata Nala and Danpur village of Rudrapur have phosphorus as 0.0016% and 0.0033%, respectively and potassium 0.0045% and 0.056%. (Figs. 1, 2, 3, 4 & Table 1, 2, 3 & 4, See appendix-I)

Light and temperature conditions have a direct effect on the population dynamics of cyanobacterial flora. Bright light favours the growth the members of order Chroococcales, whereas dim light enhances the presence of the members of order Nostocales. Similarly, a moderate temperature between 20-40°C is found suitable for the luxurious growth of cyanobacterial taxa at all the sites. In respect to electrical conductivity, the scenario is quite different as high value of electrical conductivity favours the occurrence of the members of order Nostocales, on the other hand low value of this parameter declines the presence of the members of order Nostocales. A similar finding reveals that water temperature also imparts effect on the formation of heterocystous species. It has been indicated by the researcher that high value of water temperature increases the possibility of the number of heterocystous species.

Overall study of the four research sites have brought up the conclusion that the maximum number of cyanobacterial taxa have been reported from Haldwani, which is situated at highest altitude (432 m amsl) and has minimum percentage of nitrogen in the soil as compared to the remaining sites. In contrast to this, the minimum number of cyanobacterial species has been reported from Kashipur, which is situated at lowest altitude (235 m amsl) and has maximum percentage of nitrogen in the soil among all the four research sites. This has allowed the researcher to bring out the fact that high altitude and less percentage of nitrogen in the soil increases the

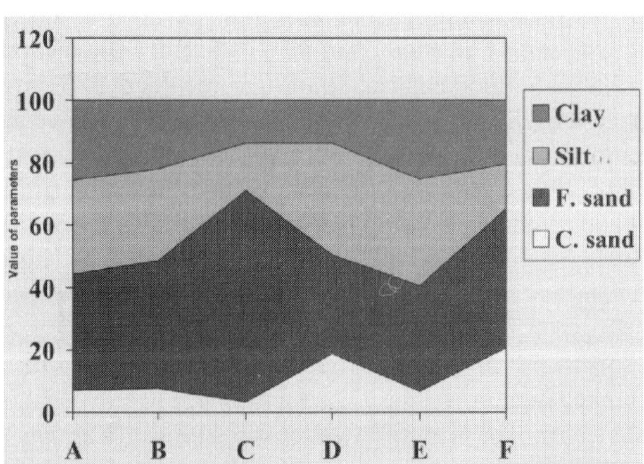

Fig. 1. Mechanical analysis showing texture of soil at different sites: A. Kashipur Awas Vikas Colony; B. Kashipur Nagarpalika; C. Rudrapur Pattharchata Nala D. Rudrapur Danpur Village; E. Ramnagar; F. Haldwani

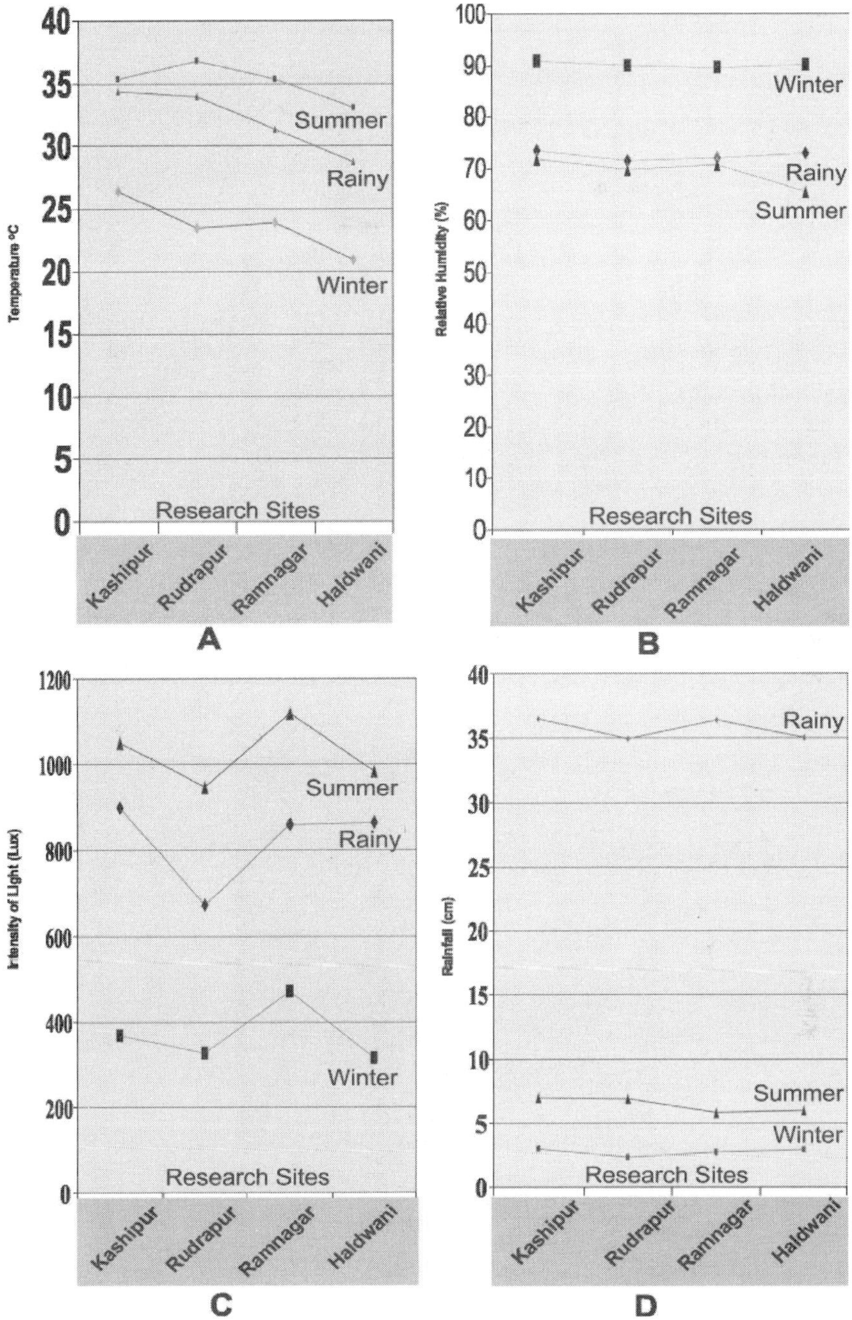

Fig. 2 (A–D): Meteorological parameters recorded at different sites.
A: Average atmospheric temperature, B: Average rainfall, C: Average intensity of light, D: Average relative humidity.

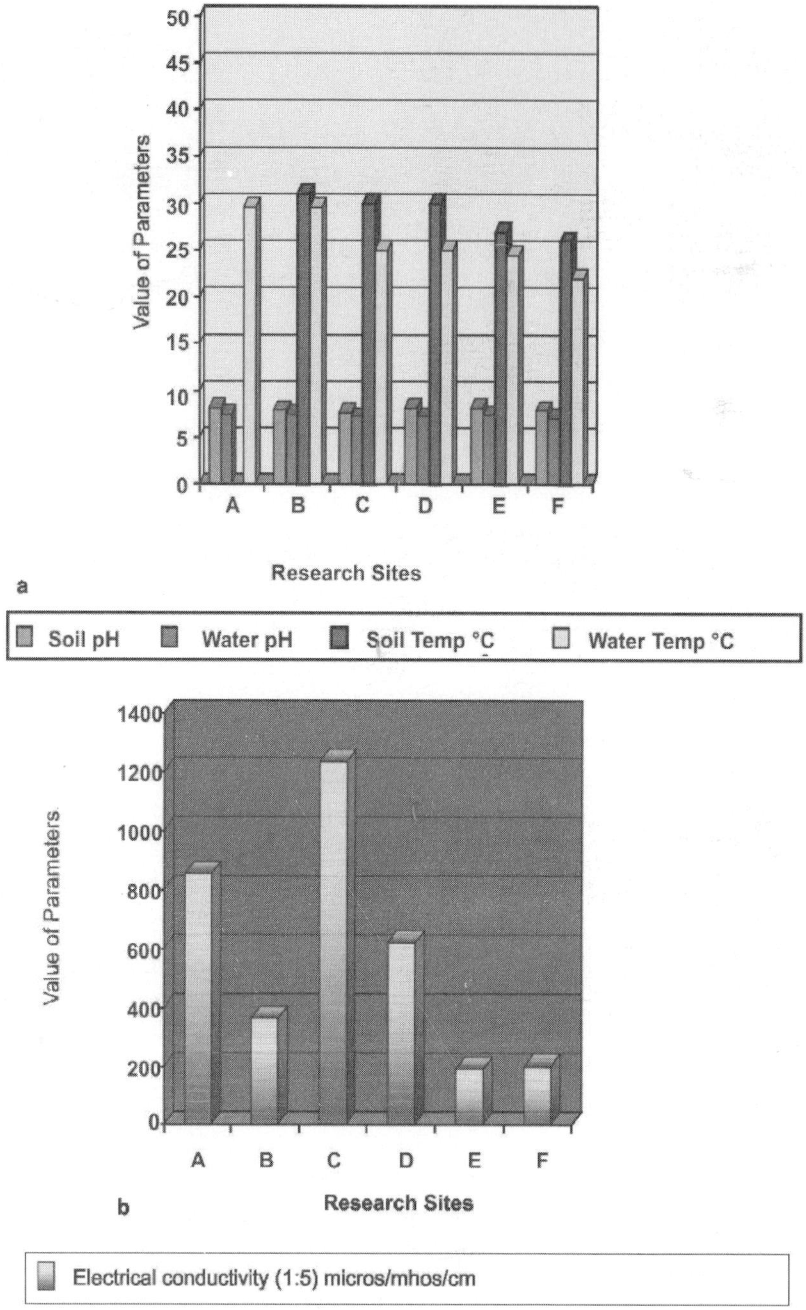

Fig. 3. Physical features at different research sites:
(a) Physical features of soil and water, (b) Physical features of soil
Index for the sites:
A. Kashipur Awas Vikas Colony, B. Kashipur Nagarpalika, C. Rudrapur Pattharchata Nala, D. Rudrapur Danpur Village, E. Ramnagar, F. Haldwani

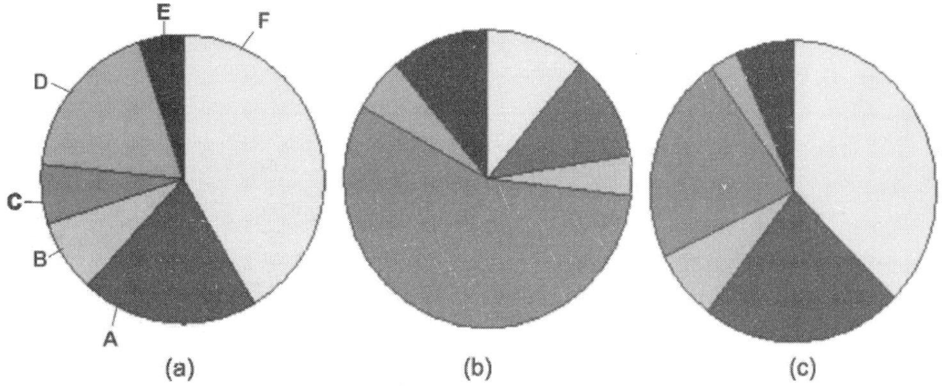

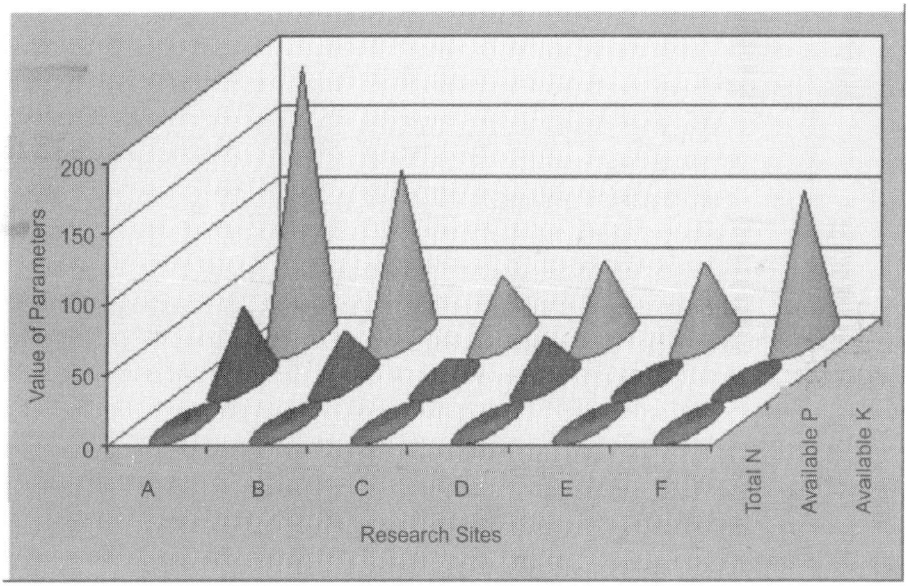

Fig. 4 (a–d). Chemical parameters of soil at different research sites:
(a) Total percentage of nitrogen per g of soil, (b) Available potassium ppm per g of soil, (c) Available phosphorus ppm per g of soil, d: Comparison of total N, P and K

Colour index Fig. (a–c):

A. Kashipur Awas Vikas Colony, B. Kashipur Nagarpalika area, C. Rudrapur Pattharchata Nala, D. Radrapur Danpur Village, E. Ramnagar, F. Haldwani

Color index Fig. (d):

■ Total Nitrogen ■ Available Phosphorus ■ Available Potassium

Table 1: Description of the research sites

S.No.		Districts	Geographical locations	Soil types	Sample collection sites
1.	Kashipur	Udham Singh Nagar	Longitude : 28°59' E Latitude : 20°13' N Altitude – 235 m amsl Lies in the Jarai and Bhabar region	Loam, clay and silt	Awas Vikas Colony, Nagarpalika, Bus Stand, Railway station
2.	Rudrapur	Udham Singh Nagar	Longitude : 70°30' E Latitude : 29°03' N Altitude : 244 m amsl	Loamy sand & sandy loam silt	Bus Stand, Danpur village Pattharchata
3.	Ram Nagar	Nainital	Longitude : 79°05' E Latitude : 29°40' N Altitude : 365 m amsl situated at the sanles of river Kosi	Silty loam	Garjiya Devi, bus stand
4.	Haldwani including Kathgodam	Nainital	Longitude : 79°31' E Latitude : 29°40' N Altitude: 423m amsl	Loamy sand and clay	Bus stand Sheetla Mata mandir, Gola river, Kaladhungi road

possibility of the occurrence of the members of this algal group.

The specimens gathered from all the four collection sites have received an identical treatment to avoid any discrepancy due to climatic hazards. The cyanobacterial flora of various research sites of foothills have been recorded separately and then combined to study the 'Cyanobacterial Flora of Foothills of Kumaon Himalayas'. All in all, the foothills of Kumaon Himalayas as a whole exhibited 170 species of 38 genera of 11 families and 4 orders. A summarized account of the species recorded during the study is presented in Table. (See Appendix)

Evidently the region of the foothills of Kumaon Himalayas is copiously rich in cyanobacterial flora but has remained neglected for a long time from the point of view of its taxonomic studies. Evidently, it needs thorough investigations to explore all possibilities of deciphering specimens with interesting features, from the academic as well as economic parameters.

ACKNOWLEDGEMENTS

The authors are highly thankful to UGC, New Delhi for financial support.

REFERENCES

Finni, T., Kononen, K., Olsonen, R. and Wallströn, K. (2001) The history of Cyanobacterial blooms in Baltic Sea. *Ambio.* **30:** 172-178

Fogg, G.E., Stewart, W.D.P., Fay, P. and Walsby, A.E. (1973). The Blue-Green Algae. Academic Press, London

vanden Hoek, C., Mann, D.J. and Johns, H.M. (Eds.) (1993). Algae: An Introduction to Phycology, 2nd ed. Cambridge Univ Press, Cambridge.

Kahru, M., Horstmann, U. and Rud, O. (1994). Satellite detection of increased Cyanobacteria bblooms in the Baltic Sea: natural fluctuation or ecosystem change? *Ambio.* **23:** 469-472

Rizvi, S.J.H. and Rizvi, V. (1992). Allelopathy: Basic and Applied Aspects. Chapman & Hall, London.

Rudi, K., Skulberg, O.M., Larsen, F. and Jakobsen, K.S. (1997). Strain characterization and classification of oxyphoto bacteria in clone cultures on the basis of 165 r – RNA sequences from the variable regions V 6, V 7 and V 8. *Appl. Environ. Microbiol.* **63:** 2593-2599.

Singh, R.N. (1961). Role of Blue-Green Algae in the Nitrogen Economy of Indian Agriculture, ICAR Publ., New Delhi.

Appendix

Order-wise distribution of cyanobacterial species as observed from different research sites

CLASS CYANOPHYCEAE			
Chroococcales **Chroococcaceae**	**Chamaesiphonales** **Cyanidiaceae**	**Nostocales** **Oscillatoriaceae**	**Stigonematales** **Nostochopsidaceae**
1. *Microcystis* (9)* 2. *Chroococcus* (10) 3. *Gloeocapsa* (4) 4. *Gloeothece* (1) 5. *Aphanocapsa* (10) 6. *Aphanothece* (1) 7. *Synechococcus* (2) 8. *Synechocystis* (1) 9. *Coelosphaerium* (1) 10. *Merismopedia* (4) 11. *Dactylococcopsis* (1)	1. *Chroococcidiopsis* (1)	1. *Arthrospira* (5) 2. *Spirulina* (8) 3. *Oscillatoria* (52) 4. *Crinalium* (1) 5. *Katagnymene* (1) 6. *Phormidium* (7) 7. *Lyngbya* (14) 8. *Schizothrix* (1) 9. *Symploca* (1)	1. *Nostochopsis* (3)
Entophysalidaceae	**Dermocarpaceae**	**Nostocaceae**	**Stigonemataceae**
12. *Chlorogloea* (2)	2. *Dermocarpa* (1) 3. *Stichosiphon* (1)	10. *Cylindrospermum* (3) 11. *Nostoc* (6) 12. *Anabaena* (3) 13. *Pseudanabaena* (1) 14. *Raphidiopsis* (1) 15. *Aulosira* (5) 16. *Hormothamnion* (1)	2. *Westiella* (1)
		Scytonemataceae	
		17. *Plectonema* (2) 18. *Pseudoscytonema* (1) 19. *Camptylonemopsis* (2)	
		Microchaetaceae	
		20. *Microchaete* (1)	
		Rivulariaceae	
		Calothrix (1)	
Families: 02	**02**	**05**	**02**
Genera: 12	**03**	**21**	**02**
Species: 46	**03**	**117**	**04**
Grand Total: Families: 11	Genera: 38	Species: 170	

Number in parenthesis (*) represents the no. of species.

Diversity of Lower Plants (2014): pp. 78-87
Editors: Rajan Kumar Gupta & Mukesh Kumar
IK International Publishing House Pvt. Ltd.

6

Cyanobacterial Biodiversity and Ecostrategist

DIVYA TOPAL* AND N.K. SHARMA

Department of Botany, Government P.G. College, Rishikesh (Uttarakhand)

ABSTRACT

Cyanobacteria obtain their energy through photosynthesis. They are not only a significant component of the marine nitrogen cycle and an important primary producer in many areas of the ocean, but are also found in habitats other than the marine environment; in particular cyanobacteria are known to occur in both freshwater and hypersaline inland lakes. Stromatolites of fossilized oxygen-producing cyanobacteria have been found from 2.8 billion years ago. The physiological properties of cyanobacteria vary from species to species. As a consequence, different "ecostrategists" inhabit different types of water bodies.

Keywords: Ecostrategist, Cyanobacteria, Eutrophic, Nitrogen fixer.

1. INTRODUCTION

Cyanobacteria are also known as blue-green algae, blue-green bacteria or cyanophyta. They belong to a phylum of bacteria that obtain their energy through photosynthesis. The name "cyanobacteria" comes from the colour of the bacteria (Greek: κυανός (kyanós) = blue). They are a significant component of the marine nitrogen cycle and an important primary producer in many areas of the ocean. They are also found in habitats other than the marine environment. Basically, cyanobacteria occur in both freshwater and hypersaline inland lakes. Stromatolites of fossilized oxygen-producing cyanobacteria have been found from 2.8 billion years ago. The ability of cyanobacteria to perform oxygenic photosynthesis is thought to have converted the early reducing atmosphere into an oxidizing one, which dramatically changed the composition of life forms on earth by provoking an explosion of biodiversity and leading to the near-extinction of oxygen-intolerant organisms. Chloroplasts in plants and eukaryotic algae have evolved from cyanobacteria via endosymbiosis.

*Communicating Author

Cyanobacteria are mostly prevalent in oceans, fresh water, bare rock and soil. Most are found in fresh water, while others in marine habitats. They occur in damp soil, or even temporarily moistened rocks in deserts. A few are endosymbionts in lichens, plants, various protists, or sponges and provide energy for the host. Some live in the fur of sloths, providing a form of camouflage.

Cyanobacteria include unicellular and colonial species. Colonies may form filaments, sheets or even hollow balls. Some filamentous colonies show the ability to differentiate into several different cell types: vegetative cells, the normal, photosynthetic cells that are formed under favourable growing conditions; akinetes, the climate-resistant spores that may form when environmental conditions become harsh; and thick-walled heterocysts, which contain the enzyme nitrogenase, vital for nitrogen fixation. Heterocysts may also form under the appropriate environmental conditions (anoxic) wherever nitrogen is necessary. Heterocyst-forming species are specialized for nitrogen fixation and are able to fix nitrogen gas, which cannot be used by plants, into ammonia (NH_3), nitrites (NO_2^-) or nitrates (NO_3^-). These are absorbed by plants and converted to protein and nucleic acids. The rice paddies of Asia, which produce about 75% of the world's rice, could not do so were it not for healthy populations of nitrogen-fixing cyanobacteria in the rice paddy fertilizer.

2. CYANOBACTERIAL BIODIVERSITY

Out of the total estimated area of 150 million sq. km of the earth, about 70.68% is occupied by oceans. However, of all the total photosynthetic productivity of 555.2 billion tons of dry weight/ year on earth, only 34.4% is contributed by the oceans. India has a vast coastline of over 7500 km. In addition, it has many lakes, ponds, puddles, backwater areas and a tropical climate. This results in abundance of natural populations of varied organisms. Cyanobacteria are widespread and abundant in most marine habitats. Their ability to grow in seawater is presumably related to a preference for alkaline conditions and an ability to tolerate high salt concentrations. The resistance, which many species show towards osmotic shock, extremes of temperature and reducing conditions, suits their existence in a variety of intertidal habitats. Desikachary suggested that probably 20% of all known cyanobacteria occur in saline conditions and a majority of them are truly marine. Thajuddin and Subramanian have made a detailed survey of marine cyanobacterial biodiversity of a continuous stretch of over 2660 km off the coastline from Tirakol of Goa state (lat.15°45′N and long. 77°33′E to Cape Comorin lat. 8°5′Nand long. 77°33′E) of Tamil Nadu and from Cape Comorinto Bhimunipattanam of Andhra Pradesh (lat. 17°55′N and long. 83°25′E) encompassing the regions such as the Arabian Sea, Indian Ocean, Paland Bay of Bengal including Andaman and Nicobar and Lakshadweep Islands. This survey included coverage of not only the shore, deeper sea but also stagnant seawater ponds and puddles, backwater and salt pans.The nature of shores in south India and Andaman and Nicobar and Lakshadweep Islands of India varied markedly in different regions. Except for the rocky spots of Kovalam, Vizhinyam Edeva, Varkale (Kerala), and Dona Paula (Goa), the shores of the Arabian Sea region (West Coast) are essentially sandy. The shores of the Bay of Bengal region are mostly sandy with few extensive rocky spots at Bhimanipattanam,Vishakapatnam, Pudimadaka (Andhra Pradesh), Kovalem and Mahabalipuram (Tamil Nadu). The shores of the Palk Strait region exhibit greater variety. Along the great lagoons, the shores are clayey and particularly in Adhirampattinam and Muthupet area, swampy and inaccessible. The shore towards the Palk Bay from Palk Strait is sandy and rich in organic matter due to deposition of dense layers of seagrass over vast stretches. However, most of the Palk Bay has a sandy shore with scattered boulders and masses of coral rocks. The shores of the Gulf of Mannar region vary from place to place. On the South Coast of Mandapam, the shores are rocky due to extensive sand stones. In Rameshwaram Islands, the shore is slightly rocky in some areas but mostly sandy. About 6–10 km away from the Mandapam and Ervadi seashore there are several small islands, viz. Kurusadai, Shingle, Pulli, Pullivasal, Nallathannithevu, Upputhannithevu and Hare Island. The shores towards Cape Comorin (Tiruchendur, Idinthakarai, etc.) are sandy with intermittent rocky spots, while at Cape Comorin it is partly rocky and partly sandy. The majority of the shores in Andaman (smooth, middle, north, little Andamans), Nicobar (Car Nicobar, Katchal island) and Lakshadweep Islands (Kavarathi, Agatti, Amini, Kadamath) are sandy with rocky and vast coral stretches in some areas.

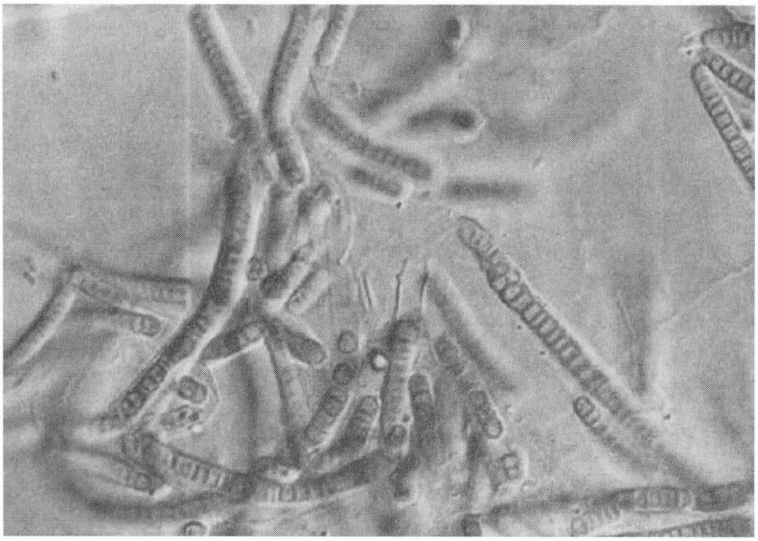

Filamentous cyanobacteria

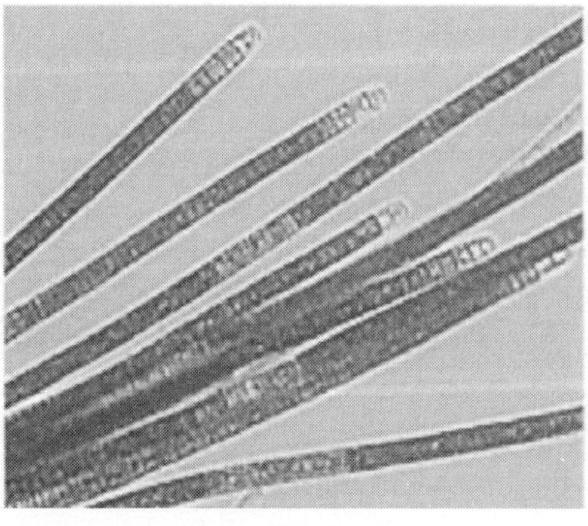

Trichodesmium

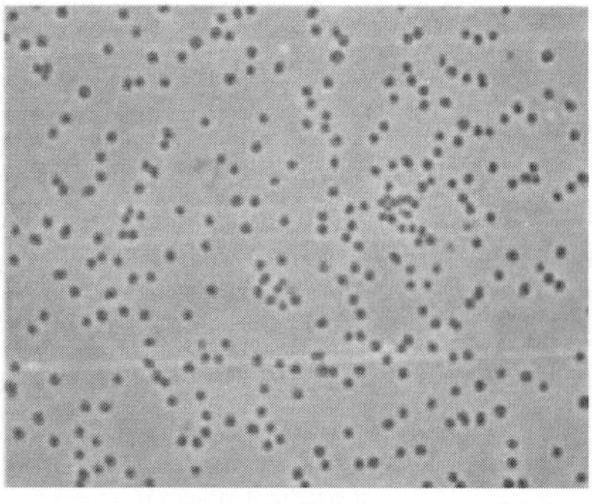

Prochlorococcus

3. BIODIVERSITY OF MARINE CYANOBACTERIA

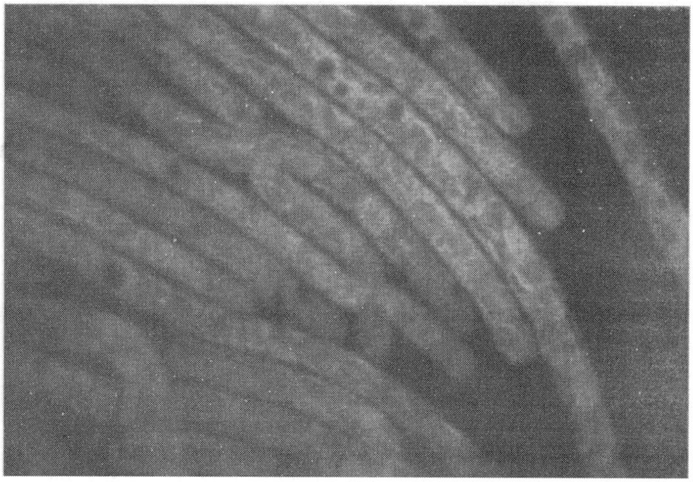

Trichodesmium

3.1 Rocky Shores

The availability of benthic cyanobacterial biodiversity is greatly influenced by the type of substrata in the intertidal area. Thajuddin and Subramanian observed that the shore in the Bay of Bengal was essentially sandy and therefore there were only 11 species of cyanobacteria whereas in the Gulf of Mannar region at several places the shores were rocky (sand stones) or coral reefs were present as many as 145 species were observed which included both benthic and planktonic forms. Cyanobacteria are apparently always present as epiliths, chasmolith and endoliths. On sandy shores, the cyanobacterial population was very poor due to rough tides, absence of substratum, and poor nutrient content of water. Benthic cyanobacteria are abundant in mangrove environments. This is on account of rich organic muddy substratum, relatively stagnant shallow water conditions, sheltered nature (hence reduced water movement), and optimum salinity conditions (15–30 ppt). 58 species of cyanobacteria belonging to 22 genera in backwaters and mangrove habitats of the southern east coast of India was reported by Thajuddin and Subramanian. They also reported 50 species of 19 genera in salt pans with salinity of over 50 ppt. Non-heterocystous forms in general and the species belonging to the family Osillatoriaceae in particular were dominant; and some could grow at even 340 ppt salinity. High sulphide content in hypersaline environment is toxic to heterocystous forms. However, this contention was subsequently disputed and the prevailing anaerobic conditions in the dark in these environments were believed to exclude the heterocystous forms. Hof and Fremy studied the flora of salt waters. Algae was divided into two physiological groups, halotolerant and halophilic. Halotolerant species are not able to grow at NaCl concentration above 3 M (175.5 ppt), e.g., *Calothrix scopularum*; other examples include salt-water forms such as, *Microcoleus chthonoplastes* and *Lyngbya aestuarii. Phormidium valderianum* BDU was shown to grow at salinity reaching 100 ppt. There was a good positive correlation between the temperature, salinity and the levels of chloride, ammonia, inorganic phosphate and total phosphorus and iron during the period of study (1988, 1989); typical negative correlation existed between wind speed, rainfall, sea level pressure and dissolved oxygen. Amongst various factors that influence the blooms of *Trichodesmium*, ammonia was more a consequence than a cause of the bloom, because of the high rates of hydrogen fixation observed in this organism. As many as six species of cyanobacteria, namely, *Ammeatoidea normanii*, *Dichothrix spiralis*, *Oscillatoria borneti*, *Pseudoanchobrosa fluminenensis*, *Radaisia violacea* and *Siphononema polonicum*, were reported for the first time from Indian coasts.

4. BIODIVERSITY OF FRESHWATER CYANOBACTERIA

Cyanobacteria, belonging to the order Chroococcales, and families Osillatoriaceae and Nostacaceae occur ordinarily as planktonic forms. Several species grow in abundance and colour the entire body of water, forming the so-called water-blooms. Huber-Pestalozzi (cited by Desikachary) lists as many as 41 genera of 251 species of cyanobacteria that occur as freshwater planktons. *Microcystis* is one of the dominant organisms that is associated with almost permanent blooms in tropical freshwaters that are exposed to constant sunshine, warmth and nutrients like phosphate, silicate, nitrates, CO_2 and lime. Cyanobacteria in freshwaters have been reported from sea level to high altitudes. They are abundant in temple tanks, ponds in the hill ranges and water lakes like Kodaikanal Lake, Ooty Lake and Yercaud Lake. Huber-Pestalozzi has listed species of genera, *Anabaena*, *Anabaenopsis*, *Aphanizominon*, *Arthrospira*, *Coelospaerium*, *Gloeotrichia*, *Lyngbya*, *Microcystis*, *Nostoc*, *Nodularia*, *Oscillatoria*, *Spirulina gomontina* and *Lyngbya* to form water blooms. The water blooms are not suitable for drinking water supplies since they choke flow in sand filters and cause mortality of fish and animals. The principal toxic species are *Aphanizomenon flos-aquae, Microcystis aeruginosa*, *Microcystis flos-aquae, Gloeotrichia echinulata* and several species of *Anabaena.* Cyanobacteria also occupy a variety of terrestrial environments. Nitrogen fixing cyanobacteria have also been reported in sugar cane and maize fields in India. Besides nitrogen fixation by heterocystous and few non-heterocystous forms, certain other cyanobacterial species such as *Lyngbya*, *Microcoleus*, *Porphyrosiphon* and *Schizothrix* often form crusts on soils. Summarizing past studies of cyanobacterial distribution in rice fields of tropical and temperate countries, it can be concluded that cyanobacteria constitute about 15% of the total algal flora in the tropics and about 2% in the temperate climate.

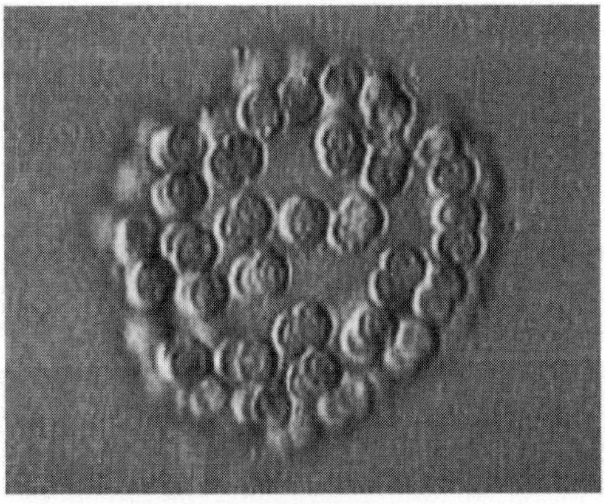

Anabaena

5. CYANOBACTERIA OF EXTREME ENVIRONMENTS

Nostoc

Cyanobacteria have been reported to grow in brines, where they form thick mats at the bottom. *Microcoleus chthonoplasts* and *Oscillatoria* species together with other cyanobacterial components as minor partners, such as *Aphanocapsa marina*, *Lyngbya aestuarii* and *Spirulina subsalsa* are found in these mats. Thajuddin and coworkers have reported as many as 89 species from the east coast and 69 species. Of a total of 36 species in 16 genera recovered from salt pans of Pudukkottai District, Tamil Nadu, 18 species exhibited varying degree of salinity tolerance (45–90 ppt). Cyanobacteria have been reported from thermal waterfall over the world. A recent account estimated that 700 taxa of nonmarine algae are present in Antarctica. The flora is dominated by species of *Anabaena*, *Aphanocapsa*, *Calothrix*, *Chroococcidiopsis*, *Gloeocapsa*, *Lyngbya*, *Mastigocladus*, *Microchaete*, *Microcoleus*, *Oscillatoria*, *Phormidium*, *Plectonema*, *Pseudoanabaena*, *Nodularia*, *Nostoc*, *Schizothrix*, *Scytonema*, *Stigonema*, *Synechococcus* and *Tolypothrix*. Extreme alkaliphiles live in soils laden with soda (natron) or in soda lakes where

Marine algal

the pH can rise to 12, but such organisms grow poorly at neutral pH. As many as 13 alkaliphilic cyanobacteria were reported to grow under alkaline conditions. Very often, soda lakes are monospecific, inhabited by *Spirulina platensis* which serves as human food (single cell protein) of high nutritional value. Cyanobacteria are often exposed to heavy metal pollution due to the disposal of industrial and domestic wastes into waterways.

6. CYANOBACTERIAL ECOSTRATEGISTS

Different species have different physiological properties of cyanobacteria. As a consequence, different "ecostrategists" inhabit different types of water bodies. A preliminary approach to describe these ecostrategists, based on ecophysiological laboratory working together with field observations (largely from north-western Europe), is described below. This information may be useful for management, as it predicts which cyanobacteria can be expected to occur under certain conditions. Further development of this approach will be possible as more data on occurrence of cyanobacteria under different growth conditions are collected from other continents.

6.1 Scum-Forming Ecostrategists

Many cyanobacteria develop large aggregates (colonies) of coccoid cells or filaments during the vegetation period which are not homogeneously distributed over the water column. *Microcystis, Anabaena* and *Aphanizomenon* are some of the important genera that show this development. The rate of photosynthesis of the colonies is high at the surface of water and the cells store large quantities of carbohydrates. Although the cells contain gas vesicles, the heavy carbohydrates acts as ballast and induce sinking within the colonies. Buoyancy regulation enables the colonies to position themselves in light conditions which are optimal for their growth. A prerequisite is that the water body is not too turbulent. During the night, all colonies may become buoyant and some of the population may be accumulated on the water surface where they can be blown together by wind, forming stable scums along downwind shores. *Microcystis* spp. may become dominant and form substantial blooms, as has been reported from Hungary, Australia, and particularly from subtropical and tropical regions. Many cyanobacteria cannot survive high light intensities over longer periods. This may limit their distribution to more turbid, eutrophic ecosystems. However, *Microcystis* species are less sensitive to high light intensities because buoyancy regulation enables them to find light conditions that are optimal for their growth. This means that the presence of *Microcystis* cannot be related strictly to the level of eutrophication. The genus is therefore found in mesotrophic, eutrophic and in hypertrophic waters. However, the amount of biomass that this species can attain depends on the level of eutrophication.

6.2 Homogeneously Dispersed Ecostrategists

This ecotype comprises filamentous species, such as *Planktothrix (Oscillatoria) agardhii* and *Limnothrix (Oscillatoria) redekei*. These filamentous species which are extremely sensitive to high light intensities and do not form colonies (Reynolds, 1987). The vertical migration by buoyancy regulation is less pronounced than their passive entrainment by water circulation due to small size of filaments. These species are therefore homogeneously dispersed throughout the epilimnion.

Distribution of phytoplankton in Lake Veluwe, The Netherlands, showing an almost permanent bloom of cyanobacteria during the years 1978-83. The dominant bloom of *Planktothrix agardhii* was only interrupted by periods of ice cover (Redrawn from RIZA, 1996).

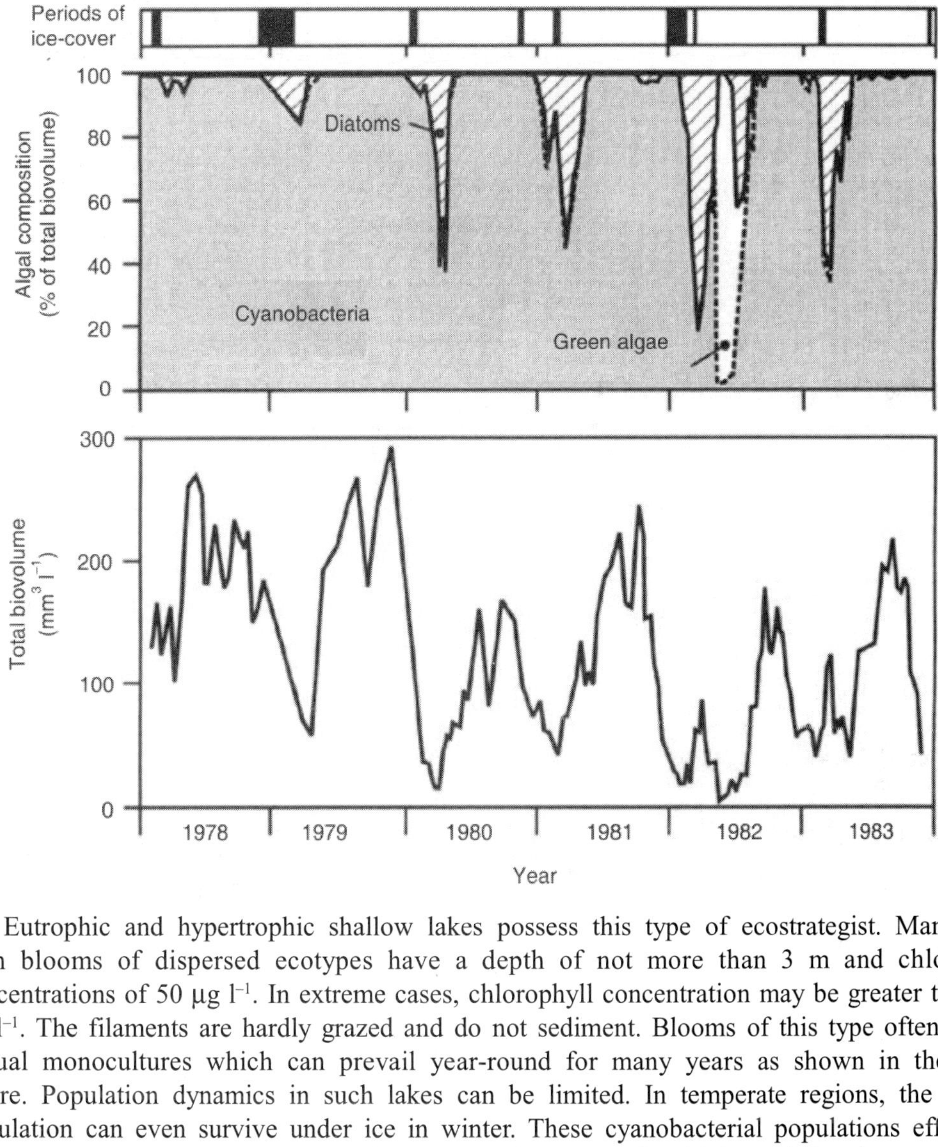

Eutrophic and hypertrophic shallow lakes possess this type of ecostrategist. Many lakes with blooms of dispersed ecotypes have a depth of not more than 3 m and chlorophyll concentrations of 50 µg l^{-1}. In extreme cases, chlorophyll concentration may be greater than 200 µg l^{-1}. The filaments are hardly grazed and do not sediment. Blooms of this type often lead to virtual monocultures which can prevail year-round for many years as shown in the above figure. Population dynamics in such lakes can be limited. In temperate regions, the autumn population can even survive under ice in winter. These cyanobacterial populations effectively suppress the growth of other phytoplankton species by causing high turbidity. As a result, the next summer population establishes itself almost without any seasonal succession between different species of phytoplankton. This high stability of the population precludes any redistribution of phosphorus and nitrogen to other components of the ecosystem and this can cause a resilience effect in lake restoration projects.

6.3 Stratifying Ecostrategists

Representatives of this ecotype develop stable summer populations in the intermediate zone of thermally stratified lakes and reservoirs known as the metalimnion. The organisms contain the red pigment phycoerythrin to absorb the green light, which is the prevailing wavelength at this depth. The most common of these species is *Planktothrix (Oscillatoria) rubescens,* but red varieties of other *Planktothrix* species can also form metalimnic populations (Aune *et al.,* 1997). The single filaments of these species hardly show any vertical migration. However, in late autumn at the end of the growing season, the cells can become buoyant and then form red surface scums (Walsby *et al.,* 1983). The niche of this type of *Planktothrix* is very limited. It needs sufficient light in the metalimnetic zone, but may be inhibited by too much light.

6.4 Nitrogen-Fixing Ecostrategists

The mass development of species capable of fixation of atmospheric nitrogen (species of the genera *Anabaena, Aphanizomenon, Cylindrospermopsis, Nodularia,* and *Nostoc*) can often be related to periodic nitrogen limitation. Examples are found in deep, as well as in shallow, systems. However, while these ecostrategists often dominate in ecosystems with low levels of inorganic dissolved nitrogen, the reverse does not necessarily apply. The lakes with clear nitrogen limitation are not dominated by nitrogen-fixing cyanobacteria. The main reason for this is the availability of low ligth, because nitrogen fixation requires high amounts of energy. A number of nitrogen fixing species can form colonies and possess gas vesicles. This means that they can regulate buoyancy, like *Microcystis,* and can form stable scums along downwind shores. Restoration measures simultaneously reduce phosphate and nitrogen loading (sewage diversion, isolation). It may also strengthen prevailing nitrogen-limiting conditions and hence the probability of large populations of nitrogen-fixing cyanobacteria.

7. SMALL COLONY-FORMING TAXA

There are many cases of large populations of the small, colony-forming genus *Aphanothece* that have been reported. Not much information is available on buoyancy regulation and scum formation by the species involved. After a decrease of *Planktothrix rubescens* populations in several water bodies, *Aphanothece* dominance has occurred. The dominance of this group is not strictly related with phosphate or nitrogen limitation, and there are no obvious relationships that can explain the sudden dominance of these cyanobacteria. They seem to dominate in an intermediate state during lake recovery after restoration measures have been taken; their ecology is unknown.

8. BENTHIC CYANOBACTERIA

Other than the planktonic ecostrategists described above, cyanobacteria may grow on the bottom sediments of water bodies which are sufficiently clear to allow light penetration to these surfaces. These benthic species may form coherent mats. Especially high rates of photosynthesis by such mats sometimes trap the photosynthetically produced oxygen. This oxygen produced is trapped as bubbles within the mats. The parts of the mats may then become sufficiently buoyant to tear loose and then rise to the surface. Awareness of these is important for monitoring and management of toxic cyanobacteria. This is because cyanotoxin problems are usually not expected in clear, oligotrophic waters. According to Gunn *et al.,* 1992 toxic benthic cyanobacteria have caused animal deaths in Scotland, where beached mats along the shore of a clear loch were scavenged by dogs.

REFERENCES

Aune, T., Ramstad, H., Skulberg, O.M., Underdal, B., Yndestad, M. and Østensvik, Ø. (1997). *Cyanotoksiner og edelkreps - toksinproduserende blågrønnalger i Steinsfjorden sommeren 1997.* Norwegian Institute for Water Research, Oslo.

Gunn, G.J., Rafferty, A.G., Rafferty, G.C., Cockburn, N., Edwards, C., Beattie, K.A. and Codd, G.A. (1992). Fatal canine neurotoxicosis attributed to blue-green algae (cyanobacteria). *Vet. Rec.* **4:** 301-302.

Riza, (1986). *Bestrijding van de Eutrofiëring van het Veluwe-Drontermeer.* Rijkswaterstaat dienst binnenwateren, Lelystad.

Reynolds, C.S. (1987). Cyanobacterial waterblooms. In: P. Callow (Ed.). *Advances in Botanical Research,* **A**cademic Press, London **13:** 17-143.

Walsby, A.E., Utkilen, H.C. and Johnsen, I.J. (1983). Buoyancy changes of a red coloured *Oscillatoria agardhii* in lake Gjersjøen, Norway. *Arch. Microbiol* **97:** 18-38.

Diversity of Lower Plants (2014): pp. 88-95
Editors: Rajan Kumar Gupta & Mukesh Kumar
IK International Publishing House Pvt. Ltd.

7

Role of Cyanobacterial Biodiversity Resources in Agricultural Sustainability

USHA PANDEY

Department of Botany, Faculty of Science and Technology,
M.G. Kashi Vidyapith, Varanasi–221 001
Email : usha_pandey28@yahoo.co.in

ABSTRACT

The use of inorganic fertilizers, during last few decades, although significantly increased the global food production, yet at a great environmental cost. In addition, excessive use of synthetic fertilizers damages agricultural sustainability and is not affordable to most of the small farm holders. This necessitates exploring most sustainable farm management practices, especially for countries like ours.

The objective of this pot-culture study is to investigate the possible use of cyanobacterial biodiversity resource in maintaining soil fertility stability in a non-paddy agroecosystem. An oil yielding crop, sunflower (*Helianthus annus* L.), was considered a test crop to establish cause-effect relationship. The chosen determinants included cow dung manure (T_3), vermicompost (T_4) and cyanobacterial biofertilizer (T_5). The results were compared with conventional and modern farm practices using unamended soil as control (T_1) and soil amended with recommended dose of NPK (T_2). The results showed that T_5 soils had superiority over rest of the treatments with respect to water holding capacity (WHC), total organic carbon (TOC), water soluble organic carbon (WSOC) and microbial biomass carbon (C_{mic}). Activities of all the four soil enzymes considered in this study remained significantly ($p < 0.01$) higher in comparison to those observed under other soil amendments. In particular, in soil amended with cyanobacterial biofertilizer, the enzyme activities increased from 1.5 to 7 fold in comparison to the NPK amended soil (T_2). The sunflower plant showed almost a synchronous trend for most of the growth variables, especially seed yield and dry matter production. The study indicates that the soils amended with cyanobacterial biofertilizer not only substantially promote the growth and yield of sunflower but also help attending soil fertility stability in croplands. This has relevance for future agricultural sustainability in India.

Keywords: Agricultural sustainability, Cyanobacteria, Biodiversity resource, Organic farming, Microbial biomass, Soil enzymes, Sunflower.

1. INTRODUCTION

The farmers in different parts of our country adopt diverse agricultural practices, most of which are based on low input subsistence farming. In our past agricultural practices, with low population pressure, we practiced crop rotation as a tool for replenishment of soil fertility under natural fallows. However, with rapid population growth, fallow periods have been shortened and continuous and intensive cropping without restoration of the soil fertility has depleted the nutrient base of most soils. The pressure of rapidly increasing population in India, the highest of any country in the world, on agricultural land has resulted in much higher nutrient outflows and the subsequent breakdown of many traditional soil fertility amendment practices. This created a vicious circle of food insecurity. Improving food production and soil resources, especially in the small farm holding sector of India, has become a big challenge. During the last few decades, tremendous use of pesticides and chemical fertilizers in the agriculture to meet the global food demand, although increased the grain yields substantially (Tilman *et al.,* 2001), deteriorated the soil fertility in many parts of the world including India. Soil fertility replenishment has been singled out as an essential condition for sustainable agriculture and soil management in India. A sustainable soil fertility replenishment strategy has the potential to supply nutrients and arrest the nutrient mining and thus, the soil fertility stability is maintained. The soil fertility stability in intense farming can only be maintained through integrated plant nutrient management, with efficient cycling of nutrients through the use of biofertilizers.

Cyanobacteria (blue-green algae), being an important component of soil microflora, help retaining soil moisture and maintaining soil fertility. These organisms are ubiquitous in distribution and sizably contribute to global nitrogen balance (Whitton, 2000). Cyanobacteria grow luxuriantly in low land paddy fields which provide an ideal environment for their vigorous growth and mat formation (Singh, 2000). Due to their strong ability to withstand environmental stresses (Goyal, 1989), cyanobacteria proliferate rapidly and establish fastly even in resource poor environments. Diazotrophic cyanobacteria contain nitrogenase, the key enzyme of nitrogen fixation, for harnessing atmospheric nitrogen. The role of these cyanobacteria in maintaining soil fertility of the paddy fields has been well substantiated and documented (Singh, 1961; Venkataraman, 1981; Singh, 2000; Whitton, 2000). Temporal fluctuation and relative abundance of cyanobacterial diversity at different crop growth stages in fields affect the total yields of crops.

In addition to their important contribution as diazotrophs and soil binders, cyanobacteria improve soil physical properties in rhizosphere making the soil environment more conducive for other microbes to grow and function. The mucilage of cyanobacteria, which often constitutes a crucial component of rhizosphere, contributes to various plant-soil interactions, microbial dynamics and nutrient turnover (Whitton, 2000). The soil contains sizable quantity of surface active materials, many of which are of cyanobacterial origin (Passioura, 1988; Pandey, 2009). These surface active materials modify surface tension of soil solution, increase hydraulic conductivity, nutrient uptake and microbial processes (Pandey, 2009). Increase in hydraulic conductivity and changed surface tension and metric potential facilitate plant water uptake especially under the condition of water deficit in soil.

The present investigation is an attempt to study the role of cyanobacterial biofertilizer, as a crop management system, in regulating soil fertility stability in a non-paddy agro-system.

2. MATERIALS AND METHODS

2.1 Experimental Design

Pot culture experiments were conducted using soil fertilizer mix filled in earthen pots of 35 cm height × 30 cm diameter in size. The experimental treatment consisted of a randomized block design using oil yielding crop as a test plant (sunflower), three organic fertilizer sources and three replicates of each set. The soil used for pot culture experiments was silt clay in texture; pH, 7.3; electrical conductivity, 0.26 d S m^{-1}; organic carbon, 0.68%; total N, 0.15%; total P, 0.02% and K, 0.12%.

Thick cyanobacterial mats were scrapped from agricultural fields. Species such as *Aulosira, Gloeocapsa*, *Nostoc*, *Scytonema* and *Oscillatoria* constituted the major component accounting for more than 90% of the total mat biomass. Uniform mats collected from agricultural fields were air dried and powdered for preparing soil mix. Experimental soil mixes were prepared using soil with cyanobacterial powder (ratio, 9 : 1), cow dung manure (ratio, 7 : 3) and commercial vermicompost (ratio, 8 : 2). To increase the validity of comparison, unamended soil was considered as control and the sets containing recommended dose of NPK were used for comparison with modern agricultural practices using synthetic fertilizers. The treatments hereafter are referred to as T_1 : unamended control; T_2 : recommended NPK; T_3 : cow dung manure (CDM) amendment; T_4 : vermicompost amendment and T_5 : soil amended with cyanobacterial biofertilizer. Seeds of sunflower (*Helianthus annus* L., MSFH-17) were sown uniformly in triplicates of each set of earthen pots containing soil mixes with uniform soil moisture levels.

2.2 Sampling and Analysis

Soil samples were collected from each experimental pot. Three bulk samples, each consisting of 500 g random cores of 2.5 cm × 15 cm size, were placed in sealed bags and brought to the laboratory for analysis. Total organic carbon (TOC) was determined by dry combustion and water-soluble organic carbon (WSOC) was estimated following dichromate oxidation titration method of Ciavatta *et al.* (1987). Microbial biomass (C_{mic}) was measured using chloroform fumigation extraction procedure as described in Vance *et al.* (1987). Activities of soil enzymes, dehydrogenase (Casida *et al.,* 1964), catalase (Beck, 1984), alkaline phosphatase (Tabatabai and Bremner, 1969) and urease (Hofmann, 1963) were determined following standard procedures.

To assess the yield, crop was finally harvested at maturation, and after 10 days of drying, seeds were removed by mild threshing. Seed yield and stalk yield, computed for all the treatments separately, were finally converted to quintal per hectare. Significant differences were assessed using analysis of variance (ANOVA). Data variabilities were expressed in terms of standard error of means (SE).

3. RESULTS AND DISCUSSION

The growing demand for sustainable cropping necessitates understanding the impact of all agricultural practices on soil fertility stability. There is a growing concern towards estimating soil properties under different organic amendments. In these lines, cyanobacterial biofertilizers have received attention globally. Table 1 shows the characteristics of the soil in the presence or absence of different fertilizer applications. In agreement with the findings of Condron *et al.* (2000), significant difference in the organic-C content was observed in soil samples having different types of organic amendments. The fertilizers considered in this study differ in nature of organic matter available for decomposition, rate of mineralization and consequently release of nutrients for plant growth. As expected, TOC was higher in T_3, T_4 and T_5 soils characterized by application of organic amendments than under control (T_1) and NPK treated (T_2) sets (Table 1). Total organic carbon (TOC) was highest in T_5 samples which showed 18.86 g of TOC kg^{-1} of soil. The water soluble organic carbon (WSOC) in soil, which is composed mainly of organic acids and water soluble carbohydrates, plays an important role in ecosystem functioning. In the present study, the WSOC content also differed significantly and WSOC value was highest in soil amended with cyanobacterial biofertilizer (530 mg kg^{-1}) and lowest in unamended control soil (438 mg kg^{-1}). This could be due to a consequence of the nature of substrate available for decomposition and moisture effects sustained by of existing organic matter (Christ and David, 1996).

Table 1: Physico-chemical and biological characteristics of experimental soils. Values are mean (n = 9) ± 1SE

Treatment	Water holding capacity (%)	Total organic carbon (TOC; g kg^{-1})	Water soluble organic carbon (WSOC; mg kg^{-1})	Microbial biomass carbon (C_{mic}; mg kg^{-1})
T_1	41.80 ± 3.64	10.5 ± 1.21	438 ± 27.50	330 ± 18.90
T_2	42.00 ± 2.87	10.5 ± 1.06	440 ± 31.70	340 ± 22.60
T_3	53.65 ± 3.56	15.5 ± 1.28	450 ± 30.65	450 ± 32.55
T_4	56.18 ± 4.15	13.8 ± 1.12	495 ± 36.15	490 ± 37.92
T_5	58.56 ± 5.04	18.86 ± 1.40	530 ± 42.11	580 ± 44.67

T_1 : control ; T_2 : soil amended with recommended dose of NPK; T_3 : soil amended with cow dung manure; T_4 : soil amended with vermicompost ; T_5 : soil amended with cyanobacterial biofertilizer.

Soil management practices are shown to have significant effect on microbial biomass (Jenkinson and Powlson, 1976). Microbial biomass carbon (C_{mic}) is often considered an useful parameter to indicate change in soil management practices that affect the turnover of organic matter (Nannipieri *et al.,* 1990). Among the five fertilizer treated management systems, untreated soil shows the lowest value of C_{mic} (330 mg kg^{-1}) compared with organically treated soils (450, 490 and 580 mg kg^{-1} for T_3, T_4 and T_5 respectively). Such differences could be expected due to differences in microbial population in soils promoted by organic fertilizers to support decomposition (Pankhurst *et al.,* 1996). In particular, soils amended with cyanobacterial biofertilizer supported highest microbial biomass (Table 1). Nitrogen-rich soft residues of cyanobacteria provide better substrate for microbes to colonize and proliferate (Pandey and Pandey, 2001). This effect was clearly evidenced in the present study indicating the significance of cyanobacterial residue maintaining soil fertility. Since enzyme activities are affected by environmental variables and farm practices (Gregorich *et al.,* 1994), these can be used as indicators of soil quality. Alkaline phosphatase and urease, both having hydrolytic activities, showed high values in T_5 soil samples, the respective values were 398 mg p-nitrophenol kg^{-1} of soil h^{-1} and 480 mg NH_4 – N kg^{-1} of soil h^{-1} (Table 2). This soil was treated with cyanobacterial mixture. The lower value of alkaline phosphatase (200 mg p-nitrophenol kg^{-1} of soil h^{-1}) and urease (98 mg NH_4-N kg^{-1} of soil h^{-1}) in NPK treated soil could be explained by phosphate and nitrogen fertilization which decreases the activity. Application of P and N fertilizers has been shown to reduce the activity of alkaline phosphatase and urease respectively (Gianfreda and Billag, 1996; Pandey and Pandey, 2009). Dehydrogenase and catalase are used as an indicator of the biological status of soil. The soil enzyme catalase, even though acting mainly as an intracellular enzyme, remains active outside the microbial cells due to an association with the soil organic matter and/or sorption on clay minerals (Stotzky, 1974). Activities of both the enzymes were high in organically amended soils, the values being highest in soil supplemented with cyanobacterial biomass (4.80 mg TPF kg^{-1} soil h^{-1} and 2.79% O_2 g^{-1} of soil min^{-1} for dehydrogenase and catalase, respectively). The higher organic – C in soil coupled with higher aggregate stability, all leading to higher aeration in cyanobacterial biofertilizer amended soil may explain the greater activity of both dehydrogenase and catalase enzymes.

Table 2: Enzyme activities in the experimental soils. Values are mean (n = 9) ± 1SE

Treatment	Alkaline phosphatase (mg p-nitrophenol kg^{-1} of soil h^{-1})	Urease (mg NH_4 – N kg^{-1} of soil h^{-1})	Dehydrogenase (mg TPF kg^{-1} of soil h^{-1})	Catalase (% O_2 g^{-1} of soil min^{-1})
T_1	249 ± 17.64	108 ± 10.05	3.60 ± 0.21	0.34 ± 0.02
T_2	200 ± 14.50	98 ± 7.85	3.20 ± 0.18	0.40 ± 0.03
T_3	325 ± 26.48	250 ± 14.80	3.80 ± 0.25	1.60 ± 0.12
T_4	380 ± 27.85	310 ± 22.40	4.50 ± 0.33	2.20 ± 0.15
T_5	398 ± 33.67	480 ± 35.48	4.80 ± 0.35	2.79 ± 0.15

T_1: control; T_2: soil amended with recommended dose of NPK; T_3: soil amended with cow dung manure; T_4: soil amended with vermicompost ; T_5 : soil amended with cyanobacterial biofertilizer.

The growth and yield of sunflower as influenced by soil amendment of various fertilizers are presented in Tables 3 and 4. Among different organic fertilizer amendments such as cow dung manure (T_3), vermicompost (T_4) and cyanobacterial biofertilizer (T_5), highest growth and yield were observed in pots treated with cyanobacterial mix. The values of overall dry matter production were 66.94, 81.65 and 90.60 g per plant respectively for T_3, T_4 and T_5 plants (Table 3). It was interesting to note that, T_5 showed superiority in this respect even over that of the recommended dose of NPK (T_2). It might be due to higher nutrient content coupled with improved physical properties of soil which, in turn, resulted in continuous supply of nutrients and better growth. Application of cyanobacterial biofertilizer improves soil physical environment particularly water holding capacity and soil aggregate formation promoting root growth and thus increases the capacity of plant for efficient absorption and utilization of nutrients and water, and consequently, a better growth and higher seed yield. Similar results were obtained by Mishra and Sharma (1997) in case of FYM. Growth in terms of plant height, head diameter and seed yield were significantly ($p < 0.01$; ANOVA) higher in pots containing cyanobacterial mix (T_5). Gill and Meelu (1992) reported a similar effect of organic manure in rice-wheat cropping system.

Table 3: Effect of soil organic amendment on growth of sunflower. Values are mean (n = 9) ± 1SE.

Treatment	Plant height (cm)	Stem diameter (cm)	Head diameter (cm)	Dry matter yield (g/plant)
T_1	110.75 ± 8.60	3.15 ± 0.24	7.79 ± 0.67	48.60 ± 3.24
T_2	156.30 ± 11.25	6.11 ± 0.45	12.65 ± 1.05	82.00 ± 5.67
T_3	130.25 ± 11.00	6.42 ± 0.38	11.25 ± 1.21	66.94 ± 5.23
T_4	142.30 ± 12.50	5.20 ± 0.45	13.15 ± 1.22	81.65 ± 6.67
T_5	150.62 ± 12.05	7.29 ± 0.49	14.20 ± 1.20	90.60 ± 7.38

T_1: control; T_2: soil amended with recommended dose of NPK; T_3: soil amended with cow dung manure; T_4: soil amended with vermicompost ; T_5 : soil amended with cyanobacterial biofertilizer.

Table 4: Effect of soil organic amendment on yield of sunflower. Values are mean (n = 9) ± 1SE.

Treatment	Seed germination index	Seed yield (Q/ha)	Stalk yield (Q/ha)	Fertility (%)
T_1	56 ± 4.27	6.15 ± 0.44	13.50 ± 1.05	75.25 ± 4.58
T_2	66 ± 4.51	16.50 ± 1.10	20.50 ± 1.67	80.00 ± 5.76
T_3	68 ± 4.50	14.40 ± 1.12	16.25 ± 1.70	82.10 ± 6.18
T_4	69 ± 6.01	15.43 ± 1.20	10.30 ± 2.11	84.06 ± 6.15
T_5	70 ± 5.65	18.20 ± 1.35	22.80 ± 2.45	90.00 ± 6.90

T_1: control; T_2: soil amended with recommended dose of NPK; T_3: soil amended with cow dung manure; T_4: soil amended with vermicompost ; T_5 : soil amended with cyanobacterial biofertilizer.

Effects of different nitrogen sources on the yield and quality of sunflower have been studied by Kavitha and Swarajya Lakshmi (2002). These authors showed that the seed size and overall yield were increased in response to nitrogen supply. Reddy and Reddy (2003) have shown the increment in the size of sunflower when cultivated under integrated nutrient management. The reason behind the improved yield of sunflower under cyanobacterial biofertilizer application may be due to adequate and more efficient supply of nutrients to germinating seeds and solubilizing action of biofertilizer on fixed forms of nutrient in soil. The seed and stalk yield were also significantly ($p < 0.01$) increased under various treatments of organic and inorganic nutrients. In comparison to others, cyanobacterial biofertilizer more efficiently improved the physical, chemical and biological properties of soil (Table 1) which enhanced the availability of nutrient and played significant role in crop growth and yield.

Continuously declining soil fertility is a fundamental limiting factor for declining per capita food production in many parts of India. Although the use of inorganic fertilizers significantly increased crop yield, excess use of such fertilizers is neither sustainable nor affordable by poor farmers. This necessitates more sustainable farm management practices to reach to farmers sooner. The results of this pot culture study, although confirmation may be required with other non-paddy agricultural crops, indicated significant differences in the biochemical characteristics of soil in five studied systems. Microbial biomass and enzyme activities are the most important variables for assessing soil fertility status. All the three organically amended systems, although showed significantly greater contents of total organic carbon and microbial biomass, soil amended with cyanobacterial biofertilizer showed superiority over the soil amended with cow dung manure and vermicompost. Similarly, the enzyme activities remained significantly higher in all the three organically amended soils with superiority to those amended with cyanobacterial biofertilizer.

The data on crop growth indicated trends almost synchronous with soil fertility status. Accordingly, cyanobacterial biofertilizer amended soil showed superiority in promoting growth and yield of sunflower. The study suggests that cyanobacterial biodiversity resources can be used as an important tool for future sustainable agriculture for reversing soil degradation and maximizing agricultural harvest without much change in the fertility stability of agricultural lands. For India, in particular, algalization of crop field is achievable in a cost-effective manner, since the country, under its vast climatic regime harbors a wide diversity of useful cyanobacteria.

REFERENCES

Beck, T. (1994). Methods and application of soil microbial analysis at the Bayerische Landensanstalt fur Bodenkulturund pflarizenbau (LBB) in Munich for the determination of some aspects of soil fertility, 5th Symp. on Soil Biology Bucharest, Rumania, Feb 1981. Rumanian National Soc. of Soil Sci., Bucharest, 13-20.

Casida, L.E., Jr., Klein, D.A. and Santoro, T. (1964). Soil dehydrogenase activity, *Soil Sci.* **98:** 371-376.

Christ, M.J. and David, M.B. (1996). Temperature and moisture effects on the production of dissolved organic carbon in a spodosol. *Soil Biol. Biochem.* **28:** 1191-1199.

Ciavatta, C., Govi, M., Antisari Vittori, L. and Sequi, P. (1991). Determination of organic carbon in aqueous extract of soil and fertilizers, *Commun. Soil Sci, Plant Anal.* **22:** 795-807.

Condron, L.M., Cameron K.C., Di, H.J., Clough, T.J., Forbes E.A., McLaren, L.G. and Silva, R.G. (2000). A comparison of soil and environmental quality under organic and conventional farming system in New Zealand, *New Zealand. J. Agric. Res.* **43:** 443-466.

Gianfreda, L. and Billag, J.M. (1996). Influence of natural and anthropogenic factors on enzyme activity in soil. In: Stotzky, G. and Billag, J.M. (Eds.). *Soil. Biochemistry,* Marcel Dekker Inc. New York **9:** 123-193.

Gill, H.S. and Meelu, O.P. (1992). Studies on the substitution of inorganic fertilizers with organic manure and their effect on soil fertility in rice-wheat rotation. *Fert. Res.* **8:** 303-314.

Goyal, S.K. (1997). Algae and the soil environment. *Phykos* **36:** 1-12.

Gregorich, E.G., Carter, M.R., Angers D.A., Monreal, C.M. and Ellert, B.H. (1994). Towards a minimum data set to assess soil organic matter quality in agricultural soil. *Can. J. Soil Sci.* **74:** 367-385.

Hofmann, E. (1963). Urease. In : *Methods of Enzyme Analysis.* Hans – Ulrich Berg meyer (Ed.). Academic Press, New York.

Jenkinson, D.S. and Powlson, D.S. (1976). The effect of biological treatments on metabolism in soil. V. A method for measuring soil biomass. *Soil Biol. Biochem.* **8:** 209-213.

Kavitha, P. and Swarajya Lakshmi, G. (2002). Effect of different sources of nitrogen on yield and quality of sunflower (*Helianthus annus* L.). *J. Oil Seeds Res.* **19(2):** 250-251.

Mishra, V.K. and Sharma, R.B. (1997). Effect of fertilizer alone and in combinations with manure or physical properties and productivity under rice based cropping system. *J. Ind. Soil. Sci.,* **45(1):** 94-98.

Nannipieri, P., Creco, S. and Ceccanti B. (1990). Ecological significance of the biological activity in soil. In: Stotzky, G. and Bollag, J.M. (Eds.). *Soil Biochemistry.* Marcel Dekker Inc., New York, **6:** 293-355.

Pandey J. and Pandey U. (2001). The influence of catchment on ecosystem properties of a tropical fresh water lake. *Biotronics,* **30:** 85-92.

Pandey U. (2009). Cyanobacterial surfectants enhance the fertility and stability of tropical soils. In: Gupta, R.K., Kumar, M. and Vyas, D. (Eds.). *Soil Microflora,* Daya Publishing House, New Delhi. pp. 102-107.

Pandey, J. and Pandey, U. (2009). Accumulation of heavy metals in dietary vegetables and cultivated soil horizon in organic farming system in relation to atmospheric deposition in a seasonally dry tropical region of India. *Environ. Mont. Assess.* **148:** 61-74.

Pankhurst, C.E., Ophel Keller, K., Doube, B.M. and Gupta V.V.S.R. (1996). Biodiversity of soil microbial communities in agricultural system. *Biodiv. Cons.* **5:** 197-209.

Passioura, J.B. (1988). Water transport in and to roots. *Ann Rev Plant Physiol. Plant Mol. Biol.,* **39:** 245-265.

Reddy, P.V.R.M. and Reddy, C.R. (2003). Integrated nutrient management in sunflower (*Helianthus annus* L.) production, *J. Oil Seeds Res.,* **18(1):** 131-133.

Singh, P.K. (2000). Biology of Azolla and blue-green algae In: Singh *et al.* (Eds.), *Biofertilizers : Blue Green Algae and Azola*, Venus Printers and Publishers, New Delhi, India.1-23.

Singh, R.N. (1961). *Role of Blue-Green Algae in Nitrogen Economy of Indian Agriculture.* Indian Council of Agricultural Research, New Delhi.

Stotzky, G. (1974). Activity, ecology and population dynamics of microorganisms in soil. In: Laskin, A. and Lechevalier, H. (Eds.). *Microbial Ecology,* CRC Press, Cleveland, Ohio, pp. 57-135.

Tabatabai, M.A. and Bremner, J.M. (1969). Use of p-nitrophenyl phosphate for assay of soil phosphatase activity. *Soil Biol. Biochem.* **1:** 301-307.

Tilman, D., Fargione, J., Wolff, B., D'Antonio, C., Dobson, A. and Howarth, R. (2001). Forecasting agricultural driven global environmental changes. *Science* **292:** 281-284.

Vance, E.D., Brookes, P.C. and Jenkinson, D.S. (1987). An extraction method for measuring soil microbial biomass carbon. *Soil Biol. Biochem.* **19:** 703-707.

Venkataraman. G.S. (1981). Blue-green Algae for Rice Production. A Manual for its Promotion. F.A.O. Bulletin No. 46. F.A.O., Rome.

Whitton, B.A. (2000). Soils and rice fields In: Whitton, B.A. and Potts, M. (Eds.) *The Ecology of Cyanobacteria*, Kluwer Academic Publishers, The Netherlands. 233-255.

Diversity of Lower Plants (2014): pp. 96-107
Editors: Rajan Kumar Gupta & Mukesh Kumar
IK International Publishing House Pvt. Ltd.

8

Algae: General Aspects, Ecological Relationships and Economic Values

ASHU*, MUKESH KUMAR** AND RAJAN KUMAR GUPTA***

****Department of Botany, Sahu Jain P.G. College, Najibabad–246 763 U.P.*
E-mail: mukesh.najibabad@gmail.com
****Department of Botany, Govt. P.G. College, Rishikesh–249 201 (Dehradun) Uttarakhand*

ABSTRACT

Algae are simple organisms that range from very small, single-celled microalgae to macroalgae that group into very large organisms such as kelp. Predominantly these are aquatic and found to occur on salt waters. Some species of algae can occur in extreme environments. Some algae even survive suspended in the atmosphere as spores or in droplets of moisture. Freshwater forms occur abundantly in ponds, lakes, slow flowing streams and water reservoirs. There are more than 300,000 species of algae in global algae culture collections. The vast majority of algae are photosynthetic, deriving energy from the sun to produce energy and biomass. They have the ability to grow in specific climatic conditions. The thallus organization of algae show a wide range of structures, i.e. from simple microscopic to multicellular forms of various shapes and sizes, on the bases of which they have been divided in several classes. A number of algae are utilized by human beings directly or indirectly at domestic and industrial levels. The most important economic products obtained from algae are associated with brown and red seaweeds, which can be utilized as food for people, and as resources for the manufacturing of industrial products. In addition, freshwater algae can cause problems when they are overly abundant. Algal blooms can cause foul tastes in the water stored in reservoirs, which may be required by nearby towns or cities as drinking water. The toxins produced by green and blue-green algae can be fetal for man and his pets. Another major problem associated with algal blooms in lakes that are receiving large inputs of nutrients through sewage disposal or the runoff fertilizers is of eutrophication, however, there are many scientific methods to handle the problems.

Keywords: Algal, classification, Thallus organization, Distribution patterns.

**Communicating Author

1. INTRODUCTION

Algae comprise a simple group of chlorophyll bearing thalloid plant body. They do not contain root, shoot, stem or leaf-like structures. They have been placed in the division thallophyta along with fungi. However, they differ from fungi in the presence of photosynthetic pigment, the chlorophyll and in their mode of nutrition. Although most of the algae are autotrophic, i.e. they synthesize there food by themselves, yet heterotrophic and holozoic forms are not uncommon. Algae are the fastest growing plants in the world and can be grown year round, unlike seasonal crops. Their farming does not require agricultural land or clean water, so it does not compete with food crops for these resources. Photosynthesis is the process by which plants utilize the energy in the sun's rays to produce energy and new plant matter (biomass). Photosynthesis is the base reaction supplying the vast majority of energy used by plants and animals on earth. In photosynthesis, energy (photons) from the sun's rays converts carbon dioxide and water to carbohydrate plus oxygen. The carbohydrate can be converted to protein or fat.

Commercially we can grow algae at a photosynthetic efficiency of approximately 5.4% under natural sunlight. General crops grow at a photosynthetic efficiency of approximately 1%. Algae can be grown much more efficiently because of the nature of the bioreactor and the removal of factors that might limit growth such as lack of nutrients or CO_2. Fast growing algae use the process of photosynthesis to harness sunlight and carbon dioxide and then convert them into carbohydrates. The cells containing these valuable materials can then be utilized for the production of various fuels such as biodiesel and ethanol, or used as ingredients for animal feed. Algae can surely make an impact on many industrial companies. The algae technology has the potential to make it more profitable without pollution.

2. GENERAL CHARACTERISTICS OF ALGAE

Algae are eukaryotic organisms, meaning their cells have nuclear material of deoxyribonucleic acid (DNA) within a discrete, membrane-bounded organelle, known as the nucleus. However, blue-green algae are prokaryotic organisms, i.e., without an organized nucleus and are more appropriately referred to as blue-green bacteria, or as cyanobacteria. The cyanobacteria are also different from the true algae in that they do not contain the photosynthetic pigment known as chlorophyll *a*, they do not have cell walls made of cellulose, and they do not store energy as starch or related polysaccharides.

Virtually all species of algae are photosynthetic. They have a relatively simple anatomy, which can range in complexity from single-celled organisms to colonial filaments, plates, or spheres, to the large, multicellular structures of the brown algae, known as thalli. Algal cell walls are generally made of cellulose, but can contain pectin, a class of hemicellulose polysaccharides that give the algae a slimy feel. The larger, multicellular algae have relatively complex tissues, which can be organized into organ-like structures that serve particular functions.

The algae, as a group have now no official existents. It is a collective term for all those chlorophyll-bearing organisms which are thalloid, pond scum, stoneworts, seaweeds and the like are collectively known as algae. They are placed in the division Thallophyta along with Fungi. The word algae derives from a Latin word 'alga' (means seaweeds). The science that deals with the study of algae is called algology. The Greek word for algae is 'phykos' and therefore their study is also called phycology. (Gr.-Phykos – seaweeds, logos – study or discourse). Linnaeus (1754) coined the term algae for first time, while giving the classification of plant kingdom.

3. SALIENT FEATURES OF ALGAE

The important distinguishing features of algae are as follows:

1. Algae are usually aquatic either marine or freshwater. Only a few algae occur in moist terrestrial habitats like tree trunks, wet rocks, moist soil, etc.
2. Their body may be unicellular, colonial, filamentous, parenchymatous or pseudoparenchymatous.
3. Algae like other aquatic plants are covered over by mucilage which protects them from epiphytic growth and decaying effect of water.
4. They do not have vascular tissues.
5. A mechanical tissue is absent.
6. The body is flexible. It helps the algae to sway with water tides without being torn.
7. Nutrition is photosynthetic.
8. Algae contain chlorophyll, carotenes and xanthophyll. Additional pigments occur in specific groups.
9. Vegetative and asexual modes of reproduction are abundant.
10. Sexual reproduction is absent in blue-green algae. Sex organs are non-jacketed.
11. Sexual reproduction involves isogamy, anisogamy and oogamy in different groups.
12. An embryo stage is absent.
13. Life cycle is of various types, i.e. haplontic diplontic or diplohaplontic.

4. DEFINITIONS OF ALGAE

To give a precise definition of algae is very difficult mainly owing to the varied nature of the plants comprising this group. Because of this Bold and Wyne (1978) have rightly stated that sometimes "even the professional botanist and biologists find algae embarrassingly elusive of definition. Some of the phycologists have defined algae as follows.

(i) F.E. Fritsch (1935)

Unless purely artificial limits are drawn the designation algae must include all holophytic organisms that fail to reach the higher level of differentiation characteristic of the archegoniate plants.

(ii) G.M. Smith (1955)

Simple plants with an autotrophic mode of nutrition are called algae.

(iii) V.J. Chapman (1962)

These plants (seaweeds of the seashore and green skeins in stagnant fresh water ponds and pools) which are among the simplest in the plant kingdom belong to the group known as algae.

(iv) G.W. Prescott (1969)

The algae are those chlorophyll-bearing organisms and their colourless relatives which are thalloid, i.e., having no true roots, stems and leaves or leaf-like organs.

(v) R.N. Singh (1974)

The algae are by and large simple plants which display a spectrum of photosynthetic pigments and evolve oxygen during the process of photosynthesis.

5. DISTRIBUTION PATTERN OF ALGAE

The algae are predominantly aquatic and are found to occur on salt waters. Freshwater forms occur abundantly in ponds, lakes, slow flowing streams and water reservoirs. They may be free swimming free floating or attained to the bottom in the shallow water. Some terrestrial algae grow in wet situations such as on damp soil, damp shaded sides of tree and walls or even rocks and thus have adapted themselves to a life in the air. They may live as epiphytes as well as epizoics. A few occur in association with fungi and form the body of lichens. Some algae are endopytic whereas a few live in symbiotic relationship with the higher plants. Most of the marine forms are seaweeds. They inhabit vast area of the ocean. The free floating and free swimming minute algae together with similar other organization constitute the plankton of the ocean and lakes.

6. CLASSIFICATION OF ALGAE BASED ON THE HABITATS

Aquatic Algae

Aquatic forms are found in freshwater or in saline water of the sea.

Freshwater Forms

These forms occur in freshwater (low salinity water) of ponds, pools, lakes, rivers, etc. Some freshwater forms like *Cladophora, Oedogonium, Ulothrix* water whereas others like *Chlamydomonas, Volvox* and *Spirogyra* occur in stagnant water.

Marine Forms

These forms occur in saline water of the sea and are represented by the members of Phaeophyceae (e.g., *Polysiphonia*).

The aquatic algae are either free floating (e.g., *Chlamydomonas, Volvox, Spirogyra*) or are attached to a substratum with the help of an attachment disc, known as holdfast (*Oedogonium, Ulothrix*) many free floating algae together with other similar organisms form colonies on the surface of water which are called water blooms or phytoplanktons.

Terrestrial Algae

Some algae are found in terrestrial habitats like soils, rocks, logs, etc. Forms like, *Vaucheria,*

Botrydium, Euglena which are found on the surface of the soil are known as saphophytes whereas many blue-green algae (e.g., *Nostoc, Anabaena* which occur under the surface of the soil are called cryptophytes.

Aerophytes

Such algae forms as are adapted for aerial mode of life and occur on the tree trunks, moist walls, flower pots and get their water and CO_2 requirements completed directly from atmosphere are called aerophytes. *Trentopohila* is found on the bank of trees in moist and humid climatic conditions.

Cryophytes

Algae growing on ice or snow are called cryophytes. These algae provide attractive colours to snow covered mountains. Green snow in Europe is due to *Chlamydomonas yellowstonensis* and species of *Mesotaenium*.

Thermophytes

Many blue-green algae (e.g., *Oscillatoria brevis*) are found in hot water springs (50-70°C). Since most hot springs are alkaline, some algologists believe that tolerance to high temperature is also associated with this factor.

Algae of Unusual Habitats

Many algae are found at various interesting places and according to their habitats may be of the following types.

Halophytic Algae

These algae are found in saline water containing high percentage of salts, e.g., *Chlamydomonas ehrenbergii.*

Lithophytic Algae

These are found on the moist surface of rocks and stones, e.g., *Batrachospermum* and *Enteromorpha.*

Epiphytic Algae

Such algal forms grow on other aquatic plants and are called epiphytes, e.g., *Oedogonium, Bulbochaete.*

Epizoic Algae

Many algae grow on the shells of molluscs, turtles and fins of fishes and are known as epizoic algae, e.g., *Cladophora* is found on snails and shells of bivalves.

Endozoic Algae

Contrary to epizoic algal forms, endozoic algae are found inside the body of aquatic animals, e.g. *Zoochlorella* is found inside *Hydra viridis*.

Parasitic Algae

Algae growing as parasites on many plants and animals, e.g., *Cephaleuros virescens* (Chlorophyaere) causes red rust of tea and coffee plantations.

Symbiotic Algae

Several members of Chlorophyccae and Cyanophyceae form symbiotic association with fungi, bryophytes, gymnosperons and angiosperms. e.g., colonies of *Nostoc* and *Anabaena* are found in symbiotic association in the thallus of *Anthoceros* and the coralloid roots of *Cycas*.

7. CLASSIFICATION OF ALGAE BASED ON THALLUS ORGANIZATION

There exists a wide range of thalli in algae particularly in vegetative plant body. The different forms show a definite range. On one hand, there are simple plants where the thallus body is microscopic and consists of only one single cell (*Chlamydomonas, Phacus*, etc.) All the morphological structures and cytological physiological genetical and the other vital and necessary activities, go on regularly in that single cell only. In many species, the cells are grouped into aggregations called colonies, e.g., *Volvox*. The most highly differentiated marine algae exhibit external differentiated marine and considerable size, e.g., *Microcystis, Sargassum*, etc.

On the basis of thallus organization, algae are divided into the following five groups:

Unicellular Forms

Simple unicellular forms are found in all groups of algae except cyanophyceae and phaeophyceae. Unicellular forms can be classified into the following four subgroups.

Rhizopodial Forms

These forms lack a rigid cell wall. They possess cytoplasmic projections which help in amoeboid movement, e.g., *Chrysamoeba.*

Flagellated Unicells

Flagellated vegetative cells are found in all groups of algae except Cyanophyceae, Phaeophyaeae and Rhodophyceae. They vary in number. and type of flagella. In Dinophyceae and Xanthophyceae there are two unequal flagella, e.g., *Euglena, Chlamydomonas.*

Spiral Filamentous Unicells

Some unicellular algae form spiral or coiled structure, e.g., *Spirulina* (Cyanophyceae).

Non-motile Unicells

There are non-motile coccoid algae which do not possess flagella eye spot, etc., meant for locomotion, e.g., *Chroococcus.*

Colonial Forms

The colonial habitat is achieved by aggregation of the products of cells divisions within a mucilaginous mass. On the basis of morphology, colonial organization may be divided into the following four types.

Coenobial

A colony with a definite shape and size and arrangement of cells is known as coenobium. Coenobia may be motile or non-motile, e.g., *Pandorina*, etc.

Palmelloid

Unlike coenobial forms in a palmelloid colony, the number of cells, their shape and size is not definite. The cells remain irregularly aggregated within a common mucilagenous matrix, e.g., *Tetraspora* (Chlorophyceae).

Dendroid

Here the colony looks like a microscopic tree. As in palmelloid colony, the number, shape and size of the cells is also indefinite in dendroid colonies, e.g., *Chrysodendron* (Chrysophyceae).

Rhizopodial

In a rhizoidal colony, the cells are united through rhizopodia, e.g., *Chrysidiastrum.*

Filamentous Forms

Filamentous forms are developed by repeated transverse division of cells. The filaments may be branched or unbranched.

Unbranched Filaments

Simple unbranched filaments are found only in a few groups of algae. Such filaments may be free floating, e.g., *Spirogyra* attached to some substratum, e.g., *Zygnema, Ulothrix* or form colony, e.g., *Nostoc.*

Branched Filaments

Branched filaments are formed by repeated transverse division of lateral outgrowth of cells. True branches which arise as lateral outgrowth may result in the following 3 types of filaments.

Simple filaments

Simple branch filaments remain attached to the substratum by a basal cell, e.g., *Cladophora.*

Heterotrichous

In this type the thallus is very much evolved and differentiated into prostrate and erect systems, e.g., *Ectocarpus*.

Pseudoparenchymatous

In many filamentous forms, one or more central or axial filaments together with their branches form a parenchymatous structure, e.g., *Batrachospermum* and *Polysiphonia*.

Siphonaceous forms

Here the thallus is made up of branched aseptate, coenocytic, tubular filaments as the nuclear divisions are not accompanied by wall formation, e.g., *Vaucheria*.

Parenchymatous forms

In these algae, cell division takes place on different sides with the result that they become parenchymatous, e.g., *Ulva*.

8. RANGE OF THALLUS ORGANIZATION IN DIFFERENT CLASSES OF ALGAE

8.1 Chlorophyceae

The thallus organization in chlorophyceae shows a wide range beginning from simple microscopic motile unicellular forms (*Chlamydomonas*) to multicellular forms flagellated (*Volvox*) or non-flagellated (*Hydrodictyon*) colonies, palmelloid form (*Tetraspora*), dendroid forms (*Prasino cladus*), filamentous forms (*Spirogyra, Ulothrix, Oedogonium*), heterotrichous (*Fritschella*), siphonous (*Bryopsis* and *Codium*) to well-developed parenchymatous type of thalli (*Ulva* and *Enteromorpha*). In genera like *Chara* and *Nitella* the plant body is so well developed and macroscopic that the plants morphologically appear like that of a small angiospermic herb. But as there is no clear differentiation of root, stem and leaves in plant body, it is thalloid in nature.

8.2 Bacillariophyceae (diatoms)

The diatom thallus is essentially a diploid unicell. The tiny unicells occur isolated often the unicellular thalli adhere by their mucilaginous sheaths to form colonies. The colonies may be filamentous or are enclosed in a common gelatinous envelope with no specialization of parts.

8.3 Phaeophyceae

The plant body in brown algae is highly complex and well organized. Unicellular colonial coccoid and unbranched filamentous thalli are absent. Except a few microscopic members of Ectocarpales, Tilopteridales and Sphacelariales, most of the brown algae are microscopic and very large in size. Because of their bulky appearance they are called giant kelps. The plant body is differentiated into a creeping prostrate system and on an erect projecting system.

In *Giffordia conifera* both the prostrate and projecting systems are almost equally developed but in *Ectocarpus irregularis* and e.g., *Elachistaeformis* the erect system is extremely reduced band-shaped or fan-shaped thalli are found in cutlerials, whereas cylindrical or flattened thalli are found in *Sporochanles*.

8.4 Rhodophyceae

In beauty the members of red algae come on the top. Genera such as *Delesseria* and *Plocamium* are really the beautiful creations of the nature. Except a few unicellular (*Porphyridium*), palmelloid (*Astarocystis*), filamentous (*Goniotrichum*) and Parenchymatous (*Bangia* and *Porphyra*) forms the majority of red algae can be grouped broadly under two types of thallus construction, i.e., uniaxial and multiaxial.

8.5 Uniaxial Construction

In such a construction the thallus contains a single axial or central filament. This central filament is generally corticated with the help of a number of well-branched laterals. All these forms a pseudoparenchymatous thallus, e.g., *Batrachospermum, Detesseria*.

8.6 Multiaxial Construction

In the multiaxial construction the thallus contains several axial or central filaments from each axial filament develop laterals which radiate out towards the margin. All these branches together from pseudoparenchymatous thallus as in *Polysiphonia*, *Helminthocladia*, etc.

8.7 Myxophyceae (Cyanophyceae)

The blue-green algae also show a wide range in their vegetative structure. It varies from unicellular genera (*Anacystis*) and along with many colonial members (*Apnanocapsa*) as well as filamentous forms (*Oscillatoria*). Some of the colonies consist of simple hollow spheres in which the cells are arranged only in a single peripheral layer. Desikachary (1959) has divided blue-green algae into two categories:

Non-filamentous forms

In non-filamentous forms coccoid and palmelloid genera have been included, e.g., *Synechococcus*, *Eucapsis*, etc.

Filamentous forms

In filamentous forms he included *Oscillatoria*, *Lyngbya*, *Scytonema*, *Calothrix*, *Rivularia*, etc.

9. FACTORS LIMITING THE PRODUCTIVITY OF ALGAE

Some species of algae can occur in extreme environments. For example, species of green algae have been observed in hot-water springs at Yellowstone National Park, in highly acidic volcanic lakes, in the extremely saline Great Salt Lake and Dead Sea, and on the surfaces of glaciers and snow. Some algae even survive suspended in the atmosphere as spores or in droplets of moisture.

These are, however, extremely stressful environmental conditions. Most algae occur in less stressful habitats, where their productivity tends to be limited by the availability of nutrients (assuming that sufficient light is available to support the photosynthetic process). In general, the productivity of freshwater algae is primarily limited by the availability of the nutrient phosphate (PO_4^{-3}), while that of marine algae is limited by nitrate (NO_3^-) or ammonium (NH_4^+). Some algal species, however, may have unusual nutrient requirements, and their productivity may be limited by certain micronutrients, such as silica, in the case of diatoms.

The structure of algal communities may also be greatly influenced by ecological interactions, such as competition and herbivory. For example, when herbivorous sea urchins are abundant, they can sometimes overgraze species of kelps in subtidal ecosystems of the west coast of North America, degrading the kelp forests. However, where sea otters (*Enhydra lutris*) are abundant this does not happen because the otters are effective predators of the urchins.

Another example of a biological influence on the structure of an algal community concerns the zebra mussel (*Dreissena polymorpha*). This is a bivalve mollusk that has been accidentally introduced by ocean-going ships to the Great Lakes of North America, where it has become an important pest because it clogs water pipes with its prolific growths, and can displace native species by competitively appropriating hard-substrate habitats. More to the present point, however, the zebra mussel is such an effective filter-feeder on phytoplankton, that its large populations in parts of the Great Lakes are apparently responsible for some of the clarification of the water that has occurred in recent years. Grazing by the zebra mussel has actually resulted in decreased standing crops of phytoplankton, even in well-fertilized waters.

10. ECONOMIC PRODUCTS OBTAINED FROM ALGAE

The most important economic products obtained from algae are associated with brown and red seaweeds, which can be utilized as food for people, and as resources for the manufacturing of industrial products. These seaweeds are mostly harvested from the wild, although increasing attention is being paid to the cultivation of large algae.

Some species of algae can be directly eaten by humans, and in eastern Asia they can be especially popular, with various species used as foods. An especially common food is the red alga known as nori in Japan (*Porphyra* spp.), which has long been eaten by coastal people of China and Japan. This alga is often used as a wrapper for other foods, such as rice or plums, or it may be cooked into a clear soup. Nori has been cultivated for centuries in eastern Asia. Another alga known as dulse or sea kale (*Rhodymenia palmata*) is consumed dried or cooked into various stews or soups. Other commonly eaten seaweeds include the sea lettuce (*Ulva lactuca*), and murlins or edible kelp (*Alaria esculenta*). Potentially, seaweeds are quite nutritious foods, because about 50% of their weight occurs as carbohydrates, with smaller concentrations of proteins and fats, and diverse micronutrients, including iodine. In practice, however, seaweeds are not very nutritious foods for humans, because we do not have the enzymes necessary to metabolize the most abundant of the complex, algal carbohydrates.

In some places, coastal livestock such as sheep and cattle, and wild ungulates such as deer, will graze algal biomass from the intertidal zone at low tide. These animals can take better advantage of the algal carbohydrates than can humans, largely because of the digestive abilities of the symbiotic microorganisms in their rumens and guts. Sometimes, algal biomass is harvested by humans and added to the fodder of livestock as a source of micronutrients.

The major economic importance of brown seaweeds, however, is as a natural resource for the manufacturing of a class of industrial chemicals known as alginates. These chemicals are extracted from algal biomass, and are used as thickening agents and as stabilizers for emulsions in the industrial preparation of foods and pharmaceuticals, and for other purposes.

Agar is another seaweed product, prepared from the mucilaginous components of the cell walls of certain red algae. Agar is used in the manufacturing of pharmaceuticals and cosmetics, as a culture medium for laboratory microorganisms, and for other purposes, including the preparation of jellied desserts and soups. Carrageenin is another, agar-like compound obtained from red algae that is widely used to stabilize emulsions in paints, pharmaceuticals, ice cream, and other products. Irish moss (*Chondrus crispus*) is a purplish alga that is a major source of carrageenin.

Researchers are investigating methods for the economic cultivation of red and brown seaweeds for the production of alginates, agar, and carrageenin. In California, use has been made of rafts anchored to float about 13 yd (12 m) below the surface, in shallow, less than 328 ft (100 m) deep water, to grow the highly productive, giant kelp *Macrocystis* as an industrial feedstock. Seaweeds are also cultivated on floating devices in coastal China, and research is investigating whether growth rates in dense plantings can be economically increased by enriching the seawater with nitrogen-containing fertilizers.

In some places, large quantities of the biomass of brown and red algae wash ashore, especially after severe storms that detach these algae from their substrates. This material, known as wrack, is an excellent substrate for composting into an organic-rich material that can greatly improve soil qualities in terms of aeration and water- and nutrient-holding capacity.

Over extremely long periods of time, the frustules of diatoms can accumulate in large quantities. This material is known as diatomaceous earth, and its small reserves are mined for use as a fine polishing substrate, as a fine filtering material, and for other industrial purposes.

11. ENVIRONMENTAL PROBLEMS ASSOCIATED WITH ALGAE

Red tides are events of great abundance (or "blooms") of red, brown, or yellow-coloured dinoflagellates of various species. These algae synthesize biochemicals, such as saxitoxin and domoic acid, which are extremely poisonous and can kill a wide range of marine animals, as well as humans who eat shellfish containing the toxins. The toxic syndromes of humans associated with dinoflagellate toxins are known as paralytic, diarrhetic, and amnesic shellfish poisoning.

The dinoflagellates involved with toxic dinoflagellate blooms are commonly of the genera *Alexandrium*, *Dincphysis*, *Nitzchia*, or *Ptychodiscus*. The algal toxins can be accumulated by filter-feeding shellfish such as clams and oysters. If these are eaten while they contain red-tide toxins, they can poison humans and animals. Even creatures of the size of large whales can die from eating fish containing large concentrations of dinoflagellate toxins.

In addition, freshwater algae can cause problems when they are overly abundant. Algal blooms can cause foul taste in the water stored in reservoirs, which may be required by nearby towns or cities as drinking water. This can be a significant problem in naturally productive, shallow-water lakes of the prairies in North America.

Another major problem associated with algal blooms in lakes that are receiving large inputs of nutrients through sewage disposal or the runoff fertilizers is of eutrophication. It can result in severe damage to the ecosystem of water when large quantities of algal biomass sink to deeper waters and consume most of the O_2 during their decomposition. The anoxic (deficiency in oxygen) conditions that develop are lethal to the animals that live in the sediment and deep waters, including most species of fish. Because the primary limiting nutrient in fresh waters is usually phosphate, inputs of this nutrient can be specifically controlled by sewage treatment, and by banning the detergents containing phosphorus. This has been done in many areas in North America, and eutrophication is now less an environmental problem than it used to be.

ACKNOWLEDGEMENTS

The authors are thankful to the Principal, Sahu Jain P.G. College, Najibabad for providing necessary facilities. Author MK and RKG are also thankful to the UGC, New Delhi for financial support.

REFERENCES

Bold, H.C. and Wynne, M.J. (1978). Introduction to the Algae, Structure and Reproduction Prentice-Hall of India, New Delhi.

Chapman, V.J. (1962). The Algae, Macmillan and Co. Ltd., New York.

Desikachary, T.V. (1959). Electron Microscopy and Algology, *Proc. Indian Acad. Sci. Sec. B*, **46:** 54-64.

Fritsch, F.E. (1935). The Structure and Reproduction of the algae, Volume 1, Cambridge University Press, Cambridge.

Linnaeus, C. (1754). Species Plantarum Vols 142: In WT Stearn 1957 Carl Linnaeus Species Plantarum Royal Society, London.

Prescott, G.W. (1969). The Algae a review, Thomas Nelson and Sons, London.

Singh, R.N. (1974). Living with algae, Presidential Address 61st Session of Indian Science Congress, Nagpur.

Smith, G.M. (1933). Fresh water Algae of the United States, McGraw-Hill Book Co. New York.

Diversity of Lower Plants (2014): pp. 108-122
Editors: Rajan Kumar Gupta & Mukesh Kumar
IK International Publishing House Pvt. Ltd.

9

Extremophilic Bacteria and Microbial Biodiversity

Jay Shankar Singh* and Rajan Kumar Gupta**

**Department of Environmental Science, BB Ambedkar (Central) University Lucknow–226 025, Uttar Pradesh*
***Department of Botany, Govt. P.G. College, Rishikesh–249 201 (Dehradun) Uttarakhand*

ABSTRACT

Using new molecular techniques, scientists are discovering more about bacteria. In their studies, microbiologists have only identified a small number of the microorganisms that exist; there are many undiscovered species, genera, families, orders, and even phyla of bacteria. Comparative ribosomal RNA sequencing is a powerful new tool that allows microbiologists to construct bacterial phylogenies. These phylogenies have revealed that there are three lineages of organisms, bacteria, archaea, and eukarya, and that far greater evolutionary differences exist between two bacterial species than between any two eukaryotic species.

Microbiologists have begun to use phylogenetic information about bacteria to construct specific ribosomal RNA sequences. These sequences can be added to extreme environments to probe for life within the often difficult to study environment. Using these rRNA probes, scientists have discovered many types of extremophiles, microorganisms that have evolved to live in certain types of extremes, such as high or low temperatures, high or low pH levels, high pressures, high salinities, and even high amounts of radiation. Studying these extremophiles may reveal how molecular stability is achieved and how enzymes function under extreme conditions. Characteristics of early life forms on the earth may also be revealed.

The different types of extremophiles have different modifications that allow them to thrive in particular extremes.

- Thermophiles are bacteria that grow optimally at temperatures between 45°C and 80°C. Hyperthermophiles have growth optima over 80°C, and some can even survive at 113°C.
- Hyperthermophiles can be found in deep sea hydrothermal vents, where a variety of temperature microhabitats exist.

*Communicating Author

- Proteins in hyperthermophiles are able to remain stable because they are more compact in their structure, they have a reduced amount of flexibility, extensive ionic bonding provides extra rigidity, and chaperonin proteins are present that refold denatured proteins into their active forms.
- DNA, which normally denatures at high temperatures, keeps its double-stranded form in hyperthermophiles due to its positive supercoiling, which allows more heat stability than the negative supercoiling in DNA of other organisms. Also, unique DNA binding proteins help maintain and stabilize the DNA.
- To prevent their cell membranes from coming apart at high temperatures, hyperthermophiles form their cell membranes as lipid monolayers instead of the lipid bilayers seen in other organisms.
- Psychrophiles are microorganisms that grow optimally at temperatures of 15°C or lower, such as those found in the Antarctic or Arctic.
- In contrast to the proteins of thermophiles, proteins in psychrophiles have structures that maximize flexibility and polarity.
- The cell membranes of psychrophiles contain more unsaturated fatty acids to keep the membrane sufficiently fluid to allow the transport of nutrients across it.
- Acidophiles grow optimally at a pH of 0.7 or lower, and they survive by keeping the acid in their environment out of their cell interiors.
- Alkaliphiles live in environments with pH levels above 10. They maintain a neutral cell interior, but instead of relying on a proton gradient to drive ATP synthesis, alkaliphiles use a sodium ion gradient.
- Halophiles are organisms that live in salty environments. They prevent the tendency of water to move out of the cells by maintaining a solute concentration inside their cells equal to or slightly above the NaCl concentration on the outside.
- Some extreme barophiles grow optimally at pressures over 500 atmospheres. The transmembrane proteins in barophiles likely are modified to still function at high pressures.
- Radiation-resistant bacteria have powerful DNA repair machinery to combat the effects of the high radiation doses on the DNA.
- Because some hyperthermophiles may have evolved relatively little in the 3.5 billion years they have been on the earth, studying them may reveal what life on early earth was like.

Studying bacterial diversity and the functions of the many unique genes possessed by bacteria has exciting potential. Continued study of extremophiles and bacterial diversity will lead to more discoveries.

Keywords: Extremophile, Environmental extremes, Psychrophiles, Halophiles.

1. INTRODUCTION

Extremophiles have different modifications that enable them to survive in particular extreme environments. Bacterial Life at High Temperatures, thermophiles have molecules and cell membranes that retain their stability at high temperatures (Horikoshi & Grant 1998).

In the Bacterial Life at Low Temperatures and Extreme pH Levels, the molecules and membranes of psychrophiles are designed to maximize flexibility. Extremophiles that grow best at pH extremes keep a neutral pH within their cells. Bacterial life in brine and other environments, Halophiles keep a balance between the solute concentration inside their cells and their salty environment. Barophiles and radiation-resistant bacteria also have unique specializations. Studies of the genetics and diversity of extremophiles may reveal more of their unique properties and characteristics of early earth (Madigan & Marrs 1997; Madigan *et al.*, 2000).

The present chapter deals about many prokaryotic organisms that inhabit "extreme environments"—habitats in which some chemical or physical variable(s) differ significantly from that found in habitats that support plant and animal life. Great strides have been made in recent years in the isolation and characterization of extremophilic prokaryotes and many of them turn out to have fascinating metabolic properties and interesting evolutionary histories. Prokaryotes that grow at very high temperatures are perhaps the most dramatic in these regards, as all cellular components need to be made heat stable and their evolutionary position is that of the least derived of all known life forms. As our knowledge of bacterial diversity improves, primarily from the introduction of molecular tools for assessing bacterial phylogeny and diversity and from new advances in isolation and laboratory culture, it is becoming clear that the bulk of evolutionary diversity on the earth does not reside in plants and animals, but instead in the invisible prokaryotic world. There is now great interest in mining the diverse genetic resources of earth's smallest cells for use in biotechnology and related areas.

1.1 A Natural Picture of the Bacterial World

Since the days almost 100 years ago when Robert Koch and his associates isolated the first pure cultures of bacteria, microbiologists worldwide have been isolating laboratory cultures of literally thousands of different bacteria. These include, of course, most of the causative agents of infectious diseases, but more important from the standpoint of the web of life on earth, many of the bacteria that carry out critical chemical reactions that form the "life support" system for plants and animals. And as diverse a group of organisms that are already known, it is now clear that microbiologists have only seen the tip of iceberg; most microorganisms that exist in nature, in particular the bacteria, have not yet been obtained in laboratory culture! Indeed, with the help of new molecular tools microbiologists have explored a variety of microbial habitats and have detected not only new *species* of bacteria, but also new *genera, families, orders* and even *phyla.* Imagine finding a new phylum of plants or animals today! The challenge for microbiologists now is to isolate these organisms, learn about their basic biology, and harness their vast genetic resources for the benefit of mankind.

Great excitement has pervaded in the field of microbial diversity in recent years because of the new found ability of microbiologists to experimentally determine the evolutionary relationships of bacteria. Such an evolutionary "Rosetta Stone" had long been sought in the field of microbiology, but not until the advent of comparative ribosomal RNA sequencing as a rapid and specific means for deducing bacterial phylogenies did microbiologists have the tool they needed to classify bacteria in a natural way, just as botanists and zoologists have done for over a century (Hugenholtz *et al.*, 1998).

Two key concepts have emerged from comparative molecular sequencing of ribosomal RNAs: (1) that cells evolved along *three* major lineages, the bacteria, the archaea, and the eukarya, instead of just two, the prokaryotes and the eukaryotes; and (2) that the evolutionary difference between a mouse and an elephant (or between *Chlorella* and *Trillium*, for the more botanically oriented) pales by comparison to the evolutionary distance between virtually any two common soil bacteria you might want to mention, like *Pseudomonas* and *Bacillus*.

The first of these conclusions, that prokaryotic life contains two major evolutionary lineages, is slowly but surely becoming mainstream thinking among microbiologists, and is even gaining support from macrobiologists as evidenced by the inclusion of this concept in recent biology textbooks. However, the second conclusion, that morphologically quite different plants or animals can be extremely closely related in a molecular evolutionary sense, has been for many a harder pill to swallow. If one steps back for a moment and considers that it is not the evolution of the mouse and the elephant, or the alga and the flowering plant, as intact entities, that molecular sequencing speaks to, but instead, the evolutionary history *of the cells that make them up*, it is easier to understand why the bulk of evolutionary change has occurred in the prokaryotic world; prokaryotes have existed for some 3.8 billion years while the mouse and the elephant have only very recently diverged (Woese 1987).

Prokaryotes ruled the earth for at least two billion years before the modern (organelle-containing) eukaryotic cell appears in the fossil record. And metazoans (multi-celled plants and animals) have only existed for some half a billion years or so. So by the time the stage was set for what botanists and zoologists consider the "evolutionary diversification of plants and animals", most of cellular evolution had already occurred. Diversification of the mouse and the elephant, for example, was simply a matter of arranging cells in different ways to yield what appears to the eye to be highly divergent organisms. But in terms of their evolutionary history, the mouse and the elephant are virtually identical organisms (Raven & Johnson 1999).

By contrast to higher organisms, prokaryotes have had the evolutionary time to show great genetic divergence. However, unlike metazoans, evolutionary change in prokaryotes is not manifest in morphological variation. For whatever reason(s), bacteria maintained a very small size and changed relatively little (compared with metazoans) in morphology through billions of years of evolutionary history. But that is not to say they did not evolve. Molecular sequencing tells us that they have indeed evolved but that the product of this evolutionary change is invisible, instead of big changes in size or shape, evolutionary change in the prokaryotes focused on metabolic diversity and the genetic capacities to explore and eventually colonize every conceivable environment on the earth, including extreme environments. Thus, we must go to the *genes* of the prokaryotes to see their true phylogenetic diversification, and with advances in nucleic acid sequencing, this world is now beginning to open up.

Using comparative ribosomal RNA sequencing microbiologists can now not only construct natural relationships of prokaryotes, but can also use phylogenetic information to construct highly specific ribosomal RNA probes as a means of identifying and tracking specific microorganisms in the environment. A natural application of this technology has been to take these tools into various extreme environments and probe for the diversity of microbial life therein. The fallout from these studies, which historically followed by many years more classical enrichment and isolation approaches, has been an awareness that extreme environments are not a place for "hangers on", but instead are habitats that flourish with microbial life, especially prokaryotes. The rest of this paper will try to introduce the reader to some of these organisms and their homes, and discuss what laboratory studies of these remarkable prokaryotes has revealed for our understanding of the physiochemical limits to life (Van *et al.*, 1998a; Van *et al.*, 1998b).

Table 1: Classes and examples of extremophiles

Parameters	Extreme	Descriptive term	Genus/species	Minimum	Optimum	Maximum
Temperature						
	High	Hyperthermophile	*Pyrolobus fumarii*	90°C	**106°C**	113°C
	Low	Psychrophile	*Polaromonas vacuolata*	0°C	**4°C**	12°C
pH						
	Low	Acidophile	*Picrophilus oshimae*	0.06	**0.7 (60°C)**	4
	High	Alkaliphile	*Natrono-bacterium magadii*	8.5	**10 (20% aCl)**	12
Pressure		Barophile	MT41 (Mariana Trench)	500 atm	**700 atm (4°C)**	> 1000 atm
Salt (NaCl)		Halophile	*Halobacterium salinarum*	15%	**25%**	32% (saturation)

2. ENVIRONMENTAL EXTREMES AND EXTREMOPHILES

Microbiological examination of extreme environments has revealed many new prokaryotes. By "extreme environment" here, it is meant an environment that humans would consider extreme or uninhabitable: extremes of heat or cold, pH, salinity, pressure, and even radiation. As previously mentioned, extreme environments are inhabited by diverse populations of microorganisms, most of which have evolved to live only in the presence of the extreme. These organisms are the **extremophiles.** Several classes of extremophiles are recognized in microbiology and laboratory cultures of representatives of each class are known. Organisms in each class are denoted by a descriptive term, usually a word with Greek or Latin roots followed by the combining form "phile", Greek for "loving". Thus, there are *thermophiles* and *hyperthermophiles* (organisms growing at high or very high temperatures, respectively), *psychrophiles* (organisms that grow best at low temperatures), *acidophiles* and *alkaliphiles* (organisms optimally adapted to acidic or basic pH values, respectively), *barophiles* (organisms that grow best under pressure), and *halophiles* (organisms that require NaCl for growth). Instead of trying to be inclusive here, as literally hundreds of different species could be included, the organisms listed in Table 1 are the current "record holders" in each of the extremophile categories. The labelled "optimum", for here it becomes clear that these organisms are not merely tolerating their lot, but that they actually do best in their punishing habitats; indeed most actually *require* their extreme(s) in order to reproduce at all!

Extremophiles are of interest to both basic and applied biology. In a basic sense these organisms hold many interesting biological secrets, such as the biochemical limits to macromolecular stability and the genetic instructions for constructing macromolecules stable to one or another extreme. But in an applied sense, these organisms have yielded an amazing array of enzymes capable of catalyzing specific biochemical reactions under extreme conditions (Adams & Kelly 1995). Such enzymes have found their way into the grist of industry in applications as diverse as laundry detergent additives (proteases, lipases) and the genetic identification of criminals. The enzymes of *Pyrococcus furiosus* and *Thermus aquatics* are extremely thermostable. Consequently the DNA Polymerase from *Pyrococcus furiosus* (Pfu DNA polymerase) and *Thermus aquaticus* (Taq DNA polymerase) can be used in the polymerase chain reaction (PCR) DNA amplification process (Ladenstein & Antranikian 1998).

Another important realization that has emerged from the study of extremophiles is that some of these organisms form the cradle of life itself. Many extremophiles, in particular the hyperthermophiles, lie close to the "universal ancestor" of all extant life on the earth. Thus, an understanding of the basic biology of these organisms is an opportunity for biologists to "look backward in time" so to speak, to a period of early life on the earth. This exciting realization has fueled much research on these organisms in order to understand the nature of primitive life forms, how the first cells "made a living" in the earth's early days, and how early organisms set the stage for the evolution of modern life forms (Raven *et al.*, 1999).

Using ribosomal RNA sequencing, scientists have described three main lineages of organisms, the two bacterial lineages containing the most diversity. Many types of extremophiles, bacteria whose members live under extreme conditions, have been identified by probing extreme environments with rRNA sequences. The growth rate of extremophiles has been presented at different temperature in Fig. 1.

3. BACTERIAL LIFE AT HIGH TEMPERATURE

Although thermophilic bacteria (organisms with growth temperature optima between 45°C and 80°C) have been known for over 80 years, **hyperthermophilic** bacteria—organisms with optima above 80°C—have only been recognized more recently. Following the pioneering work of Thomas Brock in the 1960s and 1970s, Karl Stetter and co-workers at Regensburg (Germany) have proceeded to isolate over 30 genera (over 70 species) of hyperthermophiles. Brock was the first to demonstrate, often using simple but ingenious field experiments, that bacteria were present in boiling hot springs in Yellowstone National Park (Brock 1978). By contrast, Stetter's group, whose focus has been on isolation and culture, has isolated most hyperthermophiles known today, including *Pyrolobus fumarii*, a remarkable prokaryote that can grow up to 113°C (Barns *et al.*, 1994; Blochi *et al.*, 1997).

Thermophilic microorganisms can be isolated from virtually any environment that receives intermittent heat, such as soil, compost, and the like. But hyperthermophiles thrive only in very hot and *constantly* hot environments, including hot springs, both terrestrial and undersea

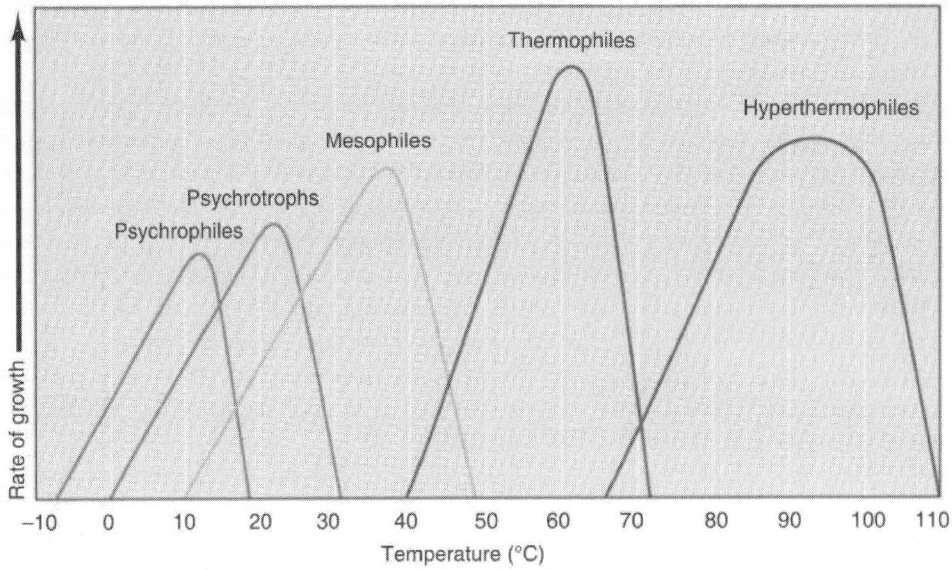

Fig. 1. Temperature range showing growth of extremophiles.

(hydrothermal vents), and active sea mounts, where volcanic lava is emitted directly onto the sea floor. It is also strongly suspected, and some supportive evidence exists, that hyperthermophiles reside deep within the earth, living a buried existence and relying on geothermal heat for their metabolic activities and reproduction (Madigan & Oren 999).

The most extreme of known hyperthermophiles, those with temperature optima *above* 100°C, have come from submarine hydrothermal vents, and examples include *P. fumarii* and the methanogen *Methanopyrus kandleri* (Kurr *et al.*, 1991). Both of these amazing prokaryotes are phylogenetically archaea and use molecular hydrogen, H_2, as their electron donor (energy source), reducing either NO_3^- (*P. fumarii*) or CO_2 (*M. kandleri*) as electron acceptors to grow by anaerobic respiration. Besides *requiring* substantial heat for growth, these bacteria can survive temperatures substantially above their upper growth temperature limits, making a conventional autoclave regimen (15 min. at 121°C) insufficient for sterilizing cultures of either species!

Both *P. fumarii* and *M. kandleri* originated from hydrothermal vent chimneys. These are precipitated iron mineral deposits that form as extremely hot water (up to 400°C) containing various minerals emerges from deep sea hydrothermal vents (note that although this water is superheated, it does not boil because of the hydrostatic pressure of the water column, usually 2000-3000 metres, that overlies these vents) (Stetter 1999). Although the water that emerges is too hot for life, the chimneys, which are about 0.5 cm thick, show a temperature gradient from about 300°C inside to 2°C outside. Because prokaryotes are so small, microenvironments differing in temperature exist across the chimney wall leading to ideal habitats for various species of heat-loving bacteria.

Using nucleic acid probe technology several morphological types of bacteria have been detected in hydrothermal vent chimney walls, suggesting that these compact thermal gradients may contain many different microbial populations in addition to the already isolated prokaryotes *P. fumarii* and *M. kandleri*. And for my botanical friends reading this, I would be remiss if I didn't point out that *P. fumarii* and *M. kandleri* are good examples, widespread in the microbial world, of primary producers totally divorced from sunlight. Besides growing at almost unbelievably high temperatures, *P. fumarii* and *M. kandleri* are also autotrophs, capable of growing in a simple anaerobic mineral salts medium supplied with CO_2 and H_2; neither sunlight nor a key product of photosynthesis, O_2, is required for either organism. Indeed, it has been hypothesized that long before the process of photosynthesis evolved, anaerobic H_2-based chemolithotrophy was the major means by which new organic material was synthesized on the earth.

4. STABILITY OF MOLECULES IN BACTERIA CELLS AT EXTREME TEMPERATURES

For an organism to grow at high temperatures, especially as high as those of the hyperthermophiles discussed here, *all* cellular components, including proteins, nucleic acids, and lipids, must be heat stable. The thermostabilities of enzymes from various hyperthermophiles, referred to as **extremozymes,** have been documented, and some have been found to remain active up to 140°C. The structural features that dictate thermal stability in proteins are not well understood but a small number of non-covalent features seem characteristic of thermostable proteins. These include a highly apolar core, which undoubtedly makes the inside of the protein "sticky" and thus more resistant to unfolding, a small surface-to-volume ratio, which confers a compact form on the protein, a reduction in glycine content that tends to remove options for flexibility and thus introduce rigidity to the molecule, and extensive ionic bonding across the protein's surface that helps the compacted protein resist unfolding at high temperature. In addition to these intrinsic stability factors special proteins called *chaperonins* are synthesized by hyperthermophiles. Chaperonins function to bind heat denatured proteins and refold them into their active form. The *thermosome* is a type of chaperonin that is widespread among hyperthermophiles capable of growth above 100°C, like *P. fumarii* and *M. kandleri* (Stetter 1996).

Several factors may combine to prevent DNA from melting in hyperthermophiles. However, the two most important features appear to be the enzyme reverse DNA gyrase, which catalyses the *positive* supercoiling of closed circular DNA (by contrast, non-hyperthermophiles contain DNA gyrase, an enzyme that supercoils DNA in a *negative* twisted fashion), and various types of DNA binding proteins, including histone-like proteins. For various physico-chemical reasons, positively supercoiled DNA is more resistant to thermal denaturation than is negatively supercoiled DNA. And the fact that reverse gyrase seems to be the *only* protein thus far found universally among hyperthermophiles (regardless of their metabolic pattern), points to an important role for it in the heat stability of DNA.

Several hyperthermophiles contain DNA binding proteins that appear to play a role in maintaining DNA in a double-stranded form at high temperature. Some of these proteins are structurally related to the core histones of eukaryotic cells and function to wind and compact the DNA into nucleosome-like structures. Others have no structural relationship to histones but when bound to DNA alter its structure in such a way as to significantly raise its melting temperature. It is likely that the combination of positive supercoiling of DNA along with proteins that prevent DNA melting are major solutions to the maintenance and integrity of DNA in hyperthermophiles.

5. MEMBRANE STABILITY IN EXTREMOPHILES

Heat can also affect membrane stability. As all biologists know, in organisms living at moderate temperatures cell membranes are constructed along the typical "lipid bilayer" model: hydrophobic residues (fatty acids) inside oppose each other and retain an affinity for one another while hydrophilic residues (glycerol phosphate) lie at the surface of the environment and the cytoplasm, respectively, maintaining contact with the aqueous phase. If one applies sufficient heat to such a membrane architecture the two leaflets of the membrane will pull apart, leading to membrane damage and cytoplasmic leakage. To prevent this from occurring at very high temperatures, hyperthermophiles have evolved a novel membrane structure. Instead of forming a membrane as a lipid *bilayer*, as just discussed, hyperthermophiles chemically bond the opposing hydrophobic residues from each layer of the membrane together. This forms a lipid *monolayer* instead of a bilayer, and prevents the membrane from melting at high temperature. Although the precise chemistry of lipid monolayer membranes can vary somewhat from species to species, they are universal among hyperthermophiles and are undoubtedly an evolutionary response to life at high temperature.

6. BACTERIAL LIFE AT LOW TEMPERATURES AND EXTREME pH LEVELS

6.1 Psychrophiles

How about life at the other end of the thermometer? Cold environments on the earth are actually much more common than hot ones. For example, the oceans, which make up over one half the earth's surface, maintain an average temperature of about 2°C. And vast land masses are intermittently cold and in some cases permanently cold, or even frozen. However, cold temperatures are no barrier to microbial life, as various microorganisms flourish in cold environments, even in ice. Many microorganisms have been isolated capable of growth at refrigerator temperatures (4-8°C). These are usually *psychrotolerant*, meaning that although they are capable of growth in the cold, they grow better at warmer temperatures, usually 25-35°C. True **psychrophiles,** defined as microorganisms that grow *best* at 15°C or lower, are usually only present in permanently cold environments like the Arctic, or in particular, the Antarctic.

A variety of microorganisms including algae and diatoms have been found in Antarctic sea ice-ocean water that remains frozen for much of the year. Sea ice is the habitat for one well characterized bacterium, *Polaromonas vacuolata*, the genus name indicating its affinity for cold temperatures (Irgens *et al.*, 1996). *P. vacuolata* grows optimally at 4°C and finds temperatures above 12°C too warm for growth! Other psychrophiles are known but because some of them appear to be very sensitive to warming, great care must be taken in their isolation and culture to prevent killing them off at temperatures as low as room temperature!

An understanding of the biochemistry and molecular biology of psychrophilic bacteria is in a much earlier stage than that of the hyperthermophiles. From what is known about the biochemistry of psychrophiles it appears that their proteins function optimally at low temperatures because they are constructed in such a way so as to maximize flexibility; this is essentially the opposite strategy from that of hyperthermophiles (see above). Moreover, proteins from psychrophiles are typically more polar and less hydrophobic than proteins from hyperthermophiles, a fact that undoubtedly also assists in their relative flexibility.

Besides keeping their enzymes functional, psychrophiles have other biological problems to contend with, transport of nutrients across the membrane being chief among them. However, just as margarine, with its higher content of unsaturated fats can stay softer than butter at cold temperatures, psychrophiles regulate the chemical composition of their membranes, including in particular the length and degree of unsaturation of fatty acids, to keep them sufficiently fluid to allow for transport processes, even at temperatures below freezing. Applications of enzymes from psychrophiles include the cold food industry, where enzymes that work at refrigerator temperatures are sometimes desirable, as well as producers of cold-water laundry detergents.

6.2 Acidophiles and Alkaliphiles

Many extremophiles have evolved to grow best at extremes of pH: these are the **acidophiles** and the **alkaliphiles.** Although extremely acidic or alkaline (below pH 3 or above pH 10) habitats are rare on the earth, in such environments one can find a variety of microorganisms thriving in chemistry the equivalent of vinegar or soda-lime. Highly acidic environments can result naturally from geochemical activities, such as from the oxidation of SO_2 and H_2S produced in hydrothermal vents and hot springs, and from the metabolic activities of certain acidophiles themselves. For example, the iron sulphide-oxidizing bacterium *Thiobacillus ferrooxidans* can generate acid by oxidizing Fe^{2+} to Fe^{3+}, the latter of which precipitates out as $Fe(OH)_3$ ($Fe^{3+} + 3H_2O \rightarrow Fe(OH)_3 + 3H^+$), or by oxidizing HS^- to SO_4^{2-} ($HS^- + 2O_2 \rightarrow SO_4^{2-} + H^+$). *T. ferrooxidans* is particularly active in surface coal mining operations where exposure to oxygen of pyrite (FeS_2) in the coal seam triggers acid production from the metabolic activities of this and related bacteria (Schleper *et al.*, 1995). Runoff from these habitats can often have a pH of less than 2, fueling conditions for further acidophile activity.

6.2.1 *Acidophiles*

The most acidophilic of all bacteria known thus far is *Picrophilus oshimae*, whose pH optimum for growth is just 0.7. *P. oshimae* is also a thermophile (temperature optimum, 60°C) so this organism must be stable to both hot *and* acidic conditions. Interestingly, acid-loving extremophiles, even those as extreme as *P. oshimae*, cannot tolerate great acidity *inside* their cells, where it would destroy such important molecules as DNA. They thus survive by keeping the acid out. The internal pH of *P. oshimae* is about pH 5 and it is the cytoplasmic membrane of this organism that keeps protons from passively entering the cell. However, studies of the *P. oshimae* membrane have shown that it can only retain its integrity in acidic solutions; above a pH of about 4, the *P. oshimae* membrane spontaneously disintegrates (Pereira & Reeve 1998).

Cultures of *P. oshimae* were isolated from an extremely acidic (< pH 1) solfatara in Italy and clearly has evolved to require these highly acidic conditions for its very existence. Major unanswered questions concerning the metabolism of *P. oshimae* and other extreme acidophiles concern how they generate a proton motive force during respiration and related issues of bioenergetics involving membrane-mediated proton translocation. Various enzymes from acidophiles have been studied and potential applications for acid-active extremozymes exist, primarily as animal-feed supplements where they function to break down inexpensive grains to more nutritionally beneficial forms directly in the animal's stomach.

6.2.2 Alkaliphiles

Extreme alkaliphiles live in soils laden with carbonate or in soda lakes where the pH can rise to as high as 13. *Natronobacterium magadii,* for example, was isolated from Lake Magadi, a soda lake located in the Rift Valley of Africa; *N. magadii* grows optimally at a pH of about 10. In the opposite scenario from the acidophiles, alkaliphiles have to contend with the problems associated with *high* pH. Above a pH of 8 or so, certain biomolecules, notably RNA, break down. Consequently, like acidophiles, alkaliphiles must maintain their cytoplasm nearer to neutrality than that of their environment. Nevertheless, any proteins found in the cell wall or in the membrane that make contact with the environment must be stable to high pH. Indeed, many such enzymes have been studied and many have found industrial applications, especially in the laundry detergent industry. Detergents that are "enzyme enriched" contain proteases and lipases (enzymes that degrade proteins or fats, respectively, in clothing stains) that function at the high pH of soapy solutions. In addition, alkali-active enzymes from thermophiles and psychrophiles have been discovered and commercialized to better target detergent additives to hot water or cold water applications, respectively.

Besides keeping their cytoplasm near neutrality, alkaliphiles have other biological problems to contend with. For example, consider the problem of membrane-mediated bioenergetics—protons extruded to the external surface of the membrane enter a sea of hydroxyl ions. Nevertheless, biochemical studies of this problem have shown that a proton motive force is indeed formed by extreme alkaliphiles and drives some of the energy-requiring reactions in the cell, such as motility and transport. But for ATP synthesis, an ion gradient of Na^+, rather than H^+, drives this key bioenergetic process. This is probably not surprising when one considers that many (but not all) extreme alkaliphiles are also extreme halophiles (see below), requiring high salt as well as high pH for metabolism and reproduction.

7. BACTERIAL LIFE IN BRINE AND OTHER ENVIRONMENTS

7.1 Halophiles

Another remarkable group of extremophiles are the halophiles—organisms adapted to grow best in salty solutions. And for extreme halophiles like *Halobacterium*, a "salty solution" means anywhere from 25% NaCl up to saturation (32% NaCl). Halophilic microorganisms abound in hypersaline lakes such as the Dead Sea, the Great Salt Lake, and solar salt evaporation ponds. Such lakes are often coloured red by the dense microbial communities of pigmented halophiles such as *Halobacterium*. Other habitats for halophilic microorganisms include highly salted foods, saline soils, and underground salt deposits. To date a very large number of halophilic bacteria have been grown in culture including members of all domains of life, including the eukarya (Oren 1999).

7.2 How Halophiles Survive in Saline Conditions

Halophiles are able to live in salty conditions by preventing dehydration of their cytoplasm. They do this by either producing large amounts of an internal organic solute or by concentrating a solute from their environment. Whatever the chemical nature of this solute, it is referred to as a "compatible solute." The archaeon *Halobacterium* concentrates in its cytoplasm large amounts of potassium (K^+, as KCl) from its environment. Dissolved KCl in cells of *Halobacterium* is present at a concentration equal to or slightly above that of the dissolved NaCl outside, and in this way cells maintain the tendency for water to enter and thereby prevent dehydration (Ventosa *et al.*, 1998). As would be expected from such a salty cytoplasm, enzymes that function inside of cells of *Halobacterium* have evolved to require this large dose of K^+ for catalytic activity. By contrast, membrane or cell wall-positioned proteins in *Halobacterium* require Na^+ and are typically stable only in the presence of high Na^+ (Javor 1989).

Extreme halophiles are sources of a variety of biomolecules that can function under salty conditions. Applications of salt-active enzymes include those that can break down viscous materials present in oil wells (oil is often found in geographic strata that contains salt) as well as enzymes that can carry out desirable transformations in highly salted foods. In addition, some halophiles that produce organic compatible solutes have been commercialized for the production of these solutes as skin care supplements (Kamekura 1998).

7.3 Halophiles Maintain High Solute Concentrations Inside their Cells

Some types of halophiles have unique chloride pumps that transport chloride from the external environment to the inside of the cells. Particular arginine residues at either end of the protein channel play roles in the uptake and release of chloride. Halophiles may also concentrate potassium ions. Halophiles concentrate solutes inside their cells to maintain osmotic balance with the external environment.

8. OTHER EXTREMOPHILES

Extremophilic microorganisms adapted to high pressure or who show no deleterious effects from exposure to high levels of radiation, are also known. **Barophiles** are microorganisms that grow best under pressure greater than 1 atmosphere. Extreme barophiles are the most interesting in this regard as they actually *require* pressure, and in some cases, extreme pressure, for growth (Harmsen *et al.*, 1997). Strain MT41, for example, a bacterium isolated from marine sediments in the Mariana Trench near the Philippines (a depth of greater than 10,000 meters), requires *at least* 500 atmospheres of pressure in order to grow and grows optimally at 700 atmospheres (and at a temperature of 4°C because strain MT41 is also a psychrophile). Because laboratory culture of extreme barophiles is rather difficult, comparatively little is known about their important biomolecules. However, although probably all macromolecules in extreme barophiles need to be biochemically tailored to high pressure to some extent, experiments with moderately barophilic bacteria, some of which can be grown without pressure, have pointed to nutrient transport proteins in the cytoplasmic membrane as key cell components requiring structural modifications in order to function at high pressure.

The bacterium *Deinococcus radiodurans* is an amazingly radiation-resistant microorganism. This remarkable organism can survive 30,000 Grays of ionizing radiation, sufficient to literally shatter its chromosome into hundreds of pieces (by contrast, a human can be killed by exposure to as little as 5 Grays). A powerful DNA repair machinery exists in cells of *D. radiodurans* that is able to piece the shattered chromosome back together and yield viable cells. Because of its remarkable radiation resistance, *Deinococcus* has been proposed as a cleanup agent for the bioremediation of toxic materials in contaminated soils that are also radioactive from the leakage of radioactive materials; these conditions exist primarily at nuclear weapons production sites.

Halophiles grow optimally in salty environments. They keep a solute concentration inside their cells at an equal or higher level than their salty environments to prevent dehydration of their cytoplasm. Pressure-resistant and radiation-resistant bacteria also have unique modifications to their environments.

9. EXTREMOPHILES IN THE EVOLUTION OF LIFE ON EARTH

A focus of research on extremophiles has centred on the hyperthermophiles. As discussed earlier, there is good reason to believe that at least some hyperthermophiles have evolved relatively little from their ancestors present on the earth over 3.5 billion years ago (Wiegel & Adams 1998). If true, an understanding of the biology of hyperthermophiles may yield a glimpse of what life was like eons ago. In this connection, the genomes of several hyperthermophiles have been sequenced and the large number of genes they contain that lack counterparts in other organisms suggests that their biological secrets have at this point only been partially revealed. As if living in boiling water isn't enough, just imagine what other tricks hyperthermophiles might be able to perform!

As previously mentioned, the excitement in microbial diversity these days comes from the fact that the evolutionary history of the prokaryotes can now be experimentally determined. Microbiologists no longer have to propose bacterial phylogeny based on speculation or "educated guesses" of what type of microbe likely preceded another; the phylogenies themselves are etched in the sequences of molecules and all one has to do is read them. Moreover, the application of molecular phylogenetic methods to natural environments has given us the exciting news that the diversity of the microbial world is enormous—indeed it is beyond our wildest expectations. Thus, in the final analysis bacterial diversity will likely dwarf that of all of the rest of biology, perhaps by several orders of magnitude. But only continued and expanded research into the diversity of microbial life in all environments, extreme and otherwise, will yield the data needed to confirm this.

It may indeed be humbling to many biologists to think that prokaryotes dominate living diversity. But within the rich genetic resources of the prokaryotes undoubtedly lies more benefit for humankind than we will extract from any other group of organisms. Antibiotics, fermentation, and biotechnology are only the beginning. The best is yet to come.

Due to their diversity and genetic uniqueness, extremophiles and bacteria in general may hold many secrets that are of evolutionary and practical significance (Woese 1987).

ACKNOWLEDGEMENTS

The author is thankful to CSIR for providing Senior Research Associateship (Pool Scientist) to Dr. Jay Shankar Singh. RKG is thankful to University Grants Commission, New Delhi, Uttarakhand Biotechnoloy Programme and Uttarakhand Council for Science and Technology (UCOST) for financial help.

REFERENCES

Adams, M.W.W. and Kelly, R.M. (1995). Enzymes from microorganisms in extreme environments, *Chemical and Engineering News* **73:** 32-42.

Barns, S.M., Fundyga, R.E. Jeffries, M.W. and Pace, N.R. (1994). "Remarkable archaeal diversity detected in a Yellowstone National Park hot spring environment," *Proc. Nat. Acad. Sci. USA,* vol. 91, pp. 1609-1613.

Blöchl, E., Rachel, R., Burggraf, S., Hafenbradl, D., Jannasch, H.W. and Stetter, K.O. (1997). "*Pyrolobus fumarii,* gen. and sp. nov. represents a novel group of archaea, extending the upper temperature for life to 113°C," *Extremophiles,* vol. 1, pp. 14-21.

Brock, T.D. (1978). *Thermophilic Microorganisms and Life at High Temperatures*, Springer, New York, pp. 465.

Harmsen, H.J.M., Prieur, D. and Jeanthon, C. (1997). "Distribution of microorganisms in deep-sea hydrothermal vent chimneys investigated by whole-cell hybridization and enrichment culture of thermophilic subpopulations," *Appl. Environ. Microbio.* vol. 63, pp. 2876-2883.

Horikoshi, K. and Grant, W.D. (1998). *Extremophiles-Microbial Life in Extreme Environments*, Wiley, New York, pp. 322.

Hugenholtz, P., Pitulle, C., Hershberger, K.L. and Pace, N.R. (1998). "Novel division level bacterial diversity in a Yellowstone hot spring," *J. Bacteriology* vol. 180, pp. 366-373.

Irgens, R.L., Gosink, J.J. and Staley, J.T. (1996). "*Polaromonas vacuolata* gen. nov. sp. nov., a psychrophilic, marine, gas vacuolate bacterium from Antarctica," *Internat. J. Systematic Bacteriology,* vol. 46, pp. 822-826.

Javor, B.J. (1989). *Hypersaline Environments-Microbiology and Biogeochemistry*, Springer-Verlag, Berlin, pp. 328.

Kamekura, M. (1998). "Diversity of extremely halophilic bacteria," *Extremophiles,* vol. 2, pp. 289-295.

Kurr, M., Huber, R., Konig, H., Jannasch, H.W., Fricke, H., Trincone, A., Kristjansson, J.K. and Stetter, K.O. (1991). "*Methanopyrus kandleri*, gen. and sp. nov. represents a novel group of hyperthermophilic methanogens, growing at 110°C," *Archives of Microbiology,* vol. 156, pp. 239-247.

Ladenstein, R., and Antranikian, G. (1998). "Proteins from hyperthermophiles: Stability and enzymatic catalysis close to the boiling point of water," *Advances in Biochemical Engineering and Biotechnology,* vol. 61, pp. 38-85.

Madigan, M.T., Marrs, B.L. (1997). "Extremophiles," *Scientific American,* vol. 276, pp. 82-87.

Madigan, M.T., Martinko, J.M. and Parker, J.(2000). *Brock Biology of Microorganisms*, 9th edition, Prentice Hall, Upper Saddle River, NJ. *In press.*

Madigan, M.T. and Oren, A. (1999). "Thermophilic and halophilic extremophiles," *Current Opinions in Microbiology, In press.*

Oren, A. (Ed.) (1999). *Microbiology and Biogeochemistry of Hypersaline Environments*, CRC Press, Boca Raton, FL, pp. 368.

Pereira, S.L. and Reeve, J.N. (1998). "Histones and nucleosomes in archaea and eukarya: a comparative analysis," *Extremophiles,* vol. 2, pp. 141-148.

Raven, P.H., Evert, R.F. and Eichhorn, S.E. (1999). *Biology of Plants*, 6th edition, W.H. Freeman and Co., New York, pp. 944.

Raven, P.H. and Johnson, G.B. (1999). *Biology,* 5th edition, WCB/McGraw-Hill, Dubuque, IA.

Schleper, C., Puhler, G., Holz, I., Gambacorta, A., Janekovic, D., Santarius, U., Klenk, H-P. and Zillig, W. (1995). "*Picrophilus* gen. nov., fam. nov., a novel aerobic heterotrophic, thermoacidophilic genus and family composing archaea capable of growth around pH 0," *J. Bacteriology,* vol. 177, pp. 7050-7059.

Stetter, K.O. (1996). "Hyperthermophilic prokaryotes," Federation of European Microbiological Societies, *Microbiology Reviews,* vol. 18, pp. 149-158.

Stetter, K.O. (1999). "Volcanoes, hydrothermal venting, and the origin of life," In: *Volcanoes and the Environment,* Edited by Marti, J. and Ernest, G.J. Cambridge University Press, Cambridge, England, In press.

van de Vossenberg, J.L.C.M., Driessen, A.J.M. and Konings, W.N. (1998a). "The essence of being extremophilic: the role of the unique archeal membrane lipids," *Extremophiles,* vol. 2, pp. 163-170.

van de Vossenberg, J.L.C.M., Driessen, A.J.M., Zillig, W. and Konings, W.N. (1998b). "Bioenergetics and cytoplasmic membrane stability of the extremely acidophilic, thermophilic archaeon, *Picrophilus oshimae,*" *Extremophiles,* vol. 2, pp. 67-74.

Ventosa, A., Nieto, J.J. and Oren, A. (1998). "Biology of moderately halophilic aerobic bacteria," *Microbiology and Molecular Biology Reviews*, vol. 62, pp. 504-544.

Wiegel, J. and Adams, M.W.W. (Eds.) (1998). *Thermophiles–The Keys to Molecular Evolution and the Origin of Life?*, Taylor and Francis, London, pp. 346.

Woese, C.R. (1987). "Bacterial evolution," *Microbiological Reviews,* vol. 51, pp. 221-271.

Diversity of Lower Plants (2014): pp. 123-134
Editors: Rajan Kumar Gupta & Mukesh Kumar
IK International Publishing House Pvt. Ltd.

10

Pathogen Biodiversity: Its Origin, Nature, Characterization and Control

Mohit Kumar. Gupta*, Rahul Kumar*, G.K. Sharma** and Mukesh Kumar***

Department of Botany, Hindu College, Moradabad–244 001
Department of Botany, Sahu Jain College, Najibabad–246 763

ABSTRACT

A pathogen is an agent which causes pathos (ailment, suffering) or damage. It includes all living organisms, often a microorganism associated with causation of diseases. Most pathogens are also parasites in that they derive the nutrients for growth from a living organism, the host or suscept. In this chapter, we will describe only phyto-pathogens among which fungi and bacteria are the most important, other groups of plant pathogens includes nematodes, viruses, viroids, parasitic algae and few angiosperms. The economic significance of plant pathogens lies in their destructive effect on yield of our crops. Pathogens are the main enemies of the crops and they cause enormous loss to the agricultural production of the world.

Keywords: Pathogen, Fungal diseases, Biological speciation.

1. INTRODUCTION

The extent of loss and damage caused by diseases varies from year to year depending upon the prevalent climatic factors, which highly influence the onset and spread of diseases. In a normal year, the damages is estimated to be approximately 10 to 25% even low, while in the years of epidemics, the entire crop is destroyed and thus the damage may extend up to 100% as happened in Tikka epidemic of groundnut in Bombay State during 1912-13 and wheat rust epidemics of 1946-47 in Peninsular India. By 1845, the epiphytotics of late blight became widespread throughout Europe where it devastated the potato fields and left famine in its wake. The famous Irish famine of 1945-46 was due largely to failure of the potato crop, which at that time constituted the staple food of over four million people of Ireland. Now this disease is also very destructive

**Communicating Author

to potato crop and making regular appearances almost throughout the plains of northern India. Although the accurate figures are not available, it is estimated that the annual loss due to plant disease in India in terms of money runs into several hundred crores of rupees.

Thus, plant pathogens are of much importance to us because they damage plants and plant products on which we depend for food textiles, housing, and medicines. The death from starvation of a quarter million people in Ireland in 1945 and much of the hunger of the underfeed millions still living in the third world, are the examples of the consequences of plant diseases. For those countries, where food production is in excess of the requirements, plant pathogens cause economic loss to farmers. They cause diseases which also result in the escalation of prices of products to consumers, especially poor men. Moreover, diseases also damage ornamental plants destroying the beauty of nature. Huge sum of money is spent in the management of plant diseases every year.

Table 1: Historical and present examples of losses caused by different diseases

Disease	Location	Remarks
Fungal Diseases		
Late blight of potato	Cool, humid areas	Annual epidemics, e.g. Irish famine (1845-1846)
Brown spot of rice	Asia	The great Bengal famine (1943)
Powdery mildew of grape	Worldwide	European epidemics (1840s-1850s)
Downy mildew of grapes	U.S., Europe	European epidemics (1870s-1880s)
Banana leaf spot or sigatoka disease	Worldwide	Great annual losses
Coffee rust	South-East Asia	Destroyed all coffee in South-East Asia (1870s-1880s), since 1970 present in South and Central America
Downy mildew of tobacco	U.S., Europe	European epidemics (1950s-1960s): epidemic in North America (1979).
Southern corn leaf blight	U.S.	Epidemic (1970)
Chestnut blight	U.S.	All American chestnut trees destroyed (1904-1940)
Dutch elm disease	U.S., Europe	All American chestnut trees destroyed (1930 to date)
Cereal rusts	Worldwide	Frequent, severe epidemics
Cereal smuts	Worldwide	Frequent losses
Ergot of rye and wheat	Worldwide	Poisonous to humans and animals.
Bacterial diseases		
Citrus canker	Asia, Africa, Brazil, U.S.	Eradicated trees in Florida (1910, 1980, 1990)
Soft rot or vegetables	Worldwide	Huge losses
Viral diseases		
Sugar cane mosaic	Worldwide	Great losses
Barley yellows dwarfs	Worldwide	On small grains
Tomato spotted wilt	Worldwide	On tomato, tobacco, peanut, etc.
Sugar beet yellows	Worldwide	Great losses
Nematodal diseases		
Root-knot	Worldwide	On vegetables and most other plants.

Table 2: Some diseases likely to cause severe losses in future

A	**Fungal diseases**	
1.	Late blight of potato and tomato	- New compatibility types of the fungus spreading worldwide.
2	Downy mildew of corn and sorghum	- Just spreading out of South-East Asia
3.	Karnal bunt of wheat	- Destructive in India, Nepal and Pakistan since 1980s, introduced into Mexico, and in the 1990s in U.S.A.
4.	Soybean rust	- Spreading from South-East Asia and Russia to Hawaii and Puerto Rico.
5.	Chrysanthemum white rust	- Important in Europe, Asia, and recently in California.
6.	Sugar cane rust	- Destructive in the Americas and elsewhere.
B	**Bacterial diseases**	
1.	Bacterial leaf blight of rice	- Destructive in Japan and India; spreading to other areas.
2	Bacterial wilt of banana	- Destructive in Americas: spreading elsewhere.
C	**Viral diseases**	
1.	African cassava mosaic	- Destructive in Africa; threatening Asia and the Americas.
2	Streak disease of maize	- Spread throughout Africa on sugar cane, corn, wheat etc.
3.	Bunchy top of banana	- Destructive in Asia, Australia, Egypt, Pacific Islands.
4.	Rice tungro	- Destructive in South-East Asia.
D	**Viroid diseases**	
1.	Cadang-cadang of coconut	- Killed more than 1.5 million trees in the Philippines.

Table 3: Estimated world crop production (FAO, 1993), and pre-harvest losses (million of tons) and percentage of world production lost to disease, insects and weeds (Oerke, 1994)

Production and losses (million of tonnes)

Crop	Actual production	Estimated loss to diseases	Potential production	Estimated loss to disease	Crop % lost to disease	Total of crop lost (%)
Cereals	1,894	906	2,800	257	9.2	9.2
Potatoes	288	137	425	93	21.8	21.8
Other root crops	603	454	1,057	176	16.7	16.7
Sugar beets	283	91	373	39	10.4	10.4
Sugar cane	1,040	1,271	2,311	444	19.2	19.2
Legumes	57	28	85	9	11.3	11.3
Vegetables	465	178	643	65	10.1	10.1
Fruits	371	113	484	61	12.6	12.6
Coffee-cocoa-tea	11	8	19	3	17.7	17.7
Oil crops(oil equi)	245	109	354	35	9.8	9.8
Fiber crops	42	19	61	7	11.0	11.0
Tobacco	8	3	11	1	12.3	12.3
Natural rubber	5	2	7	1	15.0	15.0
Average	-	-	-	-	11.8	11.8

In this chapter, we try to provide a comprehensive review of pathogenic biodiversity, evolution of diseases in plants, host parasitic relationship and new encounters between pathogens and plants. Then we describe important aspects of host ranges and host specificity. Hence, some attempts are also made to describe the concept of physiological races as the principal unit of intraspecific diversity. Emphasis is also given to the mechanisms by which diversity arises. It is essential to know the characterization of diversity, the occurrence of races and the interactions between races involved in agriculture. Identification of plant pathogen diversity and its dynamics in cropping system in different parts of the world is a matter of fundamental importance and of economic consequences. This information is essential and useful in breeding of resistant cultivars and for adopting management measures for different diseases.

2. PATHOGENICITY

Plant pathogenicity is found among most groups of microorganisms as well as in higher plants and pathogen diversity can be described in terms of taxonomy, morphology, genetics and function. Life histories display variation in modes of parasitism, of dispersal, of reproduction and of survival (Burdon, 1993).

Much has been written on the taxonomic and morphological diversity of microorganisms, including pathogens (Hawksworth, 1991; Hawksworth and Colwell, 1992). Morphology often reflects adaptive and functional strategy (Overwinkler, 1992), whereas taxonomically and morphologically distinct pathogens often have similar biophysiological strategies, expressed for instance as either specialization or opportunism found in saprophytes, necrotrophs, hemibiotrophs or biotrophs (Burdon, 1993, Mayer, 1993) displayed as adaptation to a particular habitat (Durbin, 1959) or temperature regime or similar symptoms.

Table 4: Percentage of all produce lost to disease, insects and weeds by continent or region

Continent or region	Produce lost to disease, insects and weeds (%)
Europe	25
Oceania	28
North and Central America	29
Russia and China	30
South America	33
Africa	42
Asia	43

The fungi are the largest group of plant pathogens, although only some 5% of the 7750 plant genera contain them.

Table 5: Percentage of all produce lost to disease, insects and weeds and total cost (U.S. Dollars) per year from 1988 to 1990 of major crops worldwide (Oerke *et al.*, 1994)

Continent or region	Loss (%)	Loss (billion $)
Europe	28.2	16.8
Oceania	36.2	1.9
North and Central America	31.2	23.0
Russia	40.9	22.0
South America	41.3	21.8
Africa	48.9	12.8
Asia	47.1	145.3

Many are relatively obscure and their behaviour unknown, so that the 400 genera of recognized fungal pathogens is probably an underestimate of the parasitic habit among them. For example, of some 350 species of fungi recorded from *Musa* spp. in the herbarium of the International Mycological Institute, only about 20% can be considered to be pathogens in the broad sense, although for many of these, the true parasitic nature of their association has not been investigated and most are of little current consequence in terms of damage to their host. However, changes in selection pressure, in host genotype and in agroecology can lead to the emergence of tomorrow's epidemics.

Pathogenicity is the ability to cause disease and broadly and unevenly scattered across the taxonomic spectrum of fungi. Some orders such as the erysiphales (the powdery mildews), perenosporales (the downy mildews) and uredinales (the rusts) are highly specialized pathogens, but many others also contain pathogenic species. Often within a single genus, there are species which exhibit a wide range of tropic behaviours. *Fusaruim* is a good example (Walter and Brayford, 1990). The trophic variation in this and many other genera of mitosporic fungi (*Alternaria, Cladosporium*_and *Colletotrichum*) suggest evolution of the more specialized parasitic state from saprotrophism.

Table 6: Diversity in disease symptoms in plants infected by pathogens

Type of plant disease (after Wheeler, 1969)	Typical pathogens
Post-harvest disease tuber and fruit rots; soft rots	*Penicillium, Rhizopus, Sclerotinia* spp; *Erwinia carotovora*
Damping-off and seedling blights.	*Pythium, Phytophthora, Aspergillus, Rhizoctonia, Fusarium, Helminthosporium* spp.
Root and foot rots, take-all	*Armillari, Penicillium,* Rhizopus, Sclerotinia, Erwinia Carotovora, Rhizoctonia, Pythium, Phytophthora, Aspergillus, Fusarium, Helminthosporium *Armillara mellea, Valsa eugeniae, Gaeumannomyces graminis, Corticium rolfsii, Fusarium solani, Helicobasidium, purpureum, Aphanomyces* spp.
Wilts	*Ophiostoma novo-ulmi, Fusarium oxysporum, Verticillium albo-atrum; Burkholderia solanacearum, Xanthomonas campestris pv. vasculorum, Erwinia stewartii, Striga hermonthica, Orobanche crenata.*
Downy mildews	*Peronospora, Plasmopara, Bremia* spp; *Sclerospora graminicola.*
Powdery mildews	*Erysiphe, Phyllactinia* spp; *Leveillula taurica.*
Rusts	*Puccinia, Uromyces, Melampsora* spp; *Hemileia vastatrix.*
Smuts	*Ustilago, Urocystis, Tilletia, Entyloma* spp;
Blights	*Phytophthora infestans, Ascochyta rabiei; Erwinia amylovora, Xanthomonas campestris.*
Anthracnose	*Colletorichum, Gloeosporium* spp;
Leaf spots	*Botrytis fabae; Alternaria, Septuria, Diplodia, Cercospora, Helminthosporium* spp; *Pseudomonas syringae.*
Leaf curl, witches' broom, club root, blister blight	*Plasmodiophora brassicae; Taphrina deformans; Exobasidium vexans; Crinipellis perniciosa*; tomato leaf curl geminivirus.
Galls	*Agrobacterium tumefaciens; Meloidogyne* spp;, wound tumour phytoreovirus.
Cankers and Scab	*Nectria galligena; Venturia inaequalis; Elsinoephaseoli; Spongospora subterranean; Clavibacter michiganensis.*
Mosaic, green vein-bending and yellows.	Tobacco mosaic tobamovirus, bean common mosaic potyvirus, cucumber mosaic cucumovirus, beet western yellows luteovirus, and numerous other viruses.

Studies on plant diseases indicate that interactions between plants and pathogens play an important role in the occurrence of the disease, which range from minor to epidemic (Burdon, 1987, 1993; Dobson & Crawley, 1994). In some natural plant communities, epidemics rarely occur; in others, epidemics are common. In some agroecosystems, epidemics commonly ravage crops; in others, years may go by in which disease has little, if any, effect on crop production (Bordon, 1993). There is no doubt that the evolution of disease and pathogen variability has played an important role in fostering diversity in ecosystems (Gillett, 1962). Pathogens may change the outcome of inter- and intraspecific competition, or change the genetic structure of plant populations and the diversity of the plant community (Burdon, 1987). The wild relatives of economically important crops have structural and chemical defence and resistance genes which are evidence of powerful selective forces in the past (Harper, 1990).

The coexistence of host plants and their pathogens side by side in nature indicates that the two have been evolving together. Changes in virulence of pathogen appear to be continually balanced by changes in the resistance of the host, and vice versa. In that way, a dynamic equilibrium of resistance and virulence is maintained and both, host and pathogen survive over considerable periods of time. The stepwise evolution of virulence and resistance can be explained by gene-for-gene concept, according to which for each gene that confers virulence to the pathogen, there is a corresponding gene in host that confers resistance to the host, and vice versa.

The gene-for-gene concept was first proved in cases of flax and flax rust, *Melampsora lini* by Flor in 1946. Since then, it has been shown to operate in many other rusts, smuts, powdery mildews, apple scab, late blight of potato and other fungal diseases; as well as in some diseases caused by bacteria, viruses, nematodes and higher parasitic plants.

Table 7: The distribution of plant pathogenicity in the Fungi sensu lato (after Hawksworth *et al.,* 1995)

Classes	No. of genera	No. of plant pathogenic genera
Protozoa		
Ascrasiomycota	4	0
Myxomycota (slime moulds)	4	0
plasmodiophoromycota	74	0
Chromista	15	2 (*Plasmodiophora* and *Spongospora)*
Hyphochytridiomycota	7	0
Labyrinthulomycota	13	0
Oomycota	95	19 (especially in Peronosporales and Pythiales)
Fungi sensustric to Ascomycota	3255	60 (in some 13 orders of which the most important are Dothidiales, Diaporthales, Erysiphales and Hypocrealees).
Basidiomycota	1428	249 (of which 19 are Basidiomycetes, 167 are Teliomycetes, chiefly rusts, and 63 are Ustomycetes, mostly smut fungi)
Chytridiomycota	112	3 (including *Synchytrium)*
Zygomycota	173	6 (including, *Rhizopus* and *Choanephora)*
Mitosporic fungi	2547	70
Total	7727	407

Long-term genetic interaction between host and pathogen population in many ancient plant communities results in the pathogen being held in check by both individual and population resistance (Burdon, 1987). Gene centres of plants have been recognized as rich sources of resistance (Leppik, 1970) especially to diseases against which the host and pathogen coevolved;

perhaps in a gene-for-gene relationship (Frank, 1993). Wild plants are good sources of disease resistance, but all the species of plant are not disease resistant, some of them seem to be susceptible to a fairly small number of different pathogens.

In spite of many pathogens that can infect them, numerous individuals of a single plant species survive on land every year. These remain either free of disease or show only minor symptoms. Why all the plants are not attacked by their pathogens and why are those that are attacked not usually killed by the pathogens? It is so, because plants, through coevolution or systematic breeding have acquired, in addition to genes that make them susceptible to a pathogen, one or usually many genes for resistance that protect them from infection or severe disease. When a new gene for resistance to a pathogen appears or is introduced into a plant, the plant becomes resistant to all or most of the previously existing individuals (races) of the pathogen. Such pathogens contain one and usually more than one gene for virulence, but if they do not contain the additional, new gene for virulence that is required to overcome the effect of the new resistance gene in plants, they cannot infect the plant and the plant remains resistant. Thus, even one new gene for resistance to a pathogen can protect plants that have the gene from becoming infected by all or most pre-existing races of the pathogen at least for several months or years.

With numerous host pathogen combinations, however, after a new gene for resistance to a pathogen is introduced into a crop variety and the variety planted in the fields, a new population (race) of the pathogen appears that contains a new gene to virulence that enables the pathogen to attack the crop plants containing the new gene for resistance. How did this new population of the pathogen acquired the new gene for virulence? In most cases, the new gene had already been present earlier at low levels or by mutation, only in a few pathogen individuals. New genes can arise randomly and suddenly de novo through mutations or by rearrangement of the genetic material of the pathogens through the ongoing events of genetic variability in organisms.

Table 8: Diversity in trophic specialization among crop-associated fungi

Trophic specialization	Disease/association	Pathogen examples
Symbionts	Mycorrhizae	*Glomum spp.*
Epiphytes	Inflorescence disease	*Balansia* spp.
	Sooty moulds	*Capnodium spp.*
Endophytes Biotrophs	Anthracnose of beech	*Discula umbrinella*
	Powdery mildews	*Erysiphe graminis*
	Downy mildews	*Peronospora parasitica*
		Bremia lactucae
		Peronospora parasitica
		Bremia lactuca
	Rusts	*Puccinia graminis*
		Uromyces appendiculatus
	Smuts	*Ustilago scitaminea*
		Tilletia caries.
Hemibiotrophs	Bean anthracnose	*Colletrichum lindemuthianum*
	Late blight	*Phytophthora infestans*
	Rice blast	*Pyricularia oryzae*
	Early leaf spot	*Mycosphaerella arachidis*
	Vascular wilt	*Fusarium oxysporum*
Necrotrophs	Southern leaf blight	*Cochliobolus heterostrophus*
	White mould	*Sclerotinia sclerotiorum*
	Crown rot	*Aspergillus niger*
	Southern blight	*Corticium rolfsii*

The two general mechanisms of variability, that occur in both, plants and pathogens are, mutation and recombination. Besides, these there are also some specialized mechanisms in pathogens which bring about genetic variability in them. These include heterokaryosis, parasexualism and heteroploidy among fungal pathogens and conjugation, transformation and transduction in bacterial pathogens. Genetic recombination also occurs among virus.

In most case of extreme susceptibility in the host and extreme virulence in the pathogen epidemics, are the consequence of human activities. Buddenhagen (1977) has examined the origin of epidemic diseases in tropical crops, concluding that all have one thing in common that epidemic arise from an intensified union of crop and pathogen.

Plant quarantine has played an important role in restricting the international movement of pathogen, many of which are effectively seed borne and widely distributed despite such procedures. It is evident that disease epidemics are the result of intercontinental or regional movement of a crop plant into a new environment or the movement of a pathogen into a new region where it infects a related host species. Maize streak, rice yellow mottle, groundnut rosette and black root of common bean are the examples of new-encounter diseases as a result of crop movement. Dutch elm wilt, caused by *Ophiostoma novo-ulmi*, chestnut blight in the USA caused by *Cryphonectria parasitica* and late blight of potato caused by *Phytophthora infestans* are the examples of new-encounter diseases as a result of pathogen movement.

There is great diversity in the degree of host specificity shown by plant pathogens. Those with narrow host ranges are well illustrated by biotrophic graphic fungi like the rust, smut and powdery mildew pathogens of cereals that are obligate parasites (Scott *et al.*, 1980) Some leaf spot fungi, like the Tikka disease of groundnut caused by *Cercospora arachidicola* is host-genus-specific. Necrotrophs kill tissue as they grow through it such that they are always colonizing dead substrate, are opportunistic pathogens with very wide host ranges. *Botrytis cinerea* is a facultative parasite and grows on dead remains.

There is an evolutionary progression from obligate saprophytism and facultative parasitism to obligate parasitism and, possibly mutualistic symbiosis (Tarr, 1972). Parbery (1996) suggested that biotrophic character is primitive and ancestral to necrotropism and saprophytism.

Puccinia graminis (causal organism of rust diseases) is noteworthy for their diversity of spore stages with life cycle in which, in heteroecious species, there is a host alternation. In rust fungi, diploid stages are specific to wheat (Poaceae) whereas the haploid stages are specific to *Berberis vulgaris* (Berberidaceae). These two hosts of this heteroecious rusts are very much unrelated. *Puccinia graminis* produces its spermogonia and aecia (O and I) on common barberry bush (*Berberis vulgaris*) and its uredinia and telia (II and III) on various grasses. Barberry is a dicotyledon while grasses are monocotyledon.

The phenomenon of biological specialization has been extensively studied in rust, particularly in *Puccinia graminis.* A number of biological subspecies that differ slightly in their morphology, but greatly in their ability to attack various grasses, constitute the species *Puccinia graminis.* To distinguish among these species, mycologists have added a third name to the binomial *Puccinia graminis* and designate each sub-species in the following manner:

Puccinia graminis *tritici*________ on wheat
Puccinia graminis *hordei*________ on barley
Puccinia graminis *avenea*________ on oats
Puccinia graminis *secalis*________ on rye

These above-mentioned and several other sub-species make up the species *Puccinia graminis.* But to complicate the situation, each sub-species is composed of a large number of physiological races that differ in their parasitism on various agronomic varieties of the host. *Puccinia graminis tritici,* for example, consists of over 200 such races to which wheat varieties are differentially susceptible. All these physiological races and sub-species have the common barberry bush on their alternate host.

3. BIOLOGICAL SPECIALIZATION

Biological specialization is an interesting phenomenon mycologically, but its economic importance cannot be overemphasized. The only practical method of controlling cereal rust is by an extensive breeding programme with the goal of developing rust-resistant varieties of cereals. Obviously, it is impossible to produce varieties that are resistant to all the physiological races of the parasite. Fortunately, however, there are relatively few races in a given geographical region that are violently destructive and plant breeders aim their programme at them. Importation of new races into a region of course, occurs occasionally and may seriously upset the breeding programme. Another problem is presented to the plant breeder by the constant threat of the origin of new physiological races of *Puccinia graminis* by mutation and hybridization.

Most species of *Berberis* are susceptible to the rust. *Berberis vulgaris* and *Barberis canadensis* grow wild in the United States. The Japanese barberry (*Berberis Japonica*), extensively used as an ornamental, is not susceptible. Now it is evident the eradication of barbery bushes was the logical means of breaking the life cycle of the parasite and of reducing the incidence of rust in the grain field. But it is unfortunate for plant breeders because nuclear exchanges also occur between dikaryotic mycelia on the grain host during uredinial stage.

The number of physiological races of cereals rusts in India is much less than the number encountered in cooler regions like the USA and Europe. This is apparently due to the absence of barberry bushes in the plains of India and also due to the fact that physiologic race flora on barberry is not the same as occurring on the cereals in the plains. The course of evolution of heteroecism presumably has important implications both for pathogen survival and for generation of novel diversity within species (Wilson and Henderson, 1966).

Within the pathogenic groups, variants differ in their pathogenecity. It may be host specific or non-host specific. This is only the reason that diversity exists at different levels within species of pathogens. Formae speciales are identified only physiologically or biochemically. They differ in pathogenicity or host adaptation. They are intraspecific taxa of fungi. Pathovar is synonymous term of formae speciales and used for bacteria (Holliday, 1989), *Fusarium oxysporum* has 82 formae speciales and *Xanthomonas campestris* 156 pathovars. A physiologic race may be defined as a taxon of parasites characterized by pathogenic specialization to different cultivars of one host species.

The biotrophic fungi differ in a qualitative manner in their virulence or host specific pathogenicity. The necrotropic fungi differ quantitatively or vary in virulence. The pathogens that differ in aggressiveness are called horizontal pathotypes and that vary in virulence are vertical pathotypes or physiologic race. Stakman (1926) has defined physiologic specialization as "presence of entities within morphologic species, not readily distinguishable by structure, but differing from each other physiologically" including pathogenicity, host morphology, biochemical properties, cultural variabilities, spore germination and ecological relationship. These entities have been variously named as formae speciales, biologic species, biologic forms, physiologic races, specialized races, parasitic strains, racial strains, physiologic forms. The three rusts of wheat can be easily distinguished from each other on the basis of their symptoms and structure, but each rust has distinct groups of entities of forms which cannot be distinguished from each other on morphological basis but still have different parasitic capabilities infecting only certain general species, or agricultural varieties of higher plants. *Puccinia recondita f. sp. tritici* has 228 such entities (Johnston and Browder, 1966) each of which is known as a physiologic race and is given a distinguishing number. *Sphaerotheca fuliginea* is the causal organism of powdery mildew disease on cucurbits and a serious problem of world including India, has three distinct physiologic races on cucurbits in India (Sharma and Khan, 1993). The presence of these races poses an important problem in breeding crop varieties to the disease.

Stakman and Levine (1922), Stakman *et al.* (1944, 1962) and Peterson *et al.* (1948) have described the methods of recognizing physiologic races of *P. graminis tritici* and determining the rust intensity.

Now race concept has become one of the main tools for understanding the structure and dynamics of pathogen population.

Pathogens also vary in their sensitivity to chemicals. Pathogens have developed resistance to fungicides and bacteriocides. *Pyrenophora avenae, Mycosphaerella arachidis* and *Colletotrichum kabawae* are fungal pathogens which are at risk for resistance development to fungicides. Recently chemically resistant races of powdery mildew have been isolated. Managing fungicides resistance is an essential component of effective powdery mildew control. Most powdery mildew species and their races have developed resistance to most systemic fungicides because they have single site mode of action. Thus, modification of one gene in the pathogen may be enough to enable the pathogen to resist the action of the fungicide. Most protectant fungicides are not at risk for resistance development because they have a multi-site mode of action. The powdery mildew fungus has demonstrated a high potential for developing resistance. It has already developed resistance in most parts of the world including India and Australia at least one material in the at-risk chemical class. Presence of resistant strains has been associated with control failure. Now it is important to realize that the primary goal of resistance management is to delay its development rather than to manage resistant strains after they have been selected, therefore, a management programme needs to be implemented before resistance develops to be effective. Bacterium, *Pseudomonas syringae pv. phaseolicola* have also been shown resistance to antibiotics (Russel, 1975). Thus, large-scale application of a new fungicide can provide a selective advantage for resistant mutants which may increase in the pathogen population.

Now it is seen that microbial populations have remarkable ability to adopt to changed conditions, whether to changes in the genetic structure of population of their hosts or to changes in usage of crop protection chemicals. Asexual reproduction plays an important role in virulence changes of pathogens. Asexual reproduction of *Uromuces appendiculatus* is example of large changes in virulence and emergence of new races. The origin of fungicide tolerance in *Colletotrichum kabawae* is due to mutation. Heterokaryosis, parasexuality changes the degree of virulence in pathogens. *Fusarium oxysphorum, Rhizoctonia solani* are the fungi in which these processes create variations. Heterokaryosis and parasexuality permit the segregation and recombination of genetic factors outside the sexual stage.

Population of plant pathogens in which sexual reproduction occurs in general tend to display greater diversity than asexually propagating ones. In heteroecious rust fungi the presence of the alternate host (on which the sexual stage develops) can have important implications for the joint evolution of the pathogen and its host by presenting opportunities for recombinations (Simons *et al.*, 1979). Similarly, the presence of both mating types in heterothallic fungi tends to accelerate the evolution of genetic diversity. This can be seen in the case of late blight of potato (*Phytophthora infestans*) in Europe. This disease was found only in Toluca valley of Mexico in 1842. Later the disease appeared in Belgium and then throughout the Europe.

Selection, migration and recombination may be the most important factors shaping race structure and evolution in population of sexually reproducing powdery mildews in which mutation is perhaps of limited significance.

The most widely used means of distinguishing physiologic races as 'units' of biodiversity within plant pathogen, is by comparison of reactions on a set of host cultivars carrying different sources of qualitative resistance. The races first described in *P. graminis f. sp.* tritici were identified in this manner and differential sets of host cultivars have since been used widely, with other rust, with powdery and downy mildews, and with numerous other fungi (Scott *et al.,* 1980). As the number of differential hosts is increased, so more and more races can be identified.

The existence of diversity in plant pathogens and their ability to evolve and generate new diversity in agriculture provides a continuing capacity for disease development and, as a consequence, crop loss. There is little wonder that in agriculture, pathogen diversity is seen by many as harmful, to be managed or even eliminated rather than to be conserved to generate useful disease resistance. In reality, trade-offs are needed between the two views: we need to develop cost-effective and efficient ways of managing disease which reduce food production but at the same time develop strategies to allow secure generation of diversity for disease resistance for future use in crop improvement.

Farmers have practised both direct and indirect methods of disease management since the dawn of agriculture.

REFERENCES

Burdon, J.J. (1993). The structure of pathogen populations in natural plant communities. *Ann. Rev. phytopath.* **31:** 305-323.

Burdon, J.J. (1987). Disease and Plant Population Biology. Cambridge University, Press, Cambridge.

Buddenhagen, I.W. (1977). Resistance and vulnerability of tropical crops in relation to their evolution and breeding. *Annals of the New York Acad. Sci.* **287:** 309-326.

Durbin, R.D. (1959). Factors affecting the vertical distribution of *Rhizoctonia solani* with reference to CO_2 concentration. *Am. J. Botany* **46:** 22-25.

Flor, H.H. (1956). The complementary genic systems in flax and flaxrust. *Advances in Genetics* **8:** 29-54.

Frank, S.A. (1993). Coevolutionary genetics of plants and pathogens. *Evolutionary Ecology* **7:** 45-75.

Gillett, J.B. (1962). Pest pressure, an underestimated factor in evolution. Systematics Association Publications **4:** 37-46.

Harper, J.L. (1990) Pest, pathogens and plant communities: an introduction, pp. 3-14. *In*: Burdon, J.J. and Leather, S.R. (Eds.). Pest, Pathogen and Plant Communities, Blackwell Scientific Publications, Oxford.

Hawksworth, D.L. and Colwell, R.R. (1992). Biodiversity amongst micro-organisms and its relevance. *Biodiversity and Conservation* **1:** 221-226.

Holliday, P. (1989) A dictionary of Plant Pathology. Cambridge University Press, Cambridge.

Leppik, E.E. (1970). Gene centers of plants as sources of disease resistance. *Ann. Rev. phytopath.* **8:** 323-344.

Oberwinkler, F. (1992). Biodiversity amongst filamentous fungi. *Biodiversity and Conservation* **1:** 93-311.

Russell, P.E. (1975). Variation in the virulence of some streptomycin resistant mutants of *Pseudomonas phaseolicola*. *J. Appl. Bacteriology* **39:** 175-180.

Scott, P.R., Johnson, R., Wolfe, M.S., Lowe, H.J.B. and Bennett, F.G.A. (1980). Host-specificity in cereal parasites in relation to their control. *Applied Biology* **5:** 349-393.

Stakman, E.C. (1926). Physiological specialization. Mayo Foundation Lecture, 1926.

Stakman, E.C. and Levine, M.N. (1922). The determination of biologic forms of *Puccinia graminis* on *Triticum* spp. *Minn. Agr. Exp. Sta. Tech. Bull.* **10:** 22.

Tarr, S.A.J. (1972). *The principles of Plant Pathology*. Macmillan, London.

Waller, J.M. and Brayford, D. (1990). *Fusarium* disease in the tropics. *Tropical Pest Management* **36:** 181-194.

Diversity of Lower Plants (2014): pp. 135-163
Editors: Rajan Kumar Gupta & Mukesh Kumar
IK International Publishing House Pvt. Ltd.

11

Biodiversity of Powdery Mildew Fungus: A Constant Threat to Cucurbits

G.K. Sharma*, Desh Deepak, Mohit Kumar Gupta and Manujendra Pratap Singh

Plant Pathology and Environmental Laboratory, P. G. Department of Botany, Hindu College, Moradabad–244001. India
Email: g.k.sharma1564@gmail.com

ABSTRACT

Cucurbits are attacked by a number of fungal pathogens among which Erysiphales and Peronosporales are most importent. Erysiphales are more destructive and causes powdery mildew disease on cucurbits. The present article reveals the dynamics of taxonomic biodiversity of powdery mildew disease in cucurbit crops as a previewed to our appraisal of the implications of diversity for affective management of disease in India as well as other parts of the world on cucurbitaceous crops. Powdery mildew of cucrbit is a serious disease and causes considerable loss to the number of cucurbitaceous crops grown in India. Three powdery mildew species i.e. *Sphaerotheca fuliginea, Erysiphe cichoracearum* and *Leveillula taurica* are well established causal organism of the disease on cucurbits on world-wide basis. Of these *Sphaerotheca fuliginea* and *Erysiphe cichoracearum* are more commonly encountered on cucurbits in different parts of the world.

Key words: Pathogen, *Erysiphe*, powdery mildew and cucurbits.

1. INTRODUCTION

Powdery mildew is one of the most conspicuous parasitic fungi on plants of economic importance. Numerous vegetable crops are susceptible to powdery mildew but cucurbits suffer greatly due to this serious disease.

Powdery mildew of cucurbits is a serious disease of global significance and causes considerable damage to a number of cucurbitaceous crops grown in India. Three powdery mildew species, *Sphaerotheca fuliginea* (Schlecht.) Poll., *Erysiphe cichoracearum* DC. and *Leveillula taurica* (Lev.) Arnaud are recognized as causal organisms of the disease on cucurbits

*Communicating Author

on worldwide basis. Of these, *S. fuliginea* and *E. cichoracearum* are more commonly encountered on cucurbits in different parts of the world than *L. taurica. S. fuliginea* is apparently most prevalent (Khan, 1983, 1989). All the three species are reported to exist on cucurbits in India (Khan, 1983). *S. fuliginea* and *E. cichoracearum* are known to occur on cucurbits in certain states of India like Uttar Pradesh, Bihar, Madhya Pradesh, Rajasthan and Tamil Nadu (Khan *et al.*, 1971; Khan, 1976, 1977; Dave *et al.*, 1971; Khosla *et al.*, 1974; Siradhana and Chaudhari, 1972; Sharma and Khan, 1991). In some states like Jammu and Kashmir and Punjab only *S. fuliginea* is reported to cause the disease on cucurbits (Khan *et al.*, 1974; Jhooty, 1967; Sharma, 1978). In Maharashtra, only *E. cichoracearum* is reported to cause the disease on *Coccinia indica* (Kulkarni and Mois 1976). *L. taurica* has been reported to infect cucurbits in Rajasthan and Karnataka (Mahrshi and Siradhana, 1980; Ullasa and Amin, 1981).

The cucurbitaceae or cucurbit family is a medium sized plant group and form an important and a large group of vegetables cultivated in different regions of world including India. It contains 117 genera and 825 species distributed in warmer regions of both the hemisphere. In India, 36 genera and around 100 species have been described (Seshadri, 1993). According to Indian Horticulture Database 2004, total area under cultivation of vegetables in the country is 6 millions hectares with an annual production of 84.81 million tonnes. Out of these, Uttarakhand occupies 0.7 million hectare area under cultivation with annual production of 5 million tonnes and productivity 7.2 t/ha. The crop is grown in mixed cropping of more than one kind at riverbeds and also in garden land areas. Cucurbits like cucumber, pumpkin, bottle gourd, smooth gourd, pointed gourd, bitter gourd are commonly grown in small kitchen gardens on a large scale in different parts of U.P. Most of these are seed propagated, however, pointed gourd (Parwal) is propagated vegetatively by cutting. The fruits are used as cooked vegetables, desserts, salads, pickling and as caned and preserved. Seed kernels are also edible containing high oil percentage and are good source of energy. Eastern state people consume young branches and leaves of some cucurbits as cooked green vegetables. As per data available annual production of cucurbits in the country is approximately half of the requirement as per dietary standard. Most cucurbits being perishable cannot be stored. Available cultivated area under grain production cannot be diverted for cucurbits production for obvious reasons. This emphasizes the need for increasing the production and availability from the existing area under cucurbits and ensuring that the produce remains healthy till its consumption. Powdery mildew disease is a major constraint in increasing production and availability to produce in both outdoor and indoor. Cucurbits suffer variously due to this disease (Hirata, 1966; Sitterely, 1978; Khan, 1983, 1989). According to Chakravarty (1959), out of the 108 species occurring in India, 38 are apparently endemic; Rheede (1805) and Roxburg (1832) reported that *Lageneria vulgaris* had been found to be wild in India. De Condole (1882) reported that *Citrullus vulgaris* was indigenous to tropical Africa. While Pangalo (1955) on the basis of extensive studies on Asiatic watermelons concluded that *C. vulgaris* was indigenous to India. Naudin (1859) pointed out that species of *cucumis* were African in origin except *C. anguria* and suggested that it could have been introduced into the America through slave trade. According to Meeuse (1958), *Cucumis anguria* was a cuttigen descended from a non-bitter variant (mutant) of an African wild species.

Table 1: Occurrence of powdery mildew on cucurbits in different countries of the world

Pacific	Asia	Europe	Africa	North America	South America
Japan, Formosa (Taiwan), Hawaiian Philippines, Indonesia, New Guinea, Fiji, Tonga, New Caledonia, Norfolk, Australia, New Zealand	Korea, China, Cambodia, Thailand Myanmar (Burma), Malaysia and Singapore, Nepal, India, Sri Lanka, U.S.S.R., Asian region (Siberia, Kazakh, Uzbek, Azerbaijan, Armenia, Turkmen) Afghanistan, Pakistan, Iran, Iraq, Turkey, Lebanon, Israel, Jordan, Maullritius, Cyprus, Saudi Arabia, Bangladesh	Finland, Sweden, Norway, U.S.S.R., European region (Lithuania, Latvia Ukraine), Poland, Czechoslovakia, Romania, Bulgaria, Hungry, Denmark, The Netherlands, United Kingdom, East Germany (G.D.R.), West Germany (F.R.G.), Switzerland, Greece, Yugoslavia, Italy France, Spain, Portugal, Malta, Austria, Belgium, Tunisia	Egypt, Libya, Nigeria, Morocoo, Ethiopia, Sudan, Ghana, Malawi, Sierraleone, Senegal, Kenya, Uganda, Tanzania, Mozambique, Zambia, Zimbabwe, South Africa, Modeira, Is., Ivosy Coast, Madagascar.	Canada, U.S.A., Mexico, Guatemala, El Salvador, Honduras, Nicaragua, Panama, Bermuda, Jamaica, Trinidad	Venezuela, Brazil, Peru, Bolivia, Uruguay, Argentina, Chile.

Adopted from Hirata (1986); modified and updated.

According to Naudin (1859) *Luffa cylindrica* was probably a native of Southern Asia and perhaps also of Africa, Australia, etc. Rheede (1805) observed *L. cylindrica* growing in sandy places, in wood and other localities in Malabar, while Roxburg (1832) claimed that it was wild in undivided India subcontinent. Erwin (1931) stated that there was good evidence that *Cucurbita moschata* and *C. pepo* was indigenous to South America.

A number of species of family cucurbitaceae and varieties of some taxonomic species are cultivated in different parts of the world. *Citrullus lanatus* Mansf or *Citrullus vulgaris* Schard. *Cucurbita pepo* (L), *Cucurbita moschata* Poiret, *Lageneraia leucantha* (Duch) Rusby are cultivated in many different areas in the world (Chakravarty, 1959, Khan 1972; and Sitterly, 1978).

Some cucurbits are used as baskets, as insulations, as soil filters and as alternatives of pottery and utensils. From the seeds of certain cucurbits oil is also extracted. From the hollow dried fruits of bottle gourd, *L. leucantha* some musical instruments are made. Ash-gourd, *B. hispida* is primarily used in confectionery industry. A few species of *Citrullus* and of *Cucumis* are of medicinal importance (Hill 1979).

Cucurbits constitute an important group of vegetables and raw and ripe fruits are grown all over the world. Powdery mildew disease of cucurbits is worldwide in distribution and both indoor and outdoor cucurbits suffer variously due to this disease (Hirata, 1966; Sitterly, 1978; Khan, 1983, 1989). The severity of the disease in indoor cultivations is greater than in outdoor field plots because of favourable environmental conditions such as moderate to high temperature, high soil moisture, reduced air circulation, reduced light intensity and continuous cropping (Sitterly, 1978).

2. OCCURRENCE OF POWDERY MILDEW ON CUCURBITS

Powdery mildew on cucurbits is listed to occur in all the geographical regions of the world (Hirata 1968; Amano 1986; Khan, 1983, 1989). Records of the disease on cucurbits in different countries are given in Table 1. Records of occurrence of different species of powdery mildews on cucurbits are summarized in Tables 2–5.

3. CAUSAL ORGANISMS

Teleomorphs of *Erysiphe cichoracearum* DC.; *Erysiphe communis* (Wallr.) Link.; *Erysiphe polygoni* (DC.) St. Am.; *Erysiphe polyphaga* Hammarlund; *Leveillula taurica* (Lev.) Arnaud, *Sphaerotheca fuliginea* (Schlecht.) Poll. are claimed to have been recorded on cucurbits (Ballantyne, 1975; Sitterly, 1978; Khan 1983, 1989).

There are also records of conidial stage (anamorph) of powdery mildew fungi as *Oidium* sp., which indicates that the species involved could not be determined.

The occurrence of *E. communis* and *E. polyphaga* on cucurbits has been claimed but their taxonomy and nomenclature on cucurbits are doubtful and non-recognized (Junell, 1965, 1967; Nagy, 1975). *E. polyphaga* was described from Germany (Hammarlund, 1945) and was claimed as a causal organism of cucurbit powdery mildew (Nagy 1975), however, claimed that the denomination *E. polyphaga* is not valid and *E. polyphaga* may be recognized as *E. cichoracearum*, *E. polygoni* was claimed to occur on cucurbits in Poland (Schroeter, 1983) and in Japan (Homma, 1937). Since then, these records were not substantiated by any other report. Khan (1983) opined that these may be the cases of mistaken identity. *S. fuliginea*, *E. cichoracearum* and *L. taurica* are well established causal species of the disease of cucurbits (Khan, 1983, 1989).

3.1 Leveillula Taurica

The occurrence of *L. taurica* on cucurbits is reported only in a few countries. It is reported to infect some cucurbits in Kenya, Sudan, Libya, Senegal, Mozambique, U.S.S.R., Romania, Bulgaria, France, Pakistan, Thailand, Turkey and India (Table 2). Tarr (1955) recognized *L. taurica* causing powdery mildew on *Cucurbita maxima* in Sudan. Golavin (1956) recorded *Leveillula* on vegetable marrow, cucumber and *Cucurbita* sp. in U.S.S.R., and proposed a new species *L. cucurbitacearum* for the species infecting cucurbits while splitting *L. taurica* into different species for each host family. The splitting of *L. taurica* into different species according to Table 2.

Table 2: Records of *Leveillula taurica* on cucurbits in different countries of the world

Pacific	Asia	Europe	Africa	North America	South America
	India, U.S.S.R. (Asian region), Thailand, Pakistan, Turkey	Romania, Bulgaria, U.S.S.R., (European region), France	Sudan, Kenya, Libya, Senegal, Mozambique		

Table 3: Records of *Erysiphe cichoracearum* as the causal organism of the powdery mildew of cucurbits in different countries of the world

Pacific	Asia	Europe	Africa	North America	South America
Hawaiian, Is., Figi Norfolk	Malaysia and Singapore, Nepal, Iraq, Lebanon, Cyprus, Afghanistan, Philippines	Denmark, Malta, Norway, Sweden	Senegal*, Mozambique*, Kenya*, Ghana, Ivory, Coast, Nigeria, Morocco, Ethiopia, Tanzania, Zambia, Zimbabwe, Medagascar.	Guatemala, ElSalvador, Honduras, Nicargua, Panama, Bermunda	Venezuela, Peru, Bolivia, Uruguay

* *L. taurica* also reported.

A host family has not received much recognition (Ballantyne, 1975; Khan 1983). Nasyrov (1962) also observed *L. taurica* on vegetable marrow in U.S.S.R. In Kenya, *L. taurica* was listed on cucumber (Anon., 1961). *L. taurica* was recorded on cucumber in Ramania (Docea and Fratila 1980). In other records, *L. taurica* was found associated with other cucurbit powdery mildew fungi. In U.S.S.R., Gordeeva (1961) and in Bulgaria, Elenkov *et al.* (1975) found *L. taurica* together with *E. cichoracearum* and *S. fuliginea* on greenhouse cucumbers. El–Ammari and Khan (1983) reported occurrence of *L. taurica* on greenhouse cucumbers in Libya. Molot and Lecoq (1986) found *L. taurica* infecting cucurbits in France. El-Mahjoub and Romdhami (1991) observed *S. fuliginea* on melon, *Cucumis melo var. flexuosus;* cucumber, *C. maxima* and marrows.

Palti (1988) has listed Thailand, Mozambique, Turkey and Pakistan among the countries where *L. taurica* was reported to infect cucurbits. In India, *L. taurica* in conidial stage was first reported from Jaipur in Rajasthan state (Mahrshi and Siradhana, 1980). Ullasa and Amin (1981) recognized its occurrence on squash along with *S. fuliginea* in Karnataka state.

3.2 Erysiphe Cichoracearum and Sphaerotheca Fuliginea

Early European mycologists such as Leveille (1851), Passerini (1867), Fuckel (1869) and Jackzewski (1896) diagnosed the causal organism of powdery mildew of cucurbits as *Sphaerotheca castegnei.* Later, Humphrey (1892) and Schroeter (1893) in Europe identified it as *Erysiphe cichoracearum* and *E. polygoni* respectively. Salmon (1900) observed the presence of teleomorph on *Cucurbita pepo* plants infected with powdery mildew fungus in which each ascus had two ascospores. This led Salmon to conclude that causal organism of powdery mildew was *E. cichoracearum* rather than *S. castegnei.*

Table 4: Records of *Sphaerotheca Fuliginea as the* causal organism of the powdery mildew of cucurbits in different countries of the world

Pacific	Asia	Europe	Africa	North America	South America
Formosa (Taiwan), New Zealand	Turkey*, Pakistan, Iran	Belgium, Romania, Spain	Malawi, Zimbabwe		

Table 5: Records of both *Erysiphae cichoracearum* and *Sphaerotheca Fuliginea* as the causal organisms of the powdery mildew of cucurbits in different countries of the world

Pacific	Asia	Europe	Africa	North America	South America
Japan, Australia	Korea, India*, U.S.S.R.*, Israel, Saudi Arabia, China, Bangladesh, Myanmar (Burma) Jordan.	Finland, The Netherlands, United Kingdom, Switzerland, Hungary, Austria, Czechoslovakia, West Germany (F.R.G.), Italy, France*, Bulgaria*, East Germany (G.D.R.), Tunisia, Greece	Egypt, Libya*, South Africa, Sudan	U.S.A., Canada, Mexico, Florida	Argentina, Chile, Brazil

**L. taurica* also reported

Exsiccati specimens, which had earlier been described as *S. castegnei* and observed that specimens lacked teleomorphs (perfect stage) and in each case the description was based on anamorphs (imperfect stage). This finding convinced him the correctness of his earlier identification and led him to remark, "It is of course possible that more than one species of *Erysiphe* occur on these host plants, and it would be very interesting to note if any example that perithecia of *Sphaerotheca* really exists". This emphatic statement perhaps led to the belief that the powdery mildew disease on cucurbits is caused by *E. cichoracearum*. Consequently, *E. cichoracearum* came to be known as the causal organism of the disease for long time almost throughout the world.

In the Netherlands prior to 1964, *E. cichoracearum* was believed to be the only causal organism of the disease. Dekkar (1961, 1963) reported that *E. cichoracearum* causes powdery mildew on cucumbers in the country. In 1964, Boerema and Van Kesteren (1964) identified powdery mildew of cucumber, melon and gherkins as *S. fuliginea* on the basis of anamorph characters and doubted the reliability of record of *E. cichoracearum* on cucurbits in the Netherlands. Later, Kooistra (1968) also observed anamorph of *S. fuliginea* on cucumber. Teleomorph of *S. fuliginea* was recorded from Wageningen in the country, (Anon, 1965). Recently, Schepers (1984), while studying the possible means of survival and dispersal of inoculums of powdery mildew of cucumbers recognized the pathogen as *S. fuliginea*. Tores *et..al.* (1990) and Alvarez and Tores (1995) has reported *S. fuliginea* to occur on cucurbits. Recently Perez-Garcia (2009) observed *Podosphaera fusca* (synonym *P. xanthii*) on cucurbits in Spain. In Portugal *S. fuliginea* was recognized as *Oidium* sp. and in Cyprus as *E. cichoracearum* (Hirata, 1968). In Malta, *E. cichoracearum* was reported on cucumbers (Peralta, 1946-47).

In addition to *L. taurica* (Elenkov *et al.,* 1975), both *S. fuliginea* and *E. cichoracearum* have been reported from Bulgaria (Tafradzhiiski, 1959, 1961, 1963, 1972; Lozanov and Vitanov, 1960) were the latter predominates. Recently Velkov and Masheva (2002) reported *S. fuliginea* and *E. cichoracearum* on cucumber from *Bulgaria. E. cichoracearum* was earlier recognized as powdery mildew of cucurbits in Hungary (Ubrizsy, 1946). Studies of Nagy (1970, 1972) have undoubtedly established the existence of both *E. cichoracearum* and *S. fuliginea* in Hungary. In addition to *S. fuliginea* recorded in perithecial stage (Savulescu, 1929), *L. taurica* is also reported to infect cucurbits in Romania (Docea and Fratilla, 1980). Reports of Glaeser (1970, 1973, 1974) claimed widespread occurrence of *E. cichoracearum* on cucumber in Austria. Zacha (1968) and Benada (1974) established *S. fuliginea* as the only causal organism of the disease on cucumber in Czechoslovakia although *E. cichoracearum* was earlier claimed to exist on cucurbits. Lebeda (1983), however, identified *E. cichoracearum* and *S. fuliginea* on cucurbits in Czechoslovakia from their anamorphs. A recent report by Zlochova (1988) confirmed that the two species attack cucurbits in fields and glasshouse in the country, specially in Western Slovakia. Kristkova *et al.* (1999) observed *E. cichoracearum* on most of the cucurbits in Czechoslovakia Republic. Recently Kristkov *et al.* (2002) reported both *S. fuliginea* and *E. cichoracearum* but they claimed that *E. cichoracearum* is predominating species in Czechoslovakia. Vakalounakis and Klironomou (1995) reported *S. fusca* and *E. cichoracearum* on cucumber in Greece.

In addition to *L. taurica, S. fuliginea* and *E. cichoracearum* occur on cucurbits in U.S.S.R. Both species were reported in perithecial stages on cucurbits by Deckenback and Koreneff (1927) from Crimea. Since then a number of reports have indicated the existence of both the pathogens in U.S.S.R. In some regions of the country *S. fuliginea* is reported to be confined whereas *E. cichoracearum* in other parts. In Moldavia, Armenia, Volga Basin, Crimea, Azerbaijan, Astrakhan and Bylorussia regions, existence of both the species has been recorded (Rodigin 1936; Deckenback and Koreneff, 1927; Ragimov, 1961; Nikikorova 1962; Teterevnikova-Babayan and Simonyan, 1956; Dorozhkin and Kapelyan, 1975). Simonyan (1959) recognized the causal organism of powdery mildew of cucurbits as a subspecies of *E. cichoracearum, E. cichoracearum* f.sp. *cucurbitacearum*. Rudenko (1968) reported that in Moldavin conditions *E. cichoracearum* f. sp. *cucurbitacearum* attacks all cucurbits and *S. fuliginea* f. sp. *cucurbitacearum* infects cucumber, marrow, pumpkin and muskmelon and both organisms overwinter as cleistocarps and produce perithecia annually on a number of cucurbits. Bychenko (1971), Medvendev (1974) and Voronina (1976) have also confirmed the existence of both organisms of cucurbits in U.S.S.R. from Krasnodar region.

Baimuratora (1961) identified the powdery mildew of cucurbits from Golodnaya steppe as *S. fuliginea*. Osnitskaya (1959) also determined *S. fuliginea* as causal agent of cucurbit powdery mildew near Moscow. Misiga (1966), however, identified powdery mildew on cucumber from Brastilava as *E. polyphaga*.

In East Germany, Roder (1936) reported the occurrence of *E. cichoracearum* on outdoor cucumbers near Berlin. Janke *et al.* (1972) identified powdery mildew species as *S. fuliginea* and *E. cichoracearum* on cucurbits in East Germany (G.D.R.). Bremer (1940) recognized

E. cichoracearum as the agent of cucumber mildew in Central Germany. Huttenbach (1951), for the first time, indicated the existence of *S. fuliginea* and *E. cichoracearum* in West Germany. He, however, emphasized the predominance of *E. cichoracearum*. The recent observation of *E. cichoracearum* in perithecial stage and *S. fuliginea* in conidial stage concomitantly on cucumbers by Schlosser (1972) near Bonn confirmed the existence of both organisms in West Germany. Schlosser (1976a, 1976b) based on a survey of the distribution of *S. fuliginea* and *E. cichoracearum* in the Federal Republic of Germany (West Germany) reported that *S. fuliginea* prevailed in glasshouse conditions on cucumber, and *E. cichoracearum* under field conditions. Cruger and Meyer (1976) observed the both powdery mildew agents on cucurbits. They also noted that *E. cichoracearum* was found only in outdoor conditions while *S. fuliginea* on both outdoor and glasshouse.

After Salmon's observation of perithecia of *E. cichoracearum* in 1988, Reed (1936) and Williams (1941) confirmed it as the causal organism of cucumber mildew in England. Keyworth (1959) also recorded its occurrence on cucumber and vegetable marrow; and Stone (1959) on cucumber. Zaracovitis (1965), however, in his studies on conidial characters, identified powdery mildew on cucumber collected from Berkshire as *S. fuliginea*. A recent report also indicated its existence on cucumber in U.K. (Spencer *et al.,* 1975). *E. cichoracearum* is reported to infect cucurbits particularly vegetable marrow in Switzerland (Mayor, 1947). Besides, *S. fuliginea* is also known to exist in Switzerland (Amano, 1986). Blumer (1951, 1960), however, claimed that *E. polyphaga* also infects cucumber in Switzerland. In France, Viennot-Bourgin (1958) recorded *E. cichoracearum* on *Cucumis melo* and *C. sativus* in perithecial stage. Lafon *et al.* (1966) identified *E. cichoracearum f. cucurbitae* as the organism responsible for the disease in France. Landal (1961) has also described it as *E. cichoracearum.* Recently, Molot and Lecoq (1986) recognized three species *S. fuliginea*, *E. cichoracearum* and *L. taurica* on cucurbits the first two were claimed to be of major economic importance in France. Marcelli (1949) observed both *E. cichoracearum* and *S. fuliginea* on cucurbits in Italy. *S. fuliginea* being widespread and *E. cichoracearum* only on *Cucurbita maxima*. Locci and Bisiach (1972) also observed *S. fuliginea* on cucumber in Italy. Perithecial stage of *S. fuliginea* from Turin, Italy is also on record (Blumer, 1983). Recent studies of Brunelli and Branzanti (1987) on symptoms, damage and epidemiology also indicated the occurrence of both the species on cucurbits in Italy.

Gjaerum (1956) referred *E. cichoracearum* on greenhouse cucumber in Norway. *E. cichoracearum* is reported as causal species in Sweden (Banga 1956) and Denmark (Amano, 1986). Junell (1967) obtained perithecial stage of *E. cichoracearum* in Sweden. Ristic (1985) observed that both the upper and lower surface of leaves of cucumber was attacked. The pathogen often appeared on leaves or the stem. Telemorphs were rarely formed, usually had 2-3 asci each generally with 2 ascospores indicating involvement of *E. cichoracearum* in Yugoslavia. *S. fuliginea* and *E. cichoracearum* are listed to occur on cucurbits in Finland (Hirata 1968; Khan 1983, Amano 1986).

In U.S.A. till 1963 in all reports, the disease on cucurbits was assigned to *E. cichoracearum*. Perithecia of *E. cichoracearum* were recorded on cucurbits from Massachusetts and Wisconsin by Humphries (1982) and Reed (1908) respectively. Jagger (1926) identified the powdery mildew of muskmelon in the Imperial Valley of California as *E. cichoracearum*. Middleton and Yarwood (1946) observed the appearance of *E. cichoracearum* on melons in California; Barrat (1942) at the New Hampshire Horticulture farm; Pryor and Whittaker (1942) at the U.S. Horticulture

Field Station, Lajolla, California; Walker and Weber (1949) and Conover (1957) in Florida; and Ivonoff (1957) in Mississippi. Blasquez (1970) and Jones (1974) recorded *E. cichoracearum* on cucurbits in Florida. Randall and Menzies (1956) recorded perithecia of *E. cichoracearum* for the first time in recent times, on cucumbers in the Yakima Valley, Washington in 1954. Kable and Ballantyne (1963) for the first time in U.S.A. identified the powdery mildew of cucurbits in Ithaca district, New York State on the basis of conidial characters as *S. fuliginea* and claimed that this may be prevalent in other areas of the country. Yarwood and Gardner (1964) a year later recognized that in California the powdery mildew on cucurbits was *S. fuliginea*. Since then a number of reports indicated the existence at *S. fuliginea* on cucurbits in conidial stage in different parts of U.S.A. Ellert (1966) obtained Oidium sp. resembling *S. fuliginea*. Schroeter and Provvidenti (1968) observed *S. fuliginea* on squash and cucumber at the New York State Agriculture Experiment Station. Paulus (1969) and Paulus *et al.* (1972) also observed *S. fuliginea* on cantaloupe at the University of California, Riverside and Lajolla, California. *S. fuliginea* was found on cucurbits in Georgia (Sowell and Clark, 1971; Iowa (Clark, 1975); and Wisconsin (Shanmugasundarum *et al.,* 1971). All these identifications were based on anamorph characters. Kontaxis (1977) reported the anamorphs of *S. fuliginea* on fruits of watermelon and leaves of muskmelon and cucumber. In 1979, for the first time in U.S.A., he found teleomorphs of *S. fuliginea* on squash (Kontaxis, 1979). This substantiated the earlier reports of occurrence of *S. fuliginea* in U.S.A. He remarked that the causal agent for the powdery mildew of squash, probably all cucurbits in California and possibility in the United States was *S. fuliginea* and not *E. cichoracearum*. Grand (1987a, 1987b) found teleomorphs of *S. fuliginea* on melon, squash and *Lagenaria siceraria* in at Raleigh, North Carolina. Moreno-Rico *et al.* (1991) reported cleistothecia of *S. fuliginea* in Mexico. According to them *S. fuliginea* and *E. cichoracearum,* are most important and widely known powdery mildew species. Recently, Hector *et al.* (2008) identified three powdery mildew species on cucurbits in Florida. According to them these causal species involved in disease are *Podosphaera xanthii* (syn. *Sphaerotheca fuliginea*), *Golovinomyces cucurbitacearum* (syn. *Erysiphe cichoracearum*), and *Golovinomyces orontii*. *P. xanthii* (syn. *Sphaerotheca fuliginea*) is actually more widespread than the other two species.

Early reports from Canada identified *E. cichoracearum* as causal organism of powdery mildew on cucurbits (Mckeen 1954; Fischer, 1959; Brady and Fischer, 1963). Powdery mildew on greenhouse cucumber in Essex country, Ontario was listed as *E. cichoracearum* in the report of the Ministry of Agriculture for the year 1953. Conners (1967) also reported *E. cichoracearum* on cucurbits in Canada. Jarvis and Slingsby (1984) reported the occurrence of *S. fuliginea* on cucurbits for the first time in Canada. They observed its teleomorphs on *Cucumis sativus*.

In South American countries like in other parts of the world, *E. cichoracearum* was traditionally recognized as the causal organism of the disease. A Bolivian list of diseases of temperate crops, included *E. cichoracearum* on cucurbits (Bell and Alandia, 1957). Revilla (1955) had also mentioned the powdery mildew of melon, pumpkin and cucumbers as *E. cichoracearum*. In Brazil, Rochelle (1972) reported the occurrence of *E. cichoracearum* on squash and marrow. Reifschneider (1985) has shown that *S. fuliginea* is only the causal organism of melon powdery mildew in Brazil and not *E. cichoracearum* as previously reported. *E. cichoracearum* was also recognized as causal organism of the disease in Venezuela and Uruguay. Mazzanti De Castanon *et al.* (1987) on the basis of anamorphs characters causing powdery mildew identified *S. fuliginea* on *Citrullus vulgaris*, *Cucumis melo*, *C. sativus, Lagenaria siceraria* and *Cucurbita* spp. in Argentina. Powdery mildew of *C. melo* in Peru was recognized by Mont and Villon (1975) as *E. cichoracearum*. Recently Lemus Isla *et al.* (2005) reported *S. fuliginea* on melon from Cuba. There appears to be no reports of teleomorphs of *E. cichoracearum* or *S. fuliginea* appearing on cucurbits from any country in South America. Possibly no attempt has been made to identify the causal organism taking into consideration the divergence of *S. fuliginea* as dominant species in many different parts of the world.

Although Du Toit (1948) had earlier reported that cucurbits in the Western and South Western Cape Province of South Africa suffer from powdery mildew caused by *E. cichoracearum*, Gorter (1974) established the existence of *S. fuliginea* in the country. In Egypt, powdery mildew of cucurbits was referred to as *E. cichoracearum* (Fikry, 1936; Moursi and Sirry 1956; Elarosi and Wasfly, 1971). Similar observations were made by Khadr and Abdu (1972) in Giza. El-Kazzaz (1983) surveyed various parts of Egypt and noted the presence of *S. fuliginea* on the basis of anamorph characters, specially the presence of fibrosin bodies in conidia. In Sudan, Tarr (1955, 1954) reported widespread occurrence of *S. fuliginea* along with *L. taurica* on cucurbits and other vegetables. Later, Nour (1957, 1959) also recorded the presence of *S. fuliginea* on vegetable marrow, pumpkin, cucumber and melon. He further pointed out the prevalence of this fungus in the fields during winter in Sudan. Omer (1972) has also obtained conidial stage of *S. fuliginea* on cucumbers in Sudan. Mohamed *et al.* (1995) again claimed *S. fuliginea* and *E. cichoracearum* as causal organism of powdery mildew disease on cucurbits. Omara and Taha (1995) also reported *S. fuliginea* on watermelon.

In Libya along with *L. taurica*, both *E. cichoracearum* and *S. fuliginea* are reported to infect cucurbits. Pucci (1965) regarded powdery mildew on watermelon in the country as *E. cichoracearum*. Kranz (1962) also recognized powdery mildew on certain cucurbits in the eastern region of the country, Cyrenaica, as *E. cichoracearum*. Khan (1981), for the first time, recognized the occurrence of *S. fuliginea* in conidial stage on cucumbers in the country. Later, its perfect stage was found on *C. pepo* (El-Ammari and Khan, 1985). However, *E. cichoracearum* was also recognized to occur on greenhouse cucumbers in Benghazi area (El-Ammari and Khan, 1987). Teleomorphs of *E. cichoracearum* were, however, not found. A Kenya Annual Report, Department of Agriculture 1958 enlisted *E. cichoracearum* as an agent continuously hampering the cultivation of cucurbits (Hirata, 1968). *S. fuliginea* has been recognized to cause problems to vegetables crops in Malawi (Spurling, 1973). Marocco, Ethiopia, Senegal, Uganda, Tanzania, Zambia, Zimbabwe, Madagascar and Mauritius are enlisted to have *E. cichoracearum* on cucurbits, whereas Ghana, Tunisia, Sierra Leone and Canary Island are known to have *Oidium* sp. (Amano, 1986). Khan (1983) remarked that perhaps recognition of *E. cichoracearum* in these countries is based on traditional belief rather than proper investigations.

In Australia, Blackford (1943) reported the occurrence of *E. cichoracearum* on cucumbers in Queensland. Agriculture Gazette of the New South Wales (1948) also recognized *E. cichoracearum* as cucurbit's powdery mildew. Later, Clare (1958) identified the powdery mildew of a number of cucurbits in South Eastern Queensland as *S. fuliginea*. Ballantyne (1963) successfully employed the conidial characters in identifications and established *S. fuliginea* as the only causal organism of cucurbit powdery mildew in New South Wales. McNish (1967) also recorded the widespread occurrence of *S. fuliginea* on cucurbits in Western Australia. Ballantyne (1975) after an extensive study of powdery mildew collection made in New South Wales on numerous cucurbits established the existence and predominance of *S. fuliginea* in

Australia. Letham and Priest (1989) reported *S. fuliginea* from Australia on the basis of cleistothecia. The existence of *S. fuliginea* in New Zealand has been established by the occurrence of its telemorphs on *C. pepo* (Dingley, 1959; Boesewinkel, 1976).

In several Asian countries, it was believed that *E. cichoracearum* is responsible for the disease. However, now there are also reports of *S. fuliginea* from many. In Turkey, Bremer *et al.* (1947) reported *S. fuliginea* in perithecial stage on three cucurbits. In Iran, *S. fuliginea* has been claimed to exist on all cucurbits in subtropical regions of the northern part of the country Esfandiari. (1947). Allison (1952) identified *E. cichoracearum* as powdery mildew of cucurbits in Iraq. *E. cichoracearum* was claimed to exist on cucurbits in Saudi Arabia (Anon., 1976). However, Abul-Hayja and Trabulsi (1981) observed perithecia (teleomorphs) of *S. fuliginea* on squash and cucumber and claimed that probably all cucurbits in Saudi Arabia are infected by *S. fuliginea* and not by *E. cichoracearum*. In Israel, Rayss (1947), Palti (1962) and Rudich *et al.* (1969) recognized *S. fuliginea* on cucurbits although earlier reports of powdery mildew on cucumber and vegetable marrow by Reichert *et al.* (1943) from Palestine and of cucumbers and melons by Peleg (1953) had recognized it as *E. cichoracearum*. Recently, Cohen *et al.* (1996) observed *S. fuliginea* on cucurbits in Israel. Thompson (1933) recorded *E. cichoracearum* on cucurbits in Malaysia and Singapore. Chin *et al.* (1959) from China diagnosed powdery mildew of cucumbers as *S. fuliginea*. Hashioka (1937) and Sawada (1959) also recognized *S. fuliginea* as the agent of the disease in Formosa (Taiwan). Tsay and Tung (1994) reported *S. fuliginea* on *L. siceraria* in Taiwan. In Japan, Uozumi and Yoshi (1952) emphatically recognized *S. fuliginea* as the only pathogen causing powdery mildew of cucurbits in Fukuoka region of Japan. They also observed occasional production of perithecia. Katusuki (1955) also identified *S. fuliginea* on cucumbers in Yiku Island, Kysushu, Japan. A report by Nomura (1974), however, indicated the existence of both *S. fuliginea* and *E. cichoracearum* in the country.

E. cichoracearum is mentioned to occur in Nepal, Afghanistan, Lebanon and Jordan whereas *S. fuliginea* is recognized in Pakistan. Both species are, however, known to occur in Korea (Amano, 1986). Powdery mildew of cucurbits in Philippines, Indonesia, Thailand, Cambodia, Burma and Ceylon is referred to as *Oidium* sp. (Amano, 1986), perhaps due to the lack of proper investigations in these countries.

S. fuliginea and *E. cichoracearum* are reported to infect cucurbits in different states of India. Butler (1918) reported *E. cichoracearum* on *C. cordifolia* (syn. *Cephandra indica*), *Trichosanthes dioica* and *Momordica balsamina* from the state of Bihar. He found perithecia of *E. cichoracearum* on the cucurbits. Khan (1977) on the basis of anamorph and teleomorph characters identified *S. fuliginea* on a number of cucurbits and *E. cichoracearum* on *C. grandis* (syn. *C. cordifolia*) in Bihar state. Jhooty (1967) recognized the occurrence of *S. fuliginea* on cucurbits in Punjab state. His identification was based on conidial characters. Sohi and Nayar (1969) found teleomorph of *S. fuliginea* on *C. moschata* in State of Himachal Pradesh. Khan *et al.* (1974) recognized *S. fuliginea* occurring on a number of cucurbits in the state of Jammu and Kashmir. The pathogen produced its teleomorph on some of the cucurbits. Dave *et al.* (1971) and Khosla *et al.* (1974) recorded *E. cichoracearum* and *S. fuliginea* occurring on cucurbits in the state of Madhya Pradesh. They found teleomorphs of the species. Siradhana and Chaudhari (1972) reported the occurrence of *S. fuliginea* and *E. cichoracearum* on cucurbits in the state of Rajasthan, based on teleomorph characters. Malik *et al.* (1973) identified *E. cichoracearum* on ash-gourd, (*Benincasa hispida*) from Udaipur (Rajasthan). Butler (1918), Khan and Khan (1970),

Khan *et al.* (1971), Sharma (1972, 1973) and Malik *et al.* (1973) reported *S. fuliginea* and *E. cichoracearum* infecting a number of cucurbits in the state of Uttar Pradesh. They identified the causal organisms on the basis of anamorphs as well as teleomorphs characters. Sharma and Khan (1991) reported *S. fuliginea* and *E. cichoracearum* on cucurbits from Tamil Nadu. Recently Sharma and Khan (1994) observed *S. fuliginea* and E. *cichoracearum* on cucurbits in Andhra Pradesh.

4. SYMPTOMS

Visual white fluecose, tiny, round spots are invariably present on infected parts of cucurbits. Spots were discrete on some leaves or stems. Entire foliage is covered with powdery mass, mainly composed of mycelium, conidiophores and conidia is the case of severe infection. In some cases both surfaces of leaves are found infected. Petioles and tendrils are also found infected.

Fig. 1-4. Powdery mildew infection of different cucurbits.

The fungus feeds the plant nutrients, reduces the photosynthesis and causes yellowing and sometimes the death of leaves. A severe infection may kill the entire plants, crop's yield, can be reduced because of reduced size, number of fruits. Fruits from infected plants can have low quality. Although in most of the cases fruits are free from infection but they may be malformed, sunburned and ripened prematurely or incompletely because of a loss of foliage cover caused by premature senescence of infected leaves.

5. IDENTIFICATION

Thus, powdery mildew is a common and serious disease of cucurbit crops in world. The powdery mildew species responsible for disease development are identified by using anamorph and telomorph characters.

Teleomorph characters used in identification of powdery mildew are as follow.

I. Teleomorph of *S. fuliginea*

Cleistothecia scattered to densely gregarious, brown to dark brown, globose to sub-globose, 69-114 μm in diameter. Appendages mycellioid, brown, septate, variable in number and length. Each cleistothecium contained single broadly elliptical to sub-globose ascus, measuring 57.2-90.8 × 32-64 μm. Each ascus contained 8 ellipsoidal to spherical ascospores, measuring 18-24 × 11-22 μm.

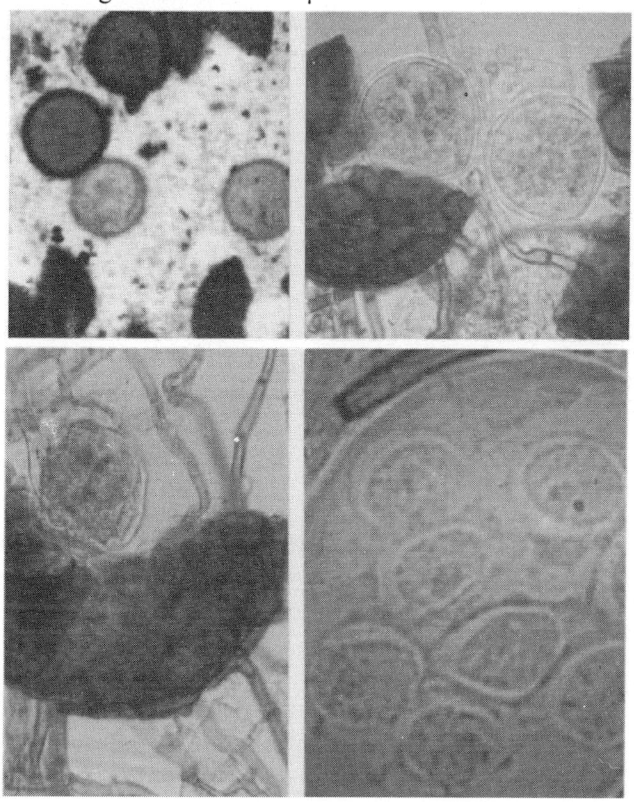

Fig. 5-8. Teleomorph of *S. fuliginea.*

II. Teleomorph of *E. cichoracearum*

Cleistothecia scattered 96-180.2 μm in diameter, appendages mycellioid brown, 15-27 oval asci measuring 76-11.1 × 27-77 μm, 2 or 3 ascospores in each ascus measuring 23-34.2 × 13.8-20.3 μm.

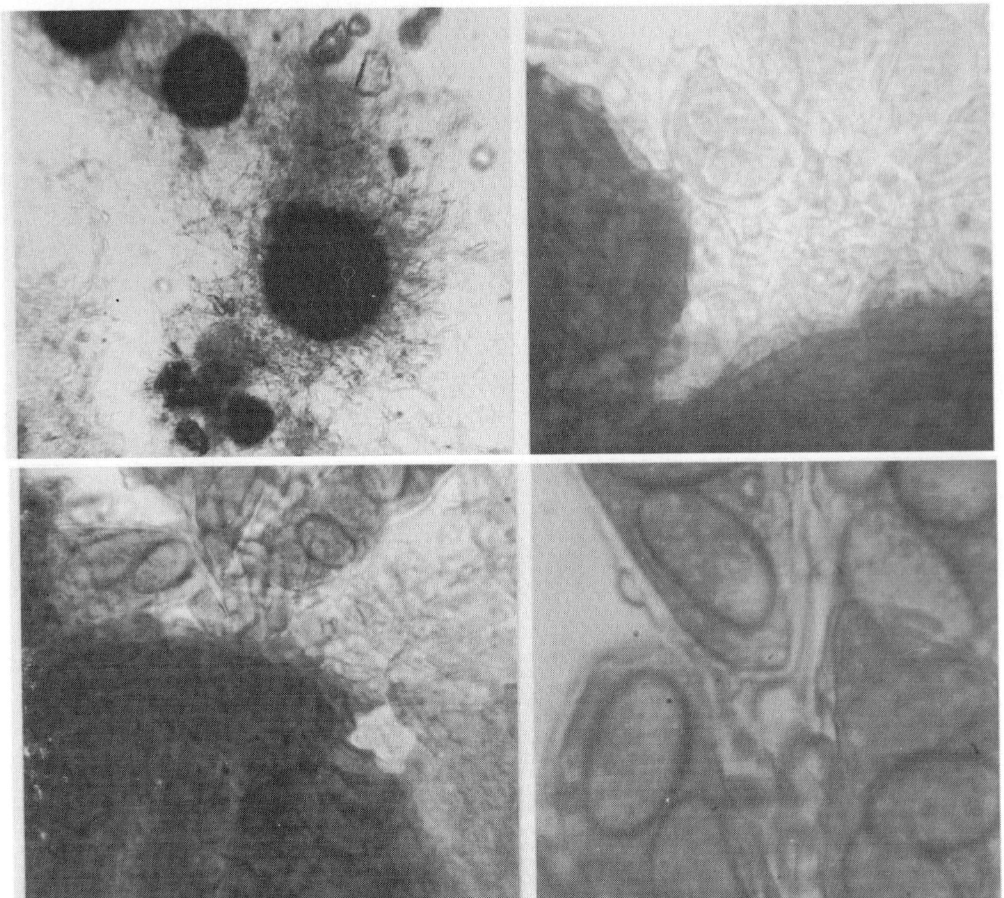

Fig. 9-12. Teleomorph of *E. cichoracearum.*

6. VALUES OF ANAMORPH CHARACTERS IN THE TAXONOMY OF POWDERY MILDEW OF CUCURBITS

Anamorph characters are attained smaller taxonomical importance as compared to teleomorph characters in the identification of powdery mildew in general. Some characters of anamorph have, however, been suggested usuable for identification of powdery fungi infecting cucurbits (Hirata, 1955; Amano, 1986; Boesewinkel, 1977, 1980; Sitterly, 1978; Yarwood, 1978 and Khan, 1983, 1989 and Khan and Sharma, 1995) have been used in diagnosis or differentiation of *S. fuliginea*, *E. cichoracearum* and *L. taurica* in the absence of their teleomorph. Taxonomic values of anamorph structures of the powdery mildew on cucurbits are briefly given below:

1. **Mycelium:** The mycelium of *E. cichoracearum* and *S. fuliginea* is ectophytic consisting of a network of colourless hyphae abundantly present on the surface of the infected parts of the hosts. In *L. taurica*, mycelium is endophytic, spreading between the mesophyll cells of the leaves. Endophytic nature of the mycelium is of great taxonomic importance for distinguishing *L. taurica* from the other species on cucurbits.
2. **Conidiophores:** The types of conidiophores observed in powdery mildew of cucurbits are either branched or unbranched; those with conidia borne singly or those borne in chains; those with fibrosin bodies or without fibrosin bodies and those with conidia of one shape or those with various shapes (Yarwood, 1978; Khan, 1983). The conidiophores of *E. cichoracearum* and *S. fuliginea* are unbranched; bearing a chain of conidia (Euoidium type) while those of *L. taurica* the conidiophore is often branched (Oidiopsis type) with two types of conidia (navicular and cylindrical). Conidiophores of *L. taurica* emerge through stomata. These features of conidiophores are helpful in differentiating *L. taurica* from *S. fuliginea* or *E. cichoracearum*.
3. **Conidia :** Shape and size of conidia are sometimes advocated to be used in taxonomy of the species.
   - **(a) Shape :** There are suggestions to use the shape of conidia to distinguish the different genera of powdery mildews involved in the development of disease on cucurbits. The shape of conidia of *S. fuliginea* is ellipsoidal or at least with the tendency towards this shape whereas those of *E. cichoracearum* are barrel-shaped with a tendency towards cylindrical forms (Khan and Sharma, 1993). *L. taurica* produces two distinct kinds of conidia, viz., cylindrical and navicular. This is a very prominent feature of conidia of *L. taurica* and can be easily distinguished from other species.
   - **(b) Size:** The size of conidia is often used in the taxonomy of powdery mildew (Allison, 1931; Clare, 1958; Boerema and Van Kesteran, 1964; Jhooty, 1967 and Nagy, 1976). Size of conidia is fairly constant for a given host in a given environment (Blumer, 1922). But the utility of this character seems doubtful as the dimensions of conidia are greatly influenced by a number of factors including relative humidity, temperature, host and its nutritional status (Yarwood, 1978; Khan, 1983; Khan and Sharma 1995).
4. **Fibrosin bodies:** Well-developed fibrosin bodies (Zopf, 1887) occur in the conidia and conidiophores of *S. fuliginea* but are absent from *E. cichoracearum* (Ballantyne, 1963; Clare, 1964; Nagy, 1970). This character is apparently useful and seems to be highly reliable for distinguishing *S. fuliginea* from *E. cichoracearum*. But there are records of *S. fuliginea*, which lack fibrosin bodies. Yarwood (1978) agreed with the utility of this character but felt that this might have been overemphasized. He has observed fibrosin bodies forming cultures of *S. fuliginea*, which did not show them in every conidium. He noticed in this culture of *S. fuliginea* 4 – 8 fibrosin bodies in each conidium produced on the upper surface of the cotyledons and true leaves in the greenhouse at about 20°C. But on the lower surface fibrosin bodies were sometimes absent. He noticed that percentage of conidia with fibrosin bodies were as low as 20% on outdoor squash in Berkley, U.S.A. He further noticed that temperature was a factor in the occurrence of fibrosin bodies in conidia. He observed that occurrence of fibrosin bodies on cucumber in greenhouse was lower at 16°C than at 20, 24, or 28°C. In other species of *Sphaerotheca*, for example, in *S. lanestris* the percentage of conidia with fibrosin bodies was as low as 30 on *Quercus*

velutina in 1971. Similarly, in Moscow (U.S.S.R.), most conidia of *S. humuli* were found without fibrosin bodies in 1975. These observations of Yarwood indicate that the presence or absence of fibrosin bodies in conidia and their number are influenced by environmental factors, age of the host as well as the species of *Sphaerotheca* involved. This creates doubt with regard to utility of fibrosin bodies and it needs further examination and thorough investigations.

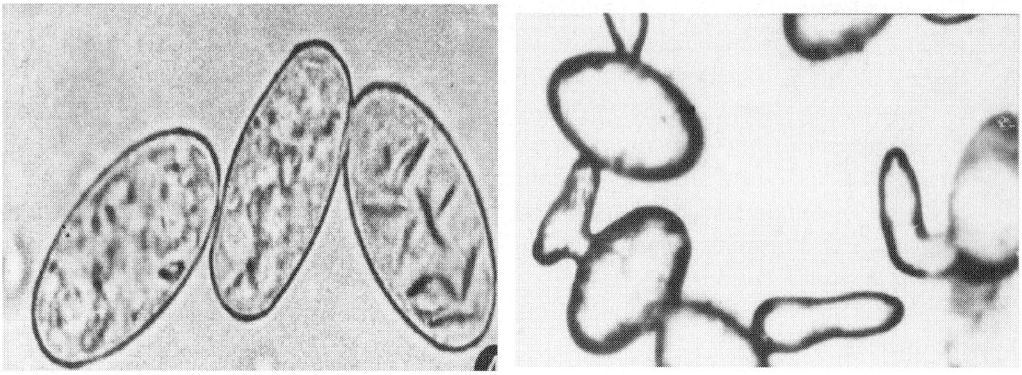

Fig. 13 & 14. Conidia of *S. fuliginea*.

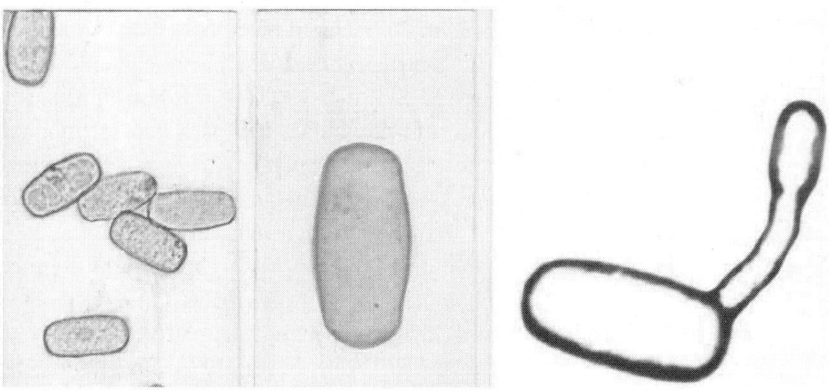

Fig. 15. Conidia of *E. cichoracearum*.

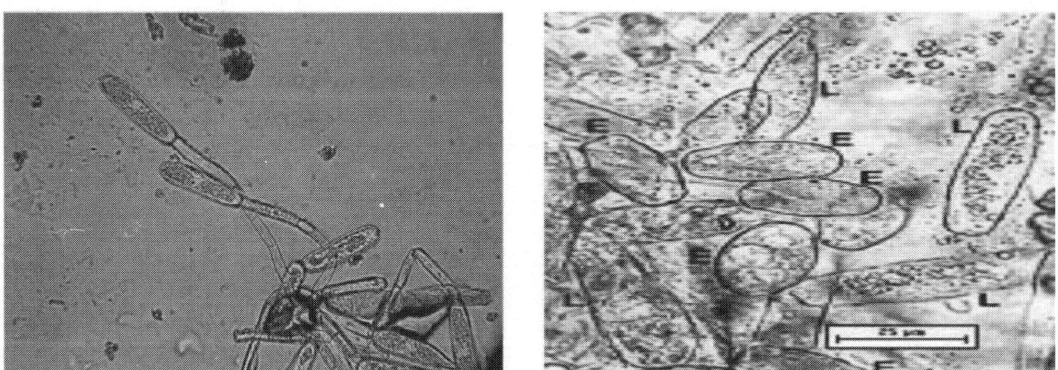

Fig. 16-17. Conidia of *L. taurica*.

5. **Germination:** Clare (1964) and Zaracovitis (1965) distinguished powdery mildew by the method of germination of the conidia. Conidia of *S. fuliginea* produced forked germ tube while those of *E. cichoracearum* produced simple germ tube with well-developed distinct apperssoria. They successfully used this as a taxonomic character.

7. IDENTIFICATION OF THE PATHOGEN IN ABSENCE OF TELEOMORPH

Identification of powdery mildews of cucurbits based on teleomorph characters is easy but they are rarely formed in nature. Then identification are based an anamorph characters. But anamorph characters of *S. fuliginea* and *E. cichoracearum* exhibit great similarities. Conidia in both are born in long chains on same kind of conidiophores. Apparently, there is no distinction in symptoms of the disease caused by them on cucurbits. This situation has caused great deal of confusion regarding identity of the species causing powdery mildew on cucurbits in different parts of the world, or in different areas or within the same area in a given country.

Sawada (1914) proposed a scheme for classification of Erysiphaceae based on conidial stage. The characters recognized as most useful in identification were formation of conidia singly or in chains, shape and dimensions of conidia and conidiophores, presence of fibrosin bodies and the type of vacuoles. Yarwood (1957) remarked that the method of germination of conidia is also a distinguishing character, though it perhaps may not be recommended for taxonomic purposes. He, however, devised a key based on conidial characters for identification of genera and species of powdery mildews in the absence of teleomorphs (Yarwood, 1973).

Boesewinkel (1977, 1980) in his synoptic key for identification of powdery mildews considered more than twelve morphological characters including presence or absence of conspicuous well-developed fibrosin bodies, the size, shape and germination pattern of conidia. Ballantyne (1975), examining a large number of samples collected from different parts in Australia, substantiated that *S. fuliginea* can be readily distinguished from *E. cichoracearum* by utilizing conidial characters.

Khan (1983, 1989) concluded that there is a great degree of agreement that the presence or absence of well-developed fibrosin bodies and mode of germination of conidia are useful criteria for distinction between *E. cichoracearum* and *S. fuliginea*.

Conidial germination characters have been successfully adopted in differentiating powdery mildew of cucurbits. Hashioka (1937) and Hirata (1955) observed that *S. fuliginea* was unique in producing forked germ tube. Zaracovitis (1965) suggested using the morphological characters of conidial germination of identifying imperfect state of powdery mildews. He observed that conidia of *E. cichoracearum* on germination on glass produced simple germ tubes with thick walled club-shaped appressoria as those of *S. fuliginea* produced no well-differentiated appreassoria and a number of conidia formed forked germ tubes. Ballantyne (1975) observed that all conidia of *S. fuliginea* do not produce forked germ tubes but percentage of forking ranges from 3-60 with more specimens in the range of 3-5 per cent.

Fibrosin bodies, also designated as Corpuscles de Zopf named after Zopf (1887) who first reported them from the conidia and conidiophores of *Podosphaera oxyacanthae*, were regarded valuable in taxonomy of powdery mildew by Foex (1912, 1925) and Bouwens (1927). Blumer (1933) and Homma (1937) considered presence or absence of fibrosin bodies valuable in distinguishing conidial stages of *E. cichoracearum* and *S. fuliginea* on cucurbits. Hashioka (1937), Uozumi and Yoshi (1952), Sawada (1959), Dingley (1959), Clare (1958, 1964), Boerema and Van Kesteren (1964), Kable and Ballantyne (1963) and Ballantyne (1963, 1975), Khan, (1982) Mazzanti de Castanona (1986), Grand, (1987), Branzanti and Brunelli (1987), Molot and Lecoq (1986), Reifschneider (1985) have recognized the significance of fibrosin bodies in differentiation of *S. fuliginea* and *E. cichoracearum*. Fibrosin bodies were also reported in *E. cichoracearum* by Klika (1922) without specifying the characteristic form. But this has not been substantiated by later reports.

Bouwens (1924), Clare (1958), Boerema and Van Kesteren (1964), Jhooty (1967) and Nagy (1979) considered that dimensions of conidia are of taxonomic importance for these species. Yarwood (1957, 1978), however, expressed doubts about the utility of dimensions of conidia in their taxonomy as these features are greatly influenced by a number of factors including relative humidity, temperature, host and nutritional status. Khan (1983) proposed that in most of the specimens the shape of conidia in *S. fuliginea* is ellipsoidal or at least with a tendency towards this shape whereas those of *E. cichoracearum* are barrel-shaped with a tendency towards cylindrical form. Similar observations have been reported by Nagy (1970). Jhooty (1967) concluded that in general conidia of *E. cichoracearum* are much longer than those of *S. fuliginea*. Nagy (1970) used the conidial dimensions as a character of differentiation and found that the length-width ratios of the two species were significantly different. Schlosser (1976b) recommended the determination of fibrosin bodies as standard procedure of differentiation of the two species. Sitterly (1978) suggested that conidial characters can be used for positive identification of the causal organism for the purpose of eliminating taxonomic confusion.

In addition to the conidial characters, Khan (1978) proposed the use of *Coccinia cordifolia* (syn. *Coccinia grandis*) and *Lagenaria leucantha* (syn. *Lagenaria siceraria*) as prospective differential hosts as aid for distinguishing between *S. fuliginea* and *E. cichoracearum*. It was claimed that *C. grandis* is infected only by *E. cichoracearum* and *L. siceraria* by *S. fuliginea*. Cross infection does not occur.

Identification of *L. taurica* in conidial stage is not difficult because of characteristic symptoms on the host, internal mycelium, branched conidiophores emerging through stomata, bearing single conidia and formation of two kinds of conidia.

In the absence of perithecia, the identity of the powdery mildew of cucurbits was established in a number of countries based on conidial measurements, germination rates, and germination in relation to moisture stress, absence and presence of fibrosin bodies. Clare (1958) in Australia; Dingley (1959) in New Zealand; Boerema and Van Kesteren (1964) and Kooistra (1968) in the Netherlands; Nagy (1970) in Hungary; Kable and Ballantyne (1963), Yarwood and Gardner (1964), Sowell and Clark (1971), Clark (1975) and Shanmugasundaram *et al.* (1971) in USA; Mazzanti de Castanon *et al.* (1987) in Argentina; Schlosser (1972) and Cruger and Meyer (1976) in West Germany; Gorter (1966) in South Africa; Viennot-Bourgin (1971) and Molot and Lecoq (1986) in France; Janke *et al.* (1977) in East Germany; Khan, (1982) in Libya, Jhooty (1967) in Punjab (India) have established the identity on the basis of anamorph characters. Some of these countries now have records of teleomorph of *S. fuliginea* of cucurbits.

8. RACES OF POWDERY MILDEW ON CUCURBITS

Powdery mildew is highly specialized parasites of plants. Existence of races within certain morphological species of powdery mildew is well established, particularly in *Erysiphe graminis*. It was shown through inoculation experiments by Neger (1902) that morphological species consist of biologically different species or races, which only infect particular plants but not closely related plants on which a different biological race occurs.

Some attempts have been made to identify the races within powdery mildews of cucurbits through their host specialization studies. According to Blumer (1933) the degree of biological specialization varies among the species and is broad in *Oidium* (*S. fuliginea* and *E. cichoracearum*) on cucurbitaceae, which infects nearly all members of this family. Alcorn (1969) reported the occurrence of four pathogenically distinct races of *S. fuliginea* on cucurbits in Queensland, Australia, all of which infected *Cucumis sativus*, *Cucurbita maxima* and *C. pepo*, but not all infected the two cultivars of *Cucumis melo* and the single cultivar of *Citrullus lanatus* var. Caffer.

In India, no indication of pathological specialization in *S. fuliginea* was noticed by Akram and Khan (1977, 1978), while screening cultivated cucurbits to five isolates of *S. fuliginea* collected from the states of Jammu and Kashmir and Uttar Pradesh.

Cross-inoculation studies of Nour (1959) between three isolated of *S. fuliginea* from various hosts were successful which indicated the lack of pathological variability within *S. fuliginea*.

Kaur and Jhooty (1985) studied the pathological specialization of six isolates of *S. fuliginea* causing powdery mildew of cucurbits in the state of Punjab (India). They noted three different pathological forms on the basis of differential reaction of cucurbitaceous hosts. They surveyed the cucurbit growing areas of Ropar, Hoshiarpur and Ludhiana districts in Punjab for the isolates of *S. fuliginea*. For testing the physiological forms, 3-4 leaf stage seedlings of *C. melo*, *C. melo* var. *utilissimus*, *C. pepo*, *Lagenaria siceraria*, *L. cylindrical* and *C. lanatus* were inoculated with different isolates of *S. fuliginea*. They noted that *C. pepo* and *L. siceraria* were infected by all the isolates; *C. melo* was resistant to isolates 5 and 6 and moderately resistant to isolate 2. *C. melo* var. *utilissimus*, *L. cylindrica* and *C. lanatus* were susceptible only to their respective isolates, i.e., isolates 5 and 6 and were resistant to the isolates 1 and 4. The above findings indicated the presence of three distinct pathological forms of *S. fuliginea* designed as form 1 on *L. cylindrica,* form 2 on *C. lanatus* and form 3 on *C. melo*, *C. melo* var. *utilissimus*, *C. pepo* and *L. siceraria*. Khan and Sharma (1993) reported three distinct races of *S. fuliginea* on cucurbits in most of the agro-ecological zones of India. Three distinct races in *S. fuliginea* are claimed to exist in U.S.A. Thomas (1978) reported a new biological race of *S. fuliginea* in USA which was designated as Race 3 while studying the reaction of five cultivars of cantaloupe. His study indicated that Race 1, Race 2 and Race 3 are present in U.S.A.

There also some records of occurrence of races of *E. cichoracearum* in cucurbits. Two distinct races in *E. cichoracearum* are differentiated in USA by Jagger and Scott (1937) and Jagger *et al.* (1938). This differentiation was made while breeding powdery mildew resistant cultivars of cucurbits. At the time of these studies, powdery mildew of cucurbits was recognized as *E. cichoracearum*. But in view of recent report of widespread occurrence of *S. fuliginea* in USA, the identity of the species selected for race differentiation by Jagger and Scott (1937) and Jaggar *et al.* (1938) as *E. cichoracearum* seems to be doubtful. It is possible that these differentiations were done in *S. fuliginea* presuming it as *E. cichoracearum*. Reifschneider *et al.* (1985) while studying the differentiation of races in *S. fuliginea* using host differentials (i.e., PMR5, PMR6, PMR45, Edisto 47) recognized the presence of Race-1 on *Cucumis melo* in the three districts of Brazil. Malot and Lecoq (1986) in their review on powdery mildew of cucurbits described three races (Race-1, Race-2 and Race-3) of *S. fuliginea* and their host differentials (i.e. PMR5, PMR6, PMR45, Edisto 47). Mohamed *et al.* (1995) observed two races (Race-1, Race-2) of *S. fuliginea* on cucurbits in Sudan. Vakalounakis and Klironomors (1995) claimed to report two races (Race-1 and Race-2) on cucurbits in Greece. Cohen *et al.* (1996, 2004) found Race-3 of *S. fuliginea* on *C. melo* in Israel. Hosoya *et al.* (1999) found the existence of 5 races in Japan. They reported Race-1 and Race-5 on melon in Japan. Velkov and Masheva (2002) reported Race-1 of *S. fuliginea* on cucurbits in Bulgaria. Cohen *et al.* (2004) observed 4 races of *S. fuliginea* from Israel. Lemus-Isla *et al.* (2005) found six races of *S. fuliginea* in Cuba. Lebeda and Sedlakova (2006) reported five races of *S. fuliginea* on cucurbits in Czechoslovakia. Bao *et. al.* (2008) identified physiological races of *S. fuliginea* on cucurbits in Sanya, Hainam.

9. MANAGEMENT OF POWDERY MILDEW

A programmed and regular observation for symptoms on cucurbit plants is necessary to detect the beginning of a powdery mildew infection, so that strategic treatment can be started according to the need. Integrated approach using a combination of several practices is found effective for the disease management.

Although the best method of disease control is prevention. Cultivating plants with genetic resistance to powdery mildew is considered effective and the best method of growing disease-free cucurbit crops. Consequently, an appropriate selection of tolerant or resistant cucurbit cultivars is the simplest way to deal with powdery mildew infections. Now powdery-mildew-resistant cultivars of different cucurbits became susceptible to disease because several races of powdery mildew fungus exist in nature (Zitter *et al.* 1996; Khan & Sharma, 1993).

For susceptible cucurbit cultivars, fungicide is the most effective means of control. Fungicides function as protectants, eradicants or both. A protectant fungicide prevents new infection from occurring whereas an eradicants can kill an existing infection. Protectant fungicide can be applied to highly susceptible plants before the disease appears. While eradicants use it earliest signs of the disease. Once mildew growth is extensive, control with any fungicide becomes more difficult.

Several least-toxic fungicides are available, including horticultural oils, neem oils, jojoba oils, sulphur, potassium bicarbonate, bicarbonate of soda (baking soda) and the biological fungicides AQ10 and Serenade with the exception of the oils, these materials are primarily preventive, although potassium bicarbonate has some eradicant activity. Oils work best as eradicants but also have some protectant activity.

Nunez-Pelinius *et al.* (2006) approved following fungicides to control powdery mildew disease under Florida environmental conditions: Flint (trioxystrobin), Nova (myclobutanil), Quadris (azoxystrobin), and Pristine (boscalid and pyraclostrobin). The misuse of these fungicides increases the risk of environmental contamination and also increases the possibility that powdery mildew fungi will develop resistance to these chemicals.

Chemically resistant races of powdery mildew have been isolated in USA and Australia. Managing fungicide resistance is an essential component of effective powdery mildew control. Most systemic fungicides are at risk for resistance development because they have single-site mode of action. Thus, modification of one gene in the pathogen may be enough to enable the pathogen to resist the action of the fungicide. Most protectant fungicides are not at risk for resistance development because they have a multi-site mode of action. The powdery mildew fungus has demonstrated a high potential for developing resistance. It has already developed resistance in Australia or elsewhere to at least one material in the at-risk chemical classes. Presence of resistant strains has been associated with control failure. It is important to realize that the primary goal of resistance management is to delay its development rather than to manage resistant strains after they have been selected; therefore, a management programme needs to be implemented before resistance develops to be effective.

The fungicide programme should include both systemic and protectant fungicides. The ideal programme consists of an alternation among effective systemic fungicides in at least 2 chemical classes and includes a protectant fungicide with the systemic fungicide in at least every other application. Protectant fungicides are important for resistance management because they control both sensitive and resistant strains of the pathogen.

10. CONCLUSIONS AND FUTURE PROSPECTS

Powdery mildew is a worldwide disease and one of the most serious threats to cucubit production. Many significant and effective breeding programmes have been started and new fungicides have developed to control the powdery mildew disease. But all these management strategies are not found effective to powdery mildew infection. Many resistant varieties of cucurbits for powdery mildew have been released and found resistant to specific fungal races. But many races of powdery mildew fungus exist in nature and this remains elusive to farmers and growers. Many fungicides are used to control the powdery mildew disease but now fungus is developing resistance to fungicides. So cucurbit powdery mildew fungus has exhibited a high potential for developing resistance to several fungicide classes in most of parts of the world including USA, Australia, India and Florida.

Although the phase of chasmothecial maturation to the production of viable sexual progeny for genetic analysis is still irresolute, the emerging molecular approaches developed for the pathogen and the forthcoming genome sequences of *B. graminis* f.sp. *hordei* and other powdery mildew fungi will be very useful for further research (Parez-Garcia *et al.*, 2009). According to them these developments will facilitate the analysis of physiological and molecular processes involved in pathogenicity and biology of powdery mildew fungus (*Podosphaera fusca*) and an understanding of the intimate molecular dialogue between host and pathogen. This will provide the basis for the design of novel chemicals effective against the different powdery mildew species involved in disease development, enabling us to face the disease and the design of disease control programmes in a more rational manner.

REFERENCES

Abul-Hayja, Z.M. and Trabulsi, I.Y. (1981). *Sphaerotheca fuliginea* on cucurbits in Saudi Arabia. *Trans. Br. Myco. Soc.* **76:** 506-507.

Allison, C.C. (1931). Powdery mildew of flax in Minnesota *Phytopathology* **24:** 305-307.

Allison, J.L. (1952). Disease of economic plants in Iraq. FAO Pl. *Prot. Bull.* **1:** 9-11. (*Rev. Appl. Mycol.* **32:** 473, 1953).

Alvarez, B. and Tores, J.A. (1995). Absence of alternative hosts of powdery mildew of cucurbitaceae on the eastern coast of Malya. Boletinde Sanidad-Vegetal, Plagas **21:** 185-193.

Amano, K. (1986). Host range and geographical distribution of the powdery mildew fungi. Japan Scientific Press, Tokyo, pp. 741.

Akram, M. and Khan, A.M. (1977). Host specialization studies of *Sphaerotheca fuliginea* (schlecht) Poll. *Acta Bot. Indica* **5:** 87-88.

Akram, M. and Khan, A.M. (1978). Studies on the cucurbits powdery mildew. III. Varietal response of some cultivated cucurbits to *Sphaerotheca fuliginea. Indian Phytopath.* **31:** 93-94.

Alcorn, J.L. (1969). Infection experiment with cucurbit powdery mildews. *Aus. J. Sci.* **31:** 296-297.

Baimuratora (Mme), G. (1961). Data on the parasitic fungus flora of the Golodnaya a steppe. *Uzbek Biol.* **3:** 8-15 (*Rev. Appl. Mycol.* 41, 18, 1962).

Ballantyne, B.J. (1963). A preliminary note on the identity of cucurbit powdery mildew. *Aust. J. Sci.* **25:** 360-361.

Ballantyne, B.J. (1975). Powdery mildew of cucurbitaceae : Identity, distribution, host range and sources of resistance. *Proc. Linn. Soc. New South Wales* **99:** 100-120.

Banga, O. (1956). International conference on the improvement of vegetable varieties at Wageningen Netherlands on August 26 and 27, 1955. *Euphytica* **5:** 18-32 (*Rev. Appl. Mycol.* **35:** 649, 1956).

Barratt, R.W. (1942). Occurrence of tomato leaf mold and of muskmelon powdery mildew in the field in New Hampshire. *Pl. Dis. Reptr.* **26:** 49.

Bell, E.H. and Alandia, S. (1957). Disease of temperate climate crops of Bolivia. *Pl. Dis. Reptr.* **41:** 646-649.

Benada, J. (1974). The susceptibility and resistance of cucumber and squash tissues to powdery mildew *Sphaerotheca fuliginea* in relation to oxidation, reduction, potential and pH. *Ceska Mycologie.* **28:** 44-53. (*Rev. Pl. Path.* **53:** 654, 1974).

Blackford, F.W. (1943). Downy mildew and powdery mildew of cucumber, *Qd. Agric. J.* **7:** 164-165.

Blasquez, C.H. (1970). Varietal resistance to three cucumber at foliar diseases in South-West Florida., *Pl. Dis. Reptr.* **54:** 52-55.

Blumer, S. (1922). Spezialisation der *Erysiphe horridula* Lev. Auf Boraginaceen. Central b1. Bakt. II **55:** 480-506.

Blumer, S. (1933). Die Erysiphaceen Mitteleuropas mit besonderer Berucksichtigung der Schweiz. *Beitrage Zur Kryptogamenflora der Schweiz.* **7(1):** 1-483.

Blumer, S. (1951). Beitrage zur kenntris der Erysiphacean Mitteilung (Contribution to the knowledge of Erysiphaceae) Note-2. *Phytopathology* **2:** 101-110 (*Rev. Appl. Mycol.* **31:** 148, 1952.)

Blumer, S. (1960). Echter Mehltau auf Susskirschan (*Prunus avium L.*) and Zwetschgen (*Prunus domestica* L.) *Phytopath. Z.* **37:** 317-320.

Boerema, G.H. and Van Kesteran, H.A. (1964). De identiteit von de echte meeldauw bij cucurbitaceae. *Neth. J. Pl. Pathol.* **70:** 33-34.

Boesewinkel, H.J. (1976). Cleistothecia of powdery mildew in New Zealand. *Trans. Br. Mycol. Soc.* **67:** 143-146.

Boesewinkel, H.J. (1977). Identification of Erysiphaceae by conidial characteristics. *Rev. de Mycol.* **41:** 493-507.

Boesewinkel, H.J. (1980). The morphology of the imperfect states of powdery mildews (Erysiphaceae). *Bot. Rev.* **46:** 167-224.

Bouwens, H. (1927). Weitere untersuchungen uber Erysipheen. Meded. Phytopath. Lab. Willie, Commelin Schotden. *Baarn* **10:** 3-31.

Bouwens, H. (1924). Untersuchungen uber Erysipheen. Meded, Phytopath. Lab. Willie, Commelin Schotden. *Baarn* **8:** 1-47.

Blumer, S. (1922). Das problem der „Bridging Species“ beiden parasitic hen Pilze. Mitt. Naturf. Ges. Bern 1992, XLV-XLVI.

Blumer, S. (1933). Die Erysiphaceem Mitteleuropas mit besonderer Beruckictigung der Schweiz. *Beitz. Kryptogamenflora der Schweiz:* **7:** 1-483.

Bao, H.Q., Xy, Y., Du Y.C., Li, J.S. and Yang, X.F. (2008). Identification of physiological races of powdery mildew on cucurbits in Sanya, Hainam. *J. Changing Vegetables* **1:** 49-51.

Bradbury, J.F. and Fischer, R.W. (1963). Control of powdery mildew of cucumbers with polybutene emulsions. *Pl. Dis. Reptr.* **47:** 1062-1064.

Omer, M.E.H. (1972). Chemical control of powdery mildews of cucumber and okra in Sudan. *Expt. Agric.* **8:** 265-270.

Osnitskaya (Mme) E.A. (1959). The biological reasons for preventive measures against powdery mildew of cucumber in net houses and forcing beds. *Erude ni Inst. Ovostch Khoz* **2:** 121-137 (*Rev. Appl. Mycol.* **41:** 118, 1962).

Omara, S.K. and Taha, M. (1995). Powdery mildew attacks commercial watermelon cultivars in Sudan. *Report cucurbit Genetic–cooperative* 18-55.

Palti, J. (1962). Prediction of powdery mildew outbreaks on cucurbits on the basis of seasonal factors and host range. *Bull. Res. Counc. Israel. Sect. D.* **10:** 236-249.

Palti, J. (1988). The *Leveillula* mildews. *Bot. Rev.* **54:** 424-535.

Pangalo, K.I. (1955). Origin and evolutionary trends in cucurbitaceous crop problemy. *Botaniki Acad. Nauk. U.S.S.R.* **2:** 329-338.

Passerini, G. (1867). Primo elenco di Fungi Purmensi. *Comment Soc. Crittog. Italy.* **2:** 472-475.

Paulus, A.O. (1969). Powdery mildew of cucurbits with systemic and non-systemic fungicides. *Pl. Dis. Reptr.* **53:** 813-816.

Paulus, A.O. Nelson, J., Shibuya, F., Whitaker, T.W., House, T., Meister, H. and Bohn, G.W. (1972). Fungicides and methods of application for the control of cautaloupe powdery mildew. *Pl. Dis. Reptr.* **56:** 935-938.

Perez-Garcia, A., Diego, R., Fernandez-Ortuno, D., Lopez-Ruiz, F., De Vicente, A. and Tores, J.N. (2009). The powdery mildew fungus *Podosphaera fusca* (synonym *Podosphaera xanthii*), a constant threat to cucurbits. *Molecular Plant Pathology.* **10(2):** 153-160.

Peleg, J. (1953). Out breaks and new records. *Israel FAO Plant Prot. Bull.* **14:** 60-61.

Peralta, G.J. (1946-1947). Appendix-6. Report by the Plant Pathologist for the year (1949–1947). *Rep. Dep. Agric. Malta Oct.* 1938 to Sept. 1946 and Agricultural Year 1946-1947. (*Rev. Appl. Mycol.* **29:** 424, 1953).

Pryor, D.E. and Whitaker, T.W. (1942). The reaction of contaloupe strains to powdery mildew. *Phytopathology* **32:** 995-1004.

Pucci, E. (1965). Lista Preliminare delia Malattie della piante ossorvate in Tripolitania. Daaaaaal–1059. Al-1964 : Sintomi Danni Lotta. Rivista di Agriculture Subtropical e Tropicale Anno. LI X No. 7-9.

Randall, T.E. and Menzies, J.D. (1956). The perithecial stage of the cucurbits mildew. *Pl. Dis. Reptr.* **40:** 255.

Rayss, T. (1947). Nouvelle contributional "etude de" la mycoflare de Palestine (quartrieme partie). *Palest. J. Bot. Jerusalum* **4:** 59-76.

Reed, G.M. (1908). Infection experiments with the mildew *Erysiphe cichoracearum* DC. *Trans. Wise. acad. Sci.* **15:** 527–Reichart, I., J. Palti and B. Kapuler. 1943. Trials for control of diseases of vegetables marroas. *Bull. Rehovoth. Agric. Sta.* **33:** 30.

Refischneider, F.J.B. (1985). Powdery mildew of melon (*Cucurmis melo*) caused by *Sphaerotheca fuliginea* in Brazil. *Plant Diseas* **69:** 1069-1070.

Reede. (1805). Malabar, III Pls. 1, 5; Royle III Himal; p. 218.

Revilla, V.A. (1955). Control del oidium del melon Zapallo, Y. pepino Bot Estac Expt. agric La Moline Loopp (*Rev. Appl. Mycol.* **36:** 164, 1956).

Ristic, S. (1985). Contribution to the study of cucumber powdery mildew, *Erysiphe cichoracearum* DC, in the glasshouse with special reference to morphological and biological characters. *Parazita Zastita Bilja.* **36:** 303-316.

Rochelle, L.A. (1972). Susceptibility of small sugar vegetable marrow to mildew. *Revista de Agriculture* **47:** 155-158.

Roder, K. (1936). Perithecia von. *Erysiphe cichoracearum* DC em Salm. au Freolandgurken (*Cucumis sativus L.*) *Angew. Bot.*, **19:** 16-163.

Rodigin, M.M. (1936). Note on the powdery mildews of cucurbits. (*Sphaerotheca fuliginea* (Sch.) Poll. and *Erysiphe cichoracearum* Fr.) *Sovetsk, Bot.*, **5:** 120-123. (*Rev. Appl. Mycol.* **16:** 364-365, 1937).

Roxburg, W. (1932). Flora India, III p. 714 as *L. clavata*; Kurg Contrb., II p. 100, Thwaites, Enum.

Rudich, J., Karchi, Z. and Esheed, N. (1969). Evidence for two races of the pathogen causing powdery mildew of muskmelon in Israel. *Israel Jnl. Agric. Res.* **19:** 41- 46.

Salmon, E.S. (1900). A monograph of the Erysiphaceae. *Mem. Torrey. Bit. Club.* **9:** 1-292.

Savulescu, T. (1929). Considerations systematiques Suz les Erysiphacees. *Ann. Sci. Acad. Hautes Etudea Bukarest.* **1:** 175.

Sawada, K. (1914). The classification of the Erysiphaceae basedon the conidial stages. *Formosa Agr. Expt. Sta. Spaec. Rep.* **9:** 1-102.

Sawada, K. (1959). Descriptive catalogue of Taiwan (Formosa) Fungi, Part XI. Special publication Coll. *Agric. Nat. Taiwan Univ. Taipei, Taiwan* **8:** 1-268.

Schepers, H.T.A.M. (1964). A pattern in the appearance of cucumber powdery mildew in Dutch glasshouses. *Neth. J. Pl. Path.* **9090:** 247-256.

Schlosser, E. (1972). Cleistothecia of *Erysiphe cichoracearum* DC ex Merat and conidia of *Sphaerotheca fuliginea* (Schl.) Poll. concomitantly on cucumber. *Meded. Fakul Landbou. Gent.* **37:** 553-558.

Schlosser, E. (1976a). Vorkommen and verbreintung der beiden arended gurkenmehlataus *Sphaerotheca fuliginea* and *Erysiphe cichoracearum*, in der Bundersrepublic Deutschland. *Fac. Land Bouw Rijks Univer Gent.* **41:** 539-544.

Schlosser, E. (1976b). Differentiation between conidia of the two species of cucurbit powdery mildew, *Sphaerotheca fuliginea* and *Erysiphe cichoracearum. Nachrichtenbfatt. des Deut-schen Pflanzen (Braunschweig)* **28:** 65-67.

Schroeder, W.T. and Providenti, R. (1968). Systemic control of powdery mildew of cucurbits with fungi 1991 applied as soil drenches and seed treatments. *Pl. Dis. Reptr.* **52:** 630-632.

Schroeter, J. (1893). In Cohn's Krypt FLora von Schlesien 3(1908) 229-247. (Original checked by Commonwealth Mycological Institute).

Shanmugasundram, S., Williams, P.H. and Peterson, C.E. (1971). Inheritance of resistance to powdery mildews in cucumber. *Phytopathology*, **61:** 1218-1221.

Sharma, P.D. (1972). Occurrence of perfect state of cucurbit powdery mildew. *Curr. Sci.* **41:** 578.

Sharma, P.D. (1973). Occurrence of perfect state of two cucurbit powdery mildews in Northern India. *Curr. Sci.* **42:** 871-872.

Sharma, A.K. (1978). Powdery mildew of some cucurbits from Jammu. *Indian Phytopath* **31:** 85-86.

Sharma, G.K. and Wajid Khan, M. (1991). Observations on occurrence and identity of powdery mildew of cucurbits in Tamil Nadu. *Indian Phytopath.* **44(1):** 45-51.

Sharma, G.K. and Wajid Khan, M. (1994). Species spectrum of cucurbit powdery mildew in Andhra Pradesh. J/ Bs-**73:** 77-79.

Siradhana, B.S. and Chaudhari, S.L. (1971). Occurrence of *Erysiphe cichoracearum* and *Sphaerotheca fuliginea* at Udiapur, Rajasthan. *Indian J. Mycol. Pl. Path.* **2:** 7-79.

Sitterly, W.R. (1978). Powdery mildews of cucurbits. Page 359–379 in D.M. Spencer Ed. "The Powdery Mildews" Academic Press, London, New York, San Francisco.

Sohi, H.S. and Nayar, S.K. (1969). Some new records of fungi from India. *Indian Phytopath.* **22:** 410-412.

*Spencer, D.M., Ebeens, M.H. and Morgan, W.M. (1975). Microbiology. Annual Report Glasshouse Crop Research Institute Sussex, U.K.

Spurling, A.T. (1973). A review of mildew problems on cucurbit vegetables crops in Malawi. *PANS* **19:** 42-45.

Stone, O.M. (1959). Annual Report 1959. Glasshouse Crop Research Institute U.K. pp. 141.

Tafradzhiiski, I. (1959). Powdery mildew on cucumbers in the greenhouse and hot industry and its control. *Grandinastwo* **1:** 22-25.

Tafradzhiiski, I. (1961). Experiments in the control of powdery mildew. (*Sphaerotheca fuliginea* and *Erysiphe cichoracearum* DC) on cucurbits. *Nauch. Trud. Vissh. Sel Kostop Inst. Vasil Kolarov Plovdiv* **10:** 239-255.

Tafradzhiiski, I. (1963). Powdery mildew on cucurbitaceae and their biology *Sphaerotheca fuliginea* and *Erysiphae cichoracearum.* Nauch. Trad. Vissn. Selkostop. *Inst. Vasil Kolarov Plovdiu.* **21:** 99-107.

Tarr. S.A.J. (1954). Diseases of economic crops in the Sudan ii, Fibres, oil seeds, coffee and tobacco. *F.A.O. Pl. Prot. Bull.* **11:** 161-165.

Tarr, S.A.J. (1955). The fungi and plant diseases of the Sudan. Commonwealth mycological Institute. *Kew*, pp. 197.

Thompson, A. (1933). Division of mycology. Annual Report for 1932. Dept. of Agric. Strait Settlement and Fed. Malay States (Reports of the res. Econ. and Agric Educ. Branches for the year 1932). Bull. 14 Gen. Ser. 53-62. (*Rev. Appl. Mycol.* **13:** 216, 1934).

Thomas, S.E. (1978). A new biological race of powdery mildew of cantaloupes. *Pl. Dis. Reptr.* **62:** 223. (*Rev. Pl. Pathol.* **11:** 467, 1978).

Tores, J.A., Canovas, J. and Relasco, M.J. (1990). A note of *Sphaerotheca fuliginea* (Shelecht ex) Poll. causal agent of powdery mildew on cucurbitaceous in the coastal areas of Malaga and Almeria. Investigation, Agraria, Production Y. *Protection Vegetable.* **5:** 3 475-479.

Ullasa, B.A. and Amin, K.S. (1981). Simultaneous occurrence of powdery mildews on *Cucurbita maxima* Dush. and *Abelmoschus esculantus* (L.) Moench from Karnataka, India. *Curr. Sci.* **50:** 238. (*Rev. Pl. Pahtol.* **60:** 462, 1981).

Ullasa, B.A. Schi, and Amin, K.S. (1979). Occurrence of Perithecial stage of *Sphaerotheca fuliginea* on muskmelon. (*Cucumis melo L.*) from Bangalore (Karnataka State) *Curr. Sci.* **48:** 696-697.

*Uozumi, T. and Yoshi, H. (1952). Some observations on the mildew fungus affecting the cucurbitaceous plants. *Ann. Phytopath. Soc. Japan.* **16:** 123-126 (*Rev. Appl. Mycol.* **32:** 466, 1953).

Vakalounakis, D.J. and Klironomors, E. (1995). Race and mating type identification of powdery mildew on cucurbits in Greece. *Plant Pathology*, **44:** 6, 1033-1038.

Viennot-Bourgin, G. (1958). Contribution a la connaissance des champigous parasites de l' *Iran. Ann. Epiphyt.* 97-270. (*Rev. Appl. Mycol.* **38:** 132, 1959).

Viennot-Bourgin, G. (1971). New rare Erysiphaceae in France. *Ann. Phytopath.* **3:** 337-352.

Velkov, N. and Masheva, S. (2002). Species and races composition of powdery mildew on cucurbits in Bulgaria. *Cucurbit Genetics cooperative.* **(25)**. 7-10.

Walker, M.N. and Weber, G.F. (1949). Disease of watermelon in Florida. *Bull. Fla. Agric. Expt. Sta.* **459:** pp. 46.

Williams, D.H. (1941). Important diseases of glasshouse plants. *Fruit Grower* **61:** 465-467. (*Rev. Appl. Mycol.* **20:** 437, 1941).

Yarwood, C.E. (1957). Powdery mildews. *Bot. Rev.* **23:** 235-300.

Yarwood, C.E. (1973). Pyrenomycetes : Erysiphales.- In the fungi: An Advanced Treatise. Academic Press, New Yrok and London. pp. 71-86.

Yarwood, C.E. (1978). History and taxonomy of powdery mildews. In the Powdery Mildew, 1-37, Editor: D. Spencer. London, Ner York: Academic Press.

Zitter, T.C., Hopkins, D.L. and Thomas, C.E. (1696). Compendium of cucurbit diseases. APS Press, St. Paul, Minn.

Zopf, W. (1887). Uber eineh neuen in haltikorper in Pflanzlichen Zellen. *Ber. Deutsch Bot. Gessel* **5:** 272-280.

Zlochova, K. (1988). Contribution to the study of the primary infection cycle of the fungus *Erysiphe cichoracearum* DX ex Merat on the family cucurbitaceous. Patologic kg fly ziologia Zborn reterator USBI, Sab, Bratislava. 397-403.

Zaracovites, C. (1965). Attempts to identify powdery mildew fungi b conidial characters. *Trans. Brit. mycol. Soc.* **48:** 553-558.

Diversity of Lower Plants (2014): pp. 164-173
Editors: Rajan Kumar Gupta & Mukesh Kumar
IK International Publishing House Pvt. Ltd.

12

Diversity of Wood Rotting Basidiomycetes of Kumaon and Garhwal Himalaya

DHIRAJ KUMAR SEHGAL*, KAUSHAL PRATAP SINGH** AND DHEERAJ MOHAN SINGHAL**

**Botanical Survey of India, 192, Kaulagarh Road, Dehradun–248 195 (Uttarakhand)*
***Botany Department, Raja Balwant Singh College, Agra–282 002 (Uttar Pradesh)*

ABSTRACT

Temperate Himalayan forests are rich in dead trees and provide adequate environment for wood rotting fungi. The macrofungal diversity is very rich due to litter decomposition. Species of wood rotting fungi of the order Aphyllophorales form an integral part of the temperate Himalayan forest ecosystems and play the primary role in the formation of forest soils by decomposition of substrates. Studies on the taxonomy and diversity of macrofungi are gaining importance, as many macrofungi are becoming extinct and facing threat of extinction because of habitat destruction. The current study deals with the diversity of macrofungi in evergreen moist temperate forest of Kumaon and Garhwal Himalaya. Surveys were conducted from July 2006 to October 2008. The macrofungi were collected, characterized, allotted accession number, preserved and all the collected specimens were photographed. A total of 197 species of aphyllophoraceous fungi belonging to 89 genera have been recorded from temperate zone of both the Kumaon and Garhwal Himalayas. Some of the important identified genera are: *Armillaria*, *Bjerkendra*, *Cystostereum*, *Dacryobolus*, *Daedalea*, *Fomes*, *Fomitopsis*, *Ganoderma*, *Jungunia Lenzites*, *Merulius*, *Phellinus*, *Phlebia*, *Inonotus*, *Phaleous Steccherinum*, *Stereum*, *Trametes*, *Trichaptum*, and *Xylobolus*.

Keywords: Diversity, Survey, Sporocarps, Wood rotting fungi, Species richness.

1. INTRODUCTION

The variety and galaxy of fungi and their natural beauty occupy prime place in the biological world and India has been a cradle for such fungi. Defining the number of fungi on earth has always been a point of discussion and several studies have focused on enumerating the world's

*Communicating Author

fungal biodiversity (Crous *et al.*, 2006). Only a fraction of total fungal wealth has been subjected to scientific scrutiny and mycologists continue to unravel the unexplored and hidden wealth. The number of existing fungi worldwide has been estimated to 1.5 million species (Hawksworth 2004). One-third of the fungal diversity of the globe exists in India and of this, only 50% are characterized until now (Manoharachary *et al.,* 2005). The number of fungi recorded in India exceeds 27,000 species, the largest biotic community after insects. The wood rotting fungi of the order Aphyllophorales constitute a cosmopolitan group, fruiting mostly on dead and wood products throughout the world under different environmental conditions. The sufficient substratum provided both by angiospermic and coniferous tree species and coupled with a great diversity in ecological habitats provides rich environmental conditions for the growth of these fungi particularly in the temperate Himalayan zones. The 'Decades of Fungi' by Berkeley (1850-54) was the first systematic information on survey and study of wood rotting mycoflora of Himalaya and this was further enhanced by Thind (1961), Bakshi (1971), Rattan (1977), Rattan and Khurana (1978) and Sharma (1995, 1999a & 1999b). The other incidental and scattered reports on these fungi from the temperate forests are also found in Bakshi and Bagchee (1950), Bagchee and Bakshi (1954), Bagchee *et al.* (1954), Thind and Adlakha (1956) Reid *et al.* (1958, 59), Rehill and Bakshi (1965, 66), Imazeki *et al.* (1966), Thind and Khara (1968, 75) Khara (1977a & 77b), Thind and Rattan (1968-73), Mass Geasteranus (1971), Rattan (1974) and Sharma (1996-97).

The temperate forests of Kumaon and Garhwal regions of Uttarakhand occupy a climatic zone with pronounced variations in seasonal temperatures. They extend approximately from 2200 m to 3000 m and form a major part of the forest vegetation in the Himalayas. The temperate broad-leaved forests in the lower parts dominated by species of *Quercus*, *Platanus*, *Aesculus*, *Juglans*, *Populus*, *Salix*, *Alnus*, *Acer*, *Pyrus*, *Lyonia*, *Viburnum*, *Elaeocarpus*, *Lindera*, *Castanopsis*, *Ilex* and *Engelhardtia* are followed by rich stands of *Pinus*, *Cedrus*, *Abies*, *Taxus*, *Cupressus*, *Larix* and *Tsuga* at higher altitudes. The dominant elements of these forests are *Abies pindrow*, *Cedrus deodara*, *Picea morinda*, *Pinus wallichiana*, *Quercus semicarpifolia*, *Rhododendron lepidotum* and *Betula utilis*.

2. MATERIALS AND METHODS

Habitat: Temperate Himalayan forests are rich in dead trees and provide adequate environment for wood rotting fungi. When trees are alive, one can hardly expect them because wood rotting fungi seldom fruit on the surface until their host trees die (Erikson, 1958). These fungi usually occur when the forest sites are disturbed to some extent. When trees are broken or damaged, the trees are exposed to wood decayers. Wind breaks and wind throws occur where trees are weakened by decay fungi or when they are shallowly rooted on exposed sites. The most favourable condition is achieved when fallen trees and slash are dumped or stacked together on the forest floor. More than 70% of the wood rotting fungi collection occurred in such type of sites.

When the trees die and defoliated, the overstory canopy is extensively opened. Increased penetration of direct sunlight down to the forest floor depletes the moisture of the forest. Substrates for fungi are abundant all over but fungal growth is quite limited because of less moisture. In such sites, the fungal fruiting becomes dramatically decreased except when the forest floor remains wet during rainy season.

Survey: Surveys are particularly sensitive to timing and location of observations. Macrofungi exhibit pattern of diversity that are related to largely to substratum and host availability (Natarajan *et al.,* 2005). Temperate regions interact with the habitat of the fungus either stimulate or retard fruiting (Lodge *et al.,* 2004). Surveying is best just after the period of rain but some macrofungi can be found any time in some locations (Susan metzler and Van metzler 1992). Difficulty was small proportions of macrofungi were visible on a single visit. Hence, repeated surveys were done during July 2006 to October 2008 in all the sampling stations. The sporocarps encountered were collected and analyzed for its identity.

Sampling: Vegetation zones and plant associations are useful criteria to use when dividing a landscape for sampling. Many macrofungi occur only in association with particular family or plants. The macrofungi generally fruit when temperature was above 15°C and RH more than 80-85%. Different species, however, exhibit different fruiting phenologies, which vary from month to month and at different altitudes (Lodge *et al.,* 1995) and regions. Thus, a particular species may fruit at different seasons cross wide geographic distances or long strong elevation gradient (Lodge *et al.,* 2004).

Equipment: Samples were collected using an axe, sharp knife, forceps, measuring tape, hand lens, pens, books, labels, camera, papers and containers. An axe was used to incise to a depth sufficient to enable identification of the host, a sharp knife for collecting sporocarps from trees. Forceps for the collection of very small sporocarps collected with the help of hand lens and containers were used for the collection of the specimens.

Collection and preservation: Sporocarps were removed from trees with a great care to avoid damage. The habitat and morphological characteristics of the macrofungi were noted (Peksen and Karaca 2003; Kaya 2005) and photographed for diagnosis during the collection (Demirel and Uzun 2002; Fadime and Mustafa 2002; Yesil and Yildiz 2004). If specimen was on the wood or litter, including substratum it was collected to facilitate the identification. The specimens were collected in the fresh paper bags for further identification in the laboratory (Afyon *et al.,* 2005). Small plastic boxes were used for woody basidiomycetes. Care was taken to avoid distortion of fleshy fungi and labelled the specimen. The collection was brought to the lab and soon preserved as dried specimens.

Characterization of macrofungi: The collected sporocarps were described for the morphological characters like colour, size, shape, odour and texture. Accurate and consistent notation of sporocarps colour, including colour changes of mature sporocarps and colours of different development stages, presence or absence of texture are important for describing macrofungi were noted. Field labels were tagged to the specimens with the collection data which includes scientific name of the fungus, date, location, habitat, a brief note on distinguishing macroscopic features, collection number, and microhabitat. A separate catalogue was maintained for each entry of the daily collection number, which facilitates the tracking of the specimens when they are sorted or identified. The sporocarps of woody substrate will remain as it is. In many species of fleshy macrofungi, the characters were lost when sporocarps were dried. Therefore, they were documented carefully before drying.

3. RESULTS

A total of 197 species of aphyllophoraceous fungi belonging to 89 genera have been recorded from temperate zone of both the Kumaon and Garhwal Himalaya (Table 1). Out of 197 species which grow on *Abies pindrow*, *Cedrus deodara*, *Picea morinda*, *Pinus wallichiana*, *Quercus semicardifolia*, the following species are more prevalent: *Daedalea incana* (Lev.) Ryv., *Fomes annosus* (Fr.) Cooke, *Fomes fomentarius* (L: Fr.) kicx., *Fomitopsis rosea* (Abt. & Schw.: Fr.) Donk, *Jungunia collabens* (Fr.) Ryv., *Lenzites betulina* (Fr.) Fr., *Lenzites striata* Swartz., *Merulius tremellosus* Schrad, *Phellinus gilvus* (Schw.: Fr.) Pat., *Phellinus pini* (Thore) Lloyd, *Phlebia roumegueri* (Bres.) Donk, *Bjerkendra adusta* (Willd.:Fr.) Karst., *Inonotus cuticularis* (Bull.:Fr.) Karst., *Phaleous schweintzii* (Fr.) Pat., *Steccherinum ochraceum* (Pers. ex Fr.) Gray, *Stereum hirsutum* (Willd. ex Fr.) Gray, *Trametes versicolor* (L: Fr.) Pilat, *Trichaptum abietinium* (L:Fr.) Quel, *Trichaptum biformis* (Fr. In Kl.) Ryv. and *Xylobolus subpiletus* (Berk. & Curt.) Biodin.

Abies pindrow form the most important and major host for these fungi. Both living and dead fallen trees support a rich aphyllophoraceous fungus flora and a rich species diversity up to the extent that over 95 species grow either exclusively on *Abies* or share it as one of their liked hosts. *Amphinema byssoides* (Fr.) Erikss., *Amylocorticium indicum* Thind and Rattan, *Antrodia seriallis* (Fr.) Erikss., *Asterostroma cervicolor* (Berk. & Curt.) Masse., *Ceraceomyces tessulatus* (Cooke) Julich, *Chaetoderma luna* (Rom.) Parm, *Coniophora betulae* Karst., *Cymatoderma sclerotiodies* (Lloyd.) Reid., *Daedalea incana* (Lev.) Ryv., *Daldinia concentrica*, *Echinodonotium japonicum* Imazeki., *Fomes annosus* (Fr.) Cooke, *Fomes pinicola* Fr., *Phellinus robustus* (Karst.) Bourd., *Fomitopsis rosea* (Abt.& Schw.: Fr.) Donk, *Lenzites repanda* (Mont.) Fr., *Merulius tremellosus* Schrad, *Phlebia hydnoidies* (Cooke & Mass.) Chirst., *Phlebia roumegueri* (Bres.) Donk, *Microporellus chocolatus* (Bose) Ryv. grow exclusively on *Abies*. Some other species which also frequently share *Abies* as one of their hosts are: *Bjerkendra adusta* (Willd.: Fr.) Karst., *Phellinus gilvus* (Schw.: Fr.) Pat., *Phanerochaete gigantea* (Fr. ex Fr) Rattan *et al*, *Skeletocutis amorpha* (Fr.) Kotl., *Inonotus cuticularis* (Bull.: Fr.) Karst., *Poria callosa* (Fr.) Cooke, *Scytinostroma odoratum* forma *Crassum* Rattan, *Serpula himantoides* (Fr.) Cunn., *Tomentella botyroides* (Schw.) Bourd.& Galz., *Tomentella cinerascens* (Karst.) Hohen and Litsch., *Tomentella muricata* (Ellis. & Everh.) Wakef., *Tomentella pilosa* (Burt.) Bourd. & Galz., *Trametes trogii* Berk., *Trechispora farnica* (Pers. ex Fr.) Liberta grow either exclusively on species of *Abies* or share them as one of the most liked hosts.

Near the timberline *Cedrus deodara* is also a common host. *Chondrostereum purpurum* (Fr.) Pouz., *Confertobasidium olivaceoalbum* (Bourd. & Galz.) Julich., *Gleophyllum carbonarium* (Berk. & Curt.) Ryv., *Hymenocheate leonina* Berk. & Curt., *Hypoderma argillaceum* (Bres.) Donk, *Hypoderma paretermissum* (Karst) Erikss, *Hypodontia alutaria* (Burt.) Erikss, *Jughunia luteoalba* (Karst) Ryv., *Metulodontia nivea* (Karst) Parm, *Phlebia subserialis* (Bourd & Galz) Donk, *Perenniporia medulla–panis* (Fr.) Donk, *Oligoporus placenta* (Fr.) Gilb., *Sistotremastrum niveocremeum* (Hoehn & Litsch) Erikss., *Stereum bicolor* (Pers. ex Fr.) Fr., *Stereum chailleti* Pers., *Subulicystidium longisporum* (Pat.) Rom., *Tomentella griseoumbrina* Litsch, *Vararia vassilevae* Parm., *Xylobolus ahmadii* (Biodin.) Biodin. also quite frequently grow on *Cedrus deodara*.

The wood rotting fungi growing frequently on *Picea morinda*, *Pinus wallichiana* and *Quercus semicardifolia* are: *Armillaria mellea* (Vahl.) Quel., *Cystostereum subabruptum* Bourd. & Galz., *Dacryobolus costratum* (Rehill & Bakshi) Rattan comb.nov., *Fomes annosus* (Fr.) Cooke, *Fomes fomentarius* (L: Fr.) kicx., *Fomitopsis rosea* (Abt. & Schw.: Fr.) Donk, *Ganoderma applanatun* (Pers.) Pat., *Lenzites betulina* (Fr.) Fr., *Merulius tremellosus* Schrad, *Bjerkendra adusta* (Willd.: Fr.) Karst., *Trametes versicolor* (L: Fr.) Pilat, *Trichaptum abietinium* (L: Fr.) Quel, *Trichaptum biformis* (Fr. In Kl.) Ryv.

Table 1

S. No.	Names	*Abies pindrow*	*Cedrus deodara*	*Picea morinda*	*Pinus wallichiana*	*Quercus semicardifolia*
1.	*Amaurodon viridis* (Alb.& Schwein: Fr.) Schroet. (Mass.) Cunn.	+	-	+	-	-
2.	*Amphinema byssoides* (Fr.) Erikss.	+	+	+	-	-
3.	*Amylocorticium indicum* Thind and Rattan	+	+	-	-	-
4.	*Amylostereum chailletii* (Pers. ex Fr.) Biodin	+	+	-	-	-
5.	*Antrodia carbonica* (Overh.) Ryv. & Gilbn.	-	-	+	-	-
6.	*Antrodia seriallis* (Fr.) Erikss.	+	-	-	+	-
7.	*Antrodiella semisupina* (Berk. & Curt.) Ryv.	-	-	+	-	-
8.	*Armillaria mellea* (Vahl.) Quel.	+	-	+	+	-
9.	*Asterostroma cervicolor* (Berk. & Curt.) Masse.	+	+	-	-	+
10.	*Asterostroma musiocolum* (Berk. & Curt.) Masse.	+	+	-	-	-
11.	*Botryohypochnus isabellinus* (Fr.) Erikss	+	+	+	-	-
12.	*Ceraceomyces fibuligera* (Thind & Rattan) Comb. Nov.	-	+	+	-	-
13.	*Ceraceomyces tessulatus* (Cooke) Julich	+	-	-	-	-
14.	*Ceratobasidium subabruptum* Rattan sp. nov.	-	+	-	-	-
15.	*Chaetoderma luna* (Rom.) Parm	+	+	-	-	-
16.	*Chondrostereum purpurum* (Fr.) Pouz.	-	+	-	-	-
17.	*Collybia velutipes* (Curt.) Fr.	-	-	+	-	-
18.	*Coltricia vallata* (Berk.) Teng.	-	-	-	+	-
19.	*Confertobasidium olivaceoalbum* (Bourd.& Galz.) Julich.	-	+	-	-	-
20.	*Coniophora arida* (Fr.) Karst.	+	+	-	-	-
21.	*Coniophora betulae* Karst.	+	+	-	-	-
22.	*Coniophora cordensis* Rattan sp. nov.	+	-	-	-	-
23.	*Coniophora dimitella* Rattan sp. nov.	+	+	-	-	-
24.	*Coniophora puteana* (Schum. ex Fr.) Karst.	+	-	-	-	-
25.	*Corticium introversum* n. sp.	-	-	+	-	-
26.	*Corticium laeve* Pers.ex Fr.	-	-	+	-	-
27.	*Corticium protentosum* Berk. & Curt.	-	-	-	-	+
28.	*Corticium sulcatum* n. sp.	-	-	+	-	-
29.	*Cristinia helvetica* (Pers.) Parm.	+	-	-	-	-
30.	*Cymatoderma sclerotiodies* (Lloyd.) Reid.	+	+	-	-	+
31.	*Cystostereum murraii* (Berk.and Curt.) Pouz.	-	+	-	-	+
32.	*Cystostereum subabruptum* Bourd. & Galz.	-	-	-	+	+
33.	*Dacryobolus costratum* (Rehill & Bakshi) Rattan comb.nov.	-	-	+	-	+
34.	*Dacryobolus karstenii* (Bres.) Over. ex Parm.	+	+	-	+	-
35.	*Daedalea flavida* Lev.	-	-	-	-	+
36.	*Daedalea incana* (Lev.) Ryv.	+	+	+	+	+
37.	*Daedaleopsis purpurea* (Cke.) Imaz.	-	-	-	-	+
38.	*Daldinia concentrica*	+	-	-	-	+
39.	*Echinodonotium japonicum* Imazeki.	+	-	-	-	+
40.	*Elammula hybrida* Fr.	+	-	-	-	-
41.	*Fomes annosus* (Fr.) Cooke	+	+	+	+	-
42.	*Fomes fomentarius* (L:Fr.) kicx.	-	+	+	+	+
43.	*Fomes pinicola* Fr.	+	-	-	-	+
44.	*Phellinus robustus* (Karst.) Bourd.	+	-	-	-	-
45.	*Fomes rufolaccatus* Bose.	+	-	-	-	-
46.	*Phellinus senex* (Nees. & Mont.) Imaz.	-	-	-	-	+
47.	*Fomitopsis rosea* (Abt. & Schw.: Fr.) Donk	+	+	+	+	+
48.	*Ganoderma applanatun* (Pers.) Pat.	-	-	+	+	+

Contd.

Table 1: Contd.

S. No.	Names	*Abies pindrow*	*Cedrus deodara*	*Picea morinda*	*Pinus wallichiana*	*Quercus semicardifolia*
49.	*Ganoderma leucophaeum* (Mont.) Pat	-	-	+	-	-
50.	*Ganoderma lucidum* (Leyss.) Karst.	-	-	+	+	+
51.	*Gloeophyllum abietinum* (Bull.: Fr.) Karst.	-	+	-	+	+
52.	*Gloeophyllum carbonarium* (Berk. & Curt.) Ryv.	-	+	-	-	-
53.	*Gloeocystidiellum lactescens* (Berk.) Biodin.	+	+	-	-	-
54.	*Gloeocystidiellum sulcatum* (Rehill and Bakshi) Biodin.	+	+	+	-	-
55.	*Heterobasidion annosum* (Fr.) Bref.	+	-	-	-	-
56.	*Hydnum coralloides* Scop.	-	+	-	-	-
57.	*Hymenocheate cacao* Berk.	-	-	-	-	+
58.	*Hymenocheate cruenta* (Pers.: Fr.) Donk	+	-	+	-	+
59.	*Hymenocheate leonina* Berk. & Curt.	-	+	-	-	+
60.	*Hymenocheate rubiginosa* (Dicks.) Lev.	-	-	-	-	+
61.	*Hymenocheate tabacinia* (Sow.: Fr.) Low.	-	-	-	-	+
62.	*Hyphoderma polonense* (Bres.) Donk	+	+	-	-	-
63.	*Hyphoderma puberum* (Fr.) Wallr.	+	-	-	-	-
64.	*Hyphoderma teutoborgenese* (BrinKm.) Erikss.	+	-	-	-	-
65.	*Hyphodontia arguta* (Fr.) Erikss.	+	+	-	-	-
66.	*Hyphodontia efibulata* Erikss.& Hjorist.	-	-	+	-	-
67.	*Hyphodontia spathulata* (Schrad. ex Fr.) Parm.	+	+	-	+	-
68.	*Hypochnicum sphaerosporum* (Hoehn. & Litsch.) Erikss.	+	-	-	-	-
69.	*Hypochnicum punctulatum* (Cooke) Erikss.	+	+	-	+	-
70.	*Hypoderma argillaceum* (Bres.) Donk	-	+	-	-	-
71.	*Hypoderma paretermissum* (Karst) Erikss	-	+	-	-	-
72.	*Hypoderma setigerum* (Fr.) Donk	-	+	-	-	-
73.	*Hypodontia alutaria* (Burt.) Erikss	-	+	-	-	-
74.	*Hypodontia longicystidiosa* Rattan sp. nov,	-	+	-	-	-
75.	*Hypodontia pallidula* (Bres.) Erikss	-	+	-	-	-
76.	*Hypodontia papillosa* (Fr.) Erikss	-	+		-	-
77.	*Inonotus drydeus* (Pers. : Fr.) Murr.	+	-	+	-	+
78.	*Inonotus flavidus* (Berk.) Ryv.	+	-	-	-	+
79.	*Inonotus diverticuloseta* Peg.	-	-	-	-	+
80.	*Jughunia luteoalba* (Karst) Ryv.	-	+	-	-	-
81.	*Jungunia collabens* (Fr.) Ryv.	+	+	+	+	-
82.	*Kavinia alboviride* (Morge.) Gilb.	+	+	-	-	-
83.	*Laetiporus sulpherus* (Bull.: Fr.) Murr.	+	-	+	-	+
84.	*Laxietextum bicolor* (Pers. ex Fr.) Lentz.	-	+	-	+	-
85.	*Lenzites adusta* Mass	-	-	-	+	-
86.	*Lenzites betulina* (Fr.)Fr.	+	+	+	+	+
87.	*Lenzites lepideus* Fr.	-	-	-	+	-
88.	*Lenzites repanda* (Mont.) Fr.	+	-	+	+	-
89.	*Lenzites sepiaria* (Wulf.) Fr.	+	+	-	+	-
90.	*Lenzites striata* Swartz.	+	+	+	+	-
91.	*Lopharia cinerascens* (Schw.) Cunn.	-	-	-	-	+
92.	*Merulius aureus* Fr.	+	-	-	+	-
93.	*Merulius tremellosus* Schrad	+	-	+	+	+
94.	*Metulodontia nivea* (Karst) Parm	-	+	-	-	-
95.	*Metulodontia qualetii* (Bourd. & Galz.) Parm	-	-	-	-	+
96.	*Morganella pyriforme*	-	+	-	-	+
97.	*Peniophora cinera* (Fr.) Cke.	-	-	+	-	-

Contd.

Table 1: Contd.

S. No.	Names	*Abies pindrow*	*Cedrus deodara*	*Picea morinda*	*Pinus wallichiana*	*Quercus semicardifolia*
98.	*Peniophora filamentosa* (Berk. & Curt.) Buds.	-	-	+	+	-
99.	*Peniophora gigantea* (Fr.) Massee	+	-	-	-	-
100.	*Peniophora roumeguerii* (Bres.) Hoehn	-	-	+	-	-
101.	*Phanerochaete filamentosa* (Berk. and Curt.) Parm	+	-	+	+	-
102.	*Phanerochaete gigantea* (Fr. ex Fr) Rattan *et al.*	+	+	-	-	-
103.	*Phellinus allardi* (Bres.) Ahmad	-	-	-	-	+
104.	*Phellinus gilvus* (Schw.: Fr.) Pat.	+	+	-	+	+
105.	*Phellinus obstusus* Berk.	-	-	-	-	+
106.	*Phellinus pectinatus* (Kl.) Quel.	-	-	+	-	-
107.	*Phellinus pini* (Thore) Lloyd	+	+	+	+	-
108.	*Phellinus xeranticus* (Berk.) Pegler	+	-	-	-	+
109.	*Phlebia albida* Post. ex Fr.	-	-	+	-	+
110.	*Phlebia hydnoidies* (Cooke & Mass.) Chirst.	+	+	-	-	-
111.	*Phlebia livida* (Fr.) Bres.	+	-	-	+	-
112.	*Phlebia roumegueri* (Bres.) Donk	+	+	+	+	-
113.	*Phlebia subserialis* (Bourd & Galz) Donk	-	+	-	-	-
114.	*Bjerkendra adusta* (Willd.: Fr.) Karst.	+	-	+	+	+
115.	*Skeletocutis amorpha* (Fr.) Kotl.	+	+	-	+	-
116.	*Polyporus biformis* Klot.	-	-	+	-	-
117.	*Inonotus rickii* (Pat.) Reid.	+	-	-	-	-
118.	*Polyporus censors* (Berk.) n. comb.Stevenson	-	-	-	-	+
119.	*Microporellus chocolatus* (Bose) Ryv.	+	-	-	+	-
120.	*Inonotus cuticularis* (Bull.: Fr.) Karst.	+	+	-	+	+
121.	*Polyporus friabilis* Bose	-	-	+	-	-
122.	*Polyporus resinosus* (Schrad.) Fr.	-	-	+	-	-
123.	*Phaleous schweintzii* (Fr.) Pat.	+	+	+	+	-
124.	*Polyporus secernibilis* Berk	-	-	+	-	-
125.	*Inonotus tomentosus* Fr.	-	+	+	-	-
126.	*Rigidoporus lineatus* (Pers.) Ryv.	-	-	+	-	-
127.	*Trametes hirsutus* (Wulf.: Fr.) Pilat.	-	+	+	+	-
128.	*Polystictus leoninus* Koltz	-	-	+	-	-
129.	*Coltricia perennis* (L.: Fr.) Murr.	-	-	+	-	-
130.	*Polystictus pergamenus* Fr.	-	-	+	-	+
131.	*Pcynoporus sanguineus* (Fr.) Murr.	-	-	-	+	-
132.	*Poria callosa* (Fr.) Cooke	+	-	-	-	-
133.	*Antrodia carbonica* (Overh) Ryv. & Gilb.	-	-	-	+	-
134.	*Oxyporus corticola* (Fr.)Ryv.	+	-	-	-	-
135.	*Junghunia nitidia* (Fr.) Ryv.	-	-	+	-	-
136.	*Perenniporia medulla –panis* (Fr.) Donk	-	+	-	-	-
137.	*Oligoporus placenta* (Fr.) Gilb.	-	+	-	-	-
138.	*Antrodia serialis* (Fr.) Donk	+	-	-	-	-
139.	*Antrodia xantha* (Fr.) Ryv.	-	-	-	+	+
140.	*Pulcherricium caeruleum* (Fr.) Parm.	-	-	+	+	-
141.	*Radulomyces molaris* Fr.) M.P. Christ.	-	-	-	+	-
142.	*Resinicium bicolor* (Fr.) Parm.	+	+	+	-	-
143.	*Schizophyllum commune* Fr.	-	-	-	-	+
144.	*Scytinostroma ochroleucu* (Bres.& Torrend) Donk	+	+	-	-	-
145.	*Scytinostroma odoratum* forma *Crassum* Rattan	+	+	-	-	-

Contd.

Table 1: Contd.

S. No.	Names	*Abies pindrow*	*Cedrus deodara*	*Picea morinda*	*Pinus wallichiana*	*Quercus semicardifolia*
146.	*Scytinostroma protentosum* (Berk. & Curt.) Donk	+	-	+	-	-
147.	*Scytinostromella heterogena* (Bourd. and Galz.) Parm.	+	+	-	-	-
148.	*Serpula himantioides* (Fr.) Cunn.	+	+	-	-	-
149.	*Serpula lachrymans* Gray	+	-	+	+	-
150.	*Sistotremastrum niveocremeum* (Hoehn & Litsch) Erikss.	-	+	-	-	-
151.	*Steccherinum ochraceum* (Pers.ex Fr.) Gray	+	+	+	-	+
152.	*Steccherinum fimbriatum* (Pers. ex Fr.) Gray	-	+	+	-	-
153.	*Stereum acanthophysatum* Rehill &Bakshi	-	+	+	-	-
154.	*Stereum bicolor* (Pers. ex Fr.) Fr.	-	+	-	-	-
155.	*Stereum chailleti* Pers.	-	+	-	-	-
156.	*Stereum fusum* (Schrad) Quel.	-	-	-	+	-
157.	*Stereum hirsutum* (Willd. ex Fr.) Gray	+	+	+	-	+
158.	*Stereum karstenii* Bres.	-	+	-	-	-
159.	*Stereum ostrea* (Blume & Nees.: Fr.) Fr.	-	-	+	-	-
160.	*Stereum ostrea* (Blume & Nees ex Fr.) Fr.	-	+	-	-	-
161.	*Stereum princes* Jungh.	+	-	-	-	+
162.	*Stereum rugosum* Pers.	-	-	-	-	+
163.	*Stereum sanguinolentum* (Alb.and Schw.) Fr.	+	+	+	-	-
164.	*Stereum scytale* Berk.	-	-	-	-	+
165.	*Subulicystidium longisporum* (Pat.) Rom.	-	+	-	-	-
166.	*Tomentella botyroides* (Schw.) Bourd.& Galz.	+	-	-	-	-
167.	*Tomentella cinerascens* (Karst.) Hohen and Litsch.	+	-	-	-	-
168.	*Tomentella coerulea* (Bres.) Hohen.& Litsch	-	+	-	-	-
169.	*Tomentella ferruginea* (Pers.) Pat.	-	+	-	-	-
170.	*Tomentella griseoumbrina* Litsch	-	+	-	-	-
171.	*Tomentella muricata* (Ellis. & Everh.) Wakef.	+	-	-	-	-
172.	*Tomentella indica* Rattan sp. nov.	+	+	+	-	-
173.	*Tomentella pilosa* (Burt.) Bourd. & Galz.	+	-	-	-	-
174.	*Tomentella stuposa* (Link.) Stalpers	+	-	-	-	-
175.	*Tomentella radiosa* (Karst.) Rick.	+	+	-	-	-
176.	*Trametes badia* (Berk.) Cooke	-	-	-	-	+
177.	*Trametes corrugata* (Pers.) Bres.	-	+	+	+	-
178.	*Trametes cubensis* Mat.	-	-	-	+	-
179.	*Trametes gibbosa* (Pers.) Fr.	-	-	-	-	+
180.	*Trametes incerta* (Currey) Cooke	-	-	-	+	-
181.	*Trametes mollis* (Willd.) Fr.	-	-	-	-	+
182.	*Trametes ochracea* (Pers.) Gilb.	-	-	-	-	+
183.	*Trametes serialis* fr.	-	-	+	+	-
184.	*Trametes trogii* Berk.	+	-	-	-	-
185.	*Trametes versicolor* (L: Fr.) Pilat	+	+	+	+	+
186.	*Trechispora alnicola* (Bourd. & Galz.) Liberta.	+	-	-	-	-
187.	*Trechispora confinis* (Bourd. & Galz.) Liberta.	+	-	-	-	+
188.	*Trechispora farnica* (Pers. ex Fr.) Liberta.	+	-	-	-	-
189.	*Trichaptum abietinium* (L: Fr.) Quel	+	+	+	+	+

Contd.

Table 1: Contd.

S. No.	Names	*Abies pindrow*	*Cedrus deodara*	*Picea morinda*	*Pinus wallichiana*	*Quercus semicardifolia*
190.	*Trichaptum biformis* (Fr. In Kl.) Ryv.	+	-	+	+	+
191.	*Tubulicrinis chaetophora* (Hohn.) Parm.	-	+	-	-	-
192.	*Varraria vassilevae* Parm.	-	+	-	-	-
193.	*Varraria pallescens* (Schw.) Rog. & Jacks.	-	+	-	-	-
194.	*Varraria rhodospora* (Wakef.) Cunn.	-	+	-	-	-
195.	*Xylobolus ahmadii* (Biodin.) Biodin.	-	+	-	-	-
196.	*Xylobolus frustulatus* (Pers. ex Fr.) Biodin.	-	+	-	-	+
197.	*Xylobolus subpiletus* (Berk. & Curt.) Biodin	+	+	+	-	+

REFERENCES

Afyon, A., Konuk, M., Yagiz, D. and Helfer, S. (2005). A study of Wood Decaying Macrofungi of the Western Black Sea region of Turkey. *Mycotaxon.* **93:** 319-322.

Bagchee, K.D. and Bakshi, B.K. (1954). Studies on Indian Thelephoraceae I. Some species of Stereum, Peniophora and Corticium. *Indian Forest Bulletin.* **166**: 1-11.

Baghchee, K.D., Puri, Y.N. and Bakshi, B.K. (1954). Principal diseases and decays of Oaks and other hardwoods in India-II. *Indian Phytopathology.* **7**: 18-42.

Bakshi, B.K. (1971). *Indian Polyporaceae.* ICAR, New Delhi, India.

Bakshi, B.K. and Bagchee, K.D. (1950). Principal diseases and decays of Oaks in India. *Indian Phytopathology.* **3**: 124-139.

Berkeley, M.J. (1850-54). Decades of fungi 25-30; *Hook J. Bot.* **2**: 1850: Decades of fungi 31-36; *Hook J. Bot.* **3**: 1851: Decades of fungi 37-40; *Hook J. Bot.* **4**: 1852; Decades of fungi 41-50; *Hook J. Bot.* **5**: 1854.

Crous, P.W. (2006). How many species of Fungi are there in tip of Africa. *Studies in Mycology.* **55:** 13.

Demirel, K. and Uzun, Y. (2002). Macrofungi of Agriprovince. *Turk. J. Bot.* **26:** 291-295.

Erikson, J. (1958). Studies in the Heterobasidiomycetes and Homobasidiomycetes –Aphyllophorales of Muddus national park in North Sweden. *Symb. Bot. Upsal.* **16:** 1-172.

Fadime, Y. and Mustafa, I. (2002). Macrofungi of Degirmangobazy (Balikesir). *Turk. J. Bot.* **26:** 161-164.

Hawksworth, D.L. (2004). Fungal diversity and its Implifications for Genetic Resource collections. *Studies in Mycology.* **50:** 19.

Imazeki, R., Kobayasi, Y. and Aoshima, K. (1966). Fungi: In *Flora of Eastern Himalayas.* pp. 611-626, (Ed.) H. Hara (Tokyo press, Tokyo, Japan).

Kaya, A. (2005). Macrofungi determined in Golbasi (Adiyaman) District. *Turk. J. Bot.* **29:** 45-50.

Khara, H.S. (1977a). Some stipitate Hydnums from North Western Himalayas. *Ind. J. Mycol. Pl. Pathol.* **7:** 127-134.

Khara, H.S. (1977b). The Hydnaceae of North Western Himalayas. *Indian Phytopatholohy* **30:** 94-95.

Lodge, D.J. (1995). A Survey of patterns of Diversity in Non-Lichinized Fungi. *Mitt. Eidgenoss. Forsch. Anst. Wald Schnee Landsch* **70(1):** 157-173.

Lodge, D.J., O'Dell, T.E. and Mueller, G.M. (2004). Approaches to Sampling Macrofungi: *Biodiversity of fungi-Inventory and Monitoring method*: 163-168.

Manoharachary, C., Sridhar, K., Singh, R., Adholeya, Suryanarayanan, T.S., Rawat, S. and Johri, B.N. (2005). Fungal Biodiversity: Distribution, Conservation and Prospecting of Fungi from India. *Curr. Sci.* **89(1):** 58-71.

Mass Geasteranus, R.A. (1971). Hydnaceous fungi of the Eastern old World. *Verhand. Kon. Nederl. Akad. Wetensch Natuurk. Ser.* **2:** 1-176.

Natarajan, K., Senthilarasuv, G., Kumaresan, V. and Taiana Riviere, (2005). Diversity in Ectomycorrhizal fungi of a Dipterocarp forest in Western Ghats. *Curr. Sci.* **88(12):** 1893-1895.

Peksen, A. and Karaca, G. (2003). Macrofungi of Samsun Province. *Turk. J. Bot.* **27:** 173-184.

Rattan, S.S. (1974). Scytinostroma in India with notes on extralimital species. *Trans. Brit. Mycol. Soc.* **63:** 1-12.

Rattan, S.S. (1977). The Resupinate Aphyllophorales of North Western Himalayas. (J. Cramer: Vaduz, Germany).

Rattan, S.S. and Khurana, I.P.S. (1978). The Clavarias of Sikkim Himalayas (J. Cramer: Vaduz, Germany).

Rehill, P.S. and Bakshi, B.K. (1965). Studies on Indian Thelephoraceae – Indian species of *Peniophora* and *Corticium. Ind. For. Bull.* **242:** 1-30.

Rehill, P.S. and Bakshi, B.K. (1966). Studies on Indian Thelephoraceae–The genus *Stereum*. *Ind. For. Bull.* **250:** 1-20.
Reid, D.A., Thind, K.S. and Adlakha, K.L. (1958). The Thelephoraceae of Mussoorie Hills. *Trans. Brit. Mycol. Soc.* **41:** 129-134.
Reid, D.A., Thind, K.S. and Chatrath, M.S. (1959). Polyporaceae of Mussoorie Hills. *Trans. Brit. Mycol. Soc.* **42:** 40-44.
Sharma, J.R. (1995). *Hymenochaetaceae of India*. Botanical Survey of India, Kolkata, India.
Sharma, J.R. (1999a). *Genera of Indian Polypores*. Botanical Survey of India, Kolkata, India.
Sharma, J.R. (1999b). *Indian Polyporoid Fungi*. Botanical Survey of India, Kolkata, India.
Sharma, J.R. (1996a). Wood rotting fungi (Aphyllophorales) from Sikkim. *Bull. Bot. Surv. Ind.* **34:** 89-99.
Sharma, J.R. (1996b). *Phellinus pachyphloeus* and its related species in India. *Bull. Bot. Surv. Ind.* **35:** 16-26.
Sharma, J.R. (1997a). Fungi. In: *Floristic Diversity and Conservation Strategies in India*. pp. 91-149, Eds. Mudgal, V. and Hajra, P.K. (Shiva Offset, Dehra Dun, India).
Sharma, J.R. (1997b). Wood rotting fungi of Temperate Timberline zone. In: *Floristic Diversity and Conservation Strategies in India*. pp. 151-194, Eds. Mudgal, V. and Hajra, P.K. (Shiva Offset, Dehra Dun, India).
Susan Metzler and Van Metzler. (1992). *Texas Mushrooms*, University of Texas press, Texas. pp. 349.
Thind, K.S. (1961). *The Clavariaceae of India*. ICAR, New Delhi, India
Thind, K.S. and Adlakha, K.L. (1956). The Thelephoraceae of Mussoorie Hills. *Indian J. Mycol. Res.* **2:** 57-64.
Thind, K.S. and Khara, H.S. (1968). The Hydnaceae of North Western Himalayas. *Indian Phytopath. Soc. Bull.* **4:** 25-33.
Thind, K.S. and Khara, H.S. (1975). The Hydnaceae of North Western Himalayas. *Indian Phytopathology*. **28:** 57-75.
Thind, K.S. and Rattan, S.S. (1968). The Thelephoraceae of North-Western Himalayas. *Indian Phytopath Soc. Bull.* **4:** 14-24.
Thind, K.S. and Rattan, S.S. (1970). The Thelephoraceae of India—the genus *Tubulicrinis* and *Hyphodontia*. *Proc. Indian Acad. Sci.* **7:** 118-131.
Thind, K.S. and Rattan, S.S. (1971). The Thelephoraceae of India—the genus *Tomentella*. *Indian Phytopathology*. **24:** 32-42.
Thind, K.S. and Rattan, S.S. (1972). Thelephoraceae of India. *Trans. Brit. Mycol. Soc.* **59:** 123-128.
Thind, K.S. and Rattan, S.S. (1973a). Thelephoraceae of India. *Indian Phytopathology*. **26:** 485-494 and 528-536.
Thind, K.S. and Rattan, S.S. (1973b). Thelephoraceae of India. *Mycologia*. **65:** 1250-1258.
Yesil, O.M. and Yildiz, A. (2004). Contributions to the Macrofungi flora of Batsman Province. *F.U. Fenve Muhendislik Bilimleri dergisi* **16(1):** 11-16.

Diversity of Lower Plants (2014): pp. 174-189
Editors: Rajan Kumar Gupta & Mukesh Kumar
IK International Publishing House Pvt. Ltd.

13

Biodiversity of Chilika Lake, East Coast of India

JNANENDRA RATH* AND SIBA PRASAD ADHIKARY**

*Department of Botany, Visva Bharati, Santiniketan, West Bengal
**Centre for Biotechnology, Visva Bharati, Santiniketan, West Bengal

ABSTRACT

Chilika is the largest brackish water lagoon in Asia covering an average area of 1065 km^2. The lagoon is an estuarine one and supports an unique assemblage of marine, brackish water and freshwater species. A distinct salinity gradient is observed in its different sectors which also varies with season. It is one of the hot spots of biodiversity in the country. Some rare, vulnerable and endangered species listed in the IUCN Red List of threatened species show at least a part of their life cycle in the lake. In Chilika, 706 species of angiospermic plants, 102 species of algae and phytoplankton, 217 species of fishes, 61 species of protozoa, 29 species of platyhelminthes, 37 species of parasitic nematodes, 31 species of polychaetes, 3 species of stomatopoda, 28 species of crustacea, 30 species of decapoda, 136 species of mollusca, 3 species of echicurans, one species of protochordata, 37 species of amphibians and reptiles, 156 species of birds and 118 species of mammals were recorded. Details of this unique ecosystem and its biodiversity has been presented.

Keywords: Biodiversity, Chilika lagoon, Fauna, Flora, Conservation measures.

1. INTRODUCTION

Chilika is the largest brackish water lagoon situated in the east coast of India between 19°28′-19°54′N and 85°06′-35′E covering the districts of Puri, Khurda and Ganjam of Orissa (Fig. 1). The pear-shaped lagoon has a maximum length of 64.3 km and average width of 20 km (Das and Samal, 1988) The water-spread area varies between 906 to 1065 km^2 during summer to

**Corresponding author: Email: adhikarysp@visva-bharati.ac.in

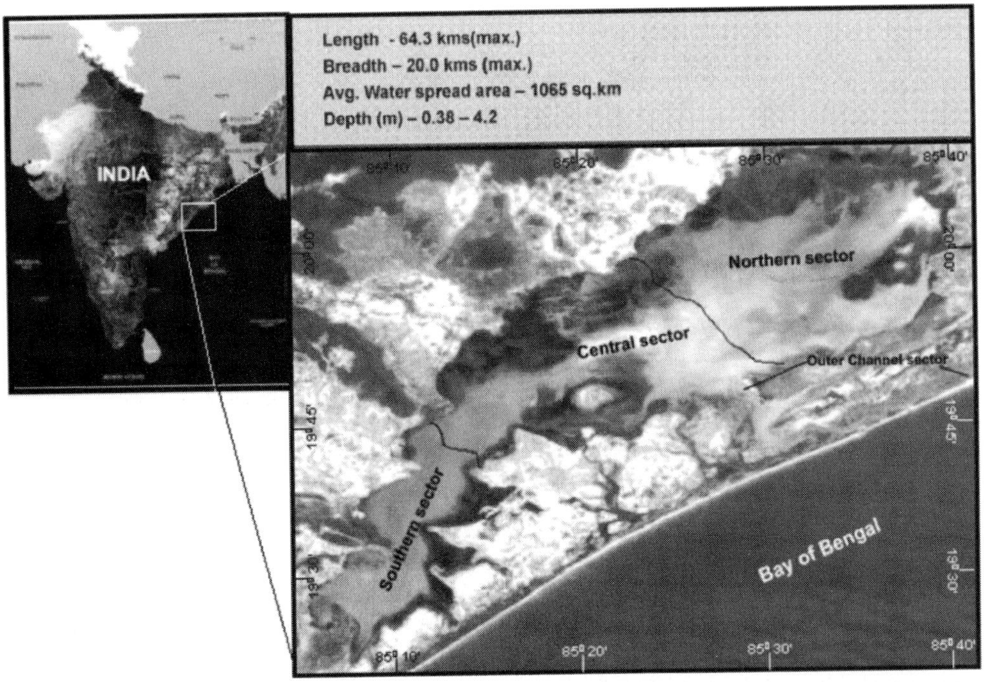

Fig. 1. Location map of Chilika lagoon in the East Coast of India. The artificial separation of the lake into four different sectors has been indicated in the map.

monsoon and the water depth, in general, fluctuates between 0.38 to 4.2 metres. Geographically the lake is circumscribed by: (i) rocky hills of eastern ghat along the western and southern margin (ii) a sixty km stretch of coastal barrier along the Bay of Bengal in the eastern side, and (iii) the deltaic plains of Mahanadi river in the northern part. Hammed in between the mountains and the sea, Chilika spreads itself out into a pear-shaped expanse of water. From about the middle line to the south, enormous rocks and emerald green islands jutting out of the water surface. Of these Badakuda, Sanakuda, Nalabana, Kalijai, Ghantasila, Chadheiguha, Arakhakuda and Kankadakuda, etc. deserve special mention. It is connected with the Bay of Bengal on its eastern side through an outlet, a 35 km long, narrow, outer channel connects the main lagoon to the Bay of Bengal near the village Motto, which cuts through liner spit that separates the lake from the sea. High tide near this inlet mouth drives in salt water through the channel during the dry months, from December to June. With the onset of the rain, the rivers fall into the Northern zone, causing freshwater currents that gradually push the seawater out. As a result of these dynamics, the inlet mouth constantly changes its position. The inlet channel is connected with Chilika at Magarmukh. The other connection with the Bay of Bengal is through Palur canal on the south-eastern side. Because of the peculiar hydrology of the lake, salinity ranges and gradients are not uniform but vary with site and season. During monsoon, the salinity is the lowest and highest salinity is observed in the late winter and summer. The varying salinity regime and depth divides the lake into four natural sectors, viz. northern sector, central sector, southern sector, and outer channel sector. The rivers Daya and Bhargavi fall into Chilika in the northern sector, which supplies major freshwater input to the lake during monsoon and post-monsoon period which contributes about 275,000 cusecs of freshwater to the northern sector. This zone is relatively shallow (0.5 to 1 m) due to sediment deposits carried by these two rivers and has a salinity ranging between 0 and 10 ppt. During the monsoon (July-September), the river Daya is in spate and the water level of the lake is the highest with an average depth of 3 metres. In this sector current generated by the river outflow is directed towards the sea. The central sector is deeper than the northern sector with medium salinity level 5 to 20 ppt and a depth of about 1.5 to 2.5 m. The southern sector is the deepest of about 2.5 to 3.5 m and has the highest salinity concentration with less fluctuation of salinity from 10 to 20 ppt. The river Rushikullya falls into the Bay of Bengal close to the Southern sector while the Palur canal connects it to the sea. The fourth zone is the main mouth of Chilika Lake along with the 35 km long channel called the outer channel (salinity up to 32 ppt). It starts from the lake at Magarmukh near village Satapada, and is connected to the sea near Arakhakuda. During the monsoon season, excess fresh water input from the rivers is discharged to the sea, while during the summer (April-June) when the water level is low, seawater flows into the lake. This unique spatial and temporal salinity gradient of Chilika Lake gives rise to a multitude of niches inhabited by a large diversity of plant and animal species hence is one of the hot spots of biodiversity of the country. The lagoon is the lifeline for more than 1.5 lakh stakeholders who live in and around the lagoon and rely upon the lake's supply of fish stock and other allied activities. Based on its rich biodiversity and socio-economic importance, Chilika has been designated as a wetland of international importance and included as one of the first two wetlands under the Ramsar Convention in 1981. It is included in the list of wetlands selected for intensive conservation and management by the Ministry of Environment and Forest, Govt. of India and also identified as a priority site for conservation and management by the National Wetland, Mangrove and Coral Reef Committee of the Ministry.

1.1 Origin and History of Chilika Lake

Chilika was a gulf of the Bay of Bengal during the Pleistocene period. The recession of coastline in the post-Pleistocene, aided by marine and fluvial activities gradually resulted in the present configuration. Whereas the wave action initiated the spit-building at the southern end and the delta-building process of the Mahanadi ensured sufficient sediment load the render. The lagoon is shallow at its north-eastern extremity. Many ancient texts mention the southern sector of Chilika as being a major harbour for maritime commerce back in the days when the kings of Kalinga was known as lord of the sea. This indicates that the area was once marine and was much deeper than it is today (Venkatratnam, 1970). Geological studies indicate the coastline added along the western shores of Chilika in the Oceane era, and that the entire north-eastern region of Chilika was under the sea. Since then the coastline has moved considerably eastwards (Chatterjee and Goswami, 1966) .

Most of the lagoons seen today were formed as a result of a worldwide rise in sea levels. There was pause in the rise in the sea as about 700 years ago, when a sandy beach might have been formed near the coast at the southern sector. As the sea rose further, this sand beach grew gradually and progressed seawards to the north east, to form what now the spit of Chilika. The partial closure of an estuary outlet to the sea possibly created a lagoon system Chilika. Recently a sub-fossil shell *Ostrea virginiana* from the southern-western edge of the spit has been dated to about 4000 years ago, which indicates that how long ago the lagoon was formed (Venkatratnam, 1970). Growth of the spit at Chilika is supposed to be due to abrupt change in the direction of coastline, long shore drift, constructive wave action, flat coastal gradient and the presence or absence of strong river and tidal currents in the different areas. The islands adjacent to the spit are believed to have been built partly by waves and partly by tidal currents as tidal deltas (Venkatratnam, 1966).

1.2 Catchment Area and Drainage Basin of Chilika

The total catchment area of Chilika is 3560 km^2 of which western catchment area of Chilika is 3,21,189 ha out of which 1,44,439 ha lies in Ganjam and 2,76,752 ha in Khurda district. The drainage basin of Chilika covers an area of over 4300 km^2. (ORSAC report, 1988), including about 1100 km^2 of the lake itself. The watershed boundaries lie between water flowing into the Mahanadi and Chilika in the north, while in the north-east area draining into the Bhargavi river make up the watershed. In the west and south-west, the watershed boundary lies between streams flowing into the Rishikullya river and those flowing into Chilika.

The catchment area of Chilika is quite heavily populated for the most part, and the population was about 2,00,000 in 2000. About 718 km^2 is covered by forests (526 km^2) and other permanent vegetation like plantation, however, 58 per cent of the vegetation is regarded as degraded. Agricultural area (mostly dryland) accounts for 2325 km^2, swamps 71 km^2 and wetland with grassy mud flats 91 km^2 occur mainly in the north-eastern part of the catchment area. About 52 km^2 are occupied by human settlements, roads, railways, etc.

1.3 Physiographic Features

Water, its quality, quantity and distribution over time is the most important factor affecting Chilika. A significant part of the fresh water and silt input to the lake comes from Mahanadi river system. The Daya and Bhargavi, which are distributaries of the Mahanadi, bring vast quantities of silt-laden water into the northern part of Chilika during the monsoon. In addition, the rivers Nuna, Ratnachira and Kani in the north and Malaguni, Dhanua and Salia and numerous smaller streams enter the lake all along the western margin (ORSAC report, 1988). The total silt added to the lake has been estimated to be 13 million tonnes every year. The lake shows an overall increase in depth of about 50 cm due to the monsoon (Sahu, 1988) and fresh water flows towards the sea, overcoming the weak tidal currents at the lake mouth. The lake drainage basin is estimated to send 1760 million cubic metres into the lake every year, while the Daya and Bhargavi rivers are believed to contribute a much larger quantity of freshwater annually.

1.4 Climate

Chilika's location on the coast in the tropical zone spares it from extreme temperatures. The mean annual temperature is about 26.6–27.7°C, with a winter minimum of 21°C and a summer maximum of about 35°C (Patnaik, 1971). It rains about 1450 mm per year at Chilika, and the rainfall over the area varies from 1200-1600 mm. Most of the rain (80 per cent) occurs during the monsoon season between June-September. The lake area proper experiences mean wind speed ranging from 5 kmph (winter) and 10-15 kmph in the summer. However, the coastal areas experience higher average wind speed up to 25 kmph.

1.5 Salinity Fluctuation in Chilika Lagoon

Chilika undergoes a cyclical variation in salinity throughout the year, with different patterns seen in different sectors. It is this periodicity in salinity that allows fresh water as well as marine and estuarine species to thrive in the lake.

During monsoons (July to September) River Daya is in spate and water level in the lake is at its peak (about 3 m depth). In the northern sector the main currents are generated by the river in flow and directed towards the sea. They are strong enough to overcome tidal currents in the mouth. Therefore, during monsoon, the northern and central sectors show low salinity (1.8-2.0 ppt) and in the mouth region 3.5 ppt. In the southern sector, salinity remains higher (6-10 ppt) as the fresh water input is much lower in this part of the lake. (Fig. 2). In the

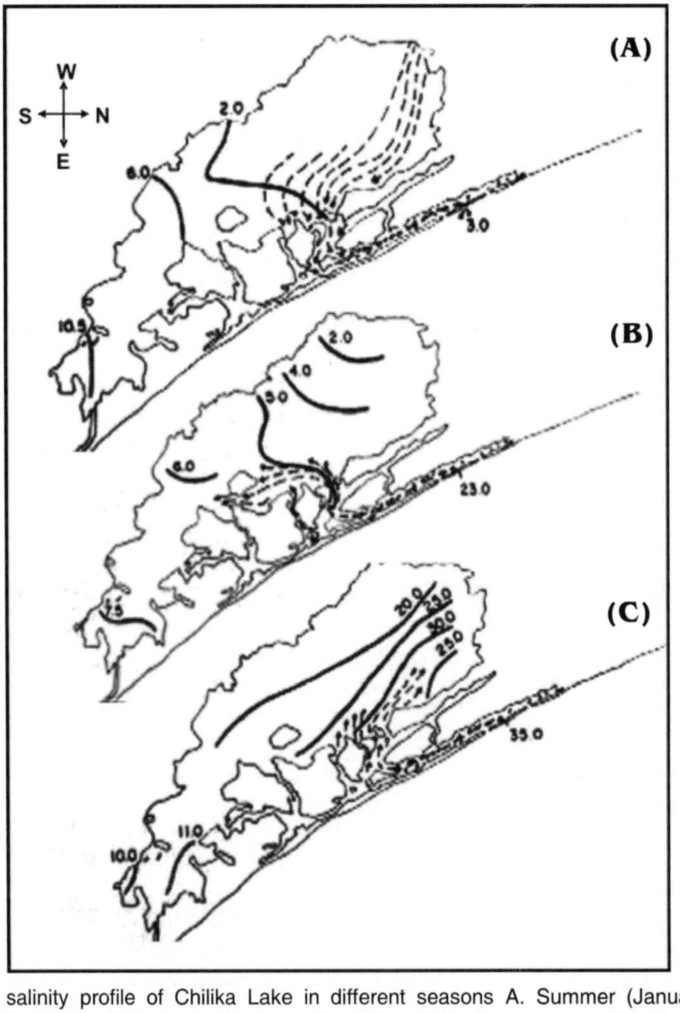

Fig. 2. Surface salinity profile of Chilika Lake in different seasons A. Summer (January-March), B. Rainy (April- June), C. Winter (July-September). = *Surface salinity %,* ——— = *Direction of water current.*

post-monsoons (October to December) the salinity remains low in the northern sector (1.8–4.5), increases slightly in the central sector and in the mouth (5 ppt) but decreases in the southern sector (7-8 ppt) as a result of slow mixing with the rest of the lake (Ramanadhan *et al.*, 1964; Banerjee and Roychoudhury 1966; Mohanty, 1975). The outer channel shows the lowest salinity as freshwater from the body of the lake continues to flow through it (Sarkar, 1977).

During winter (January-March), almost no fresh water enters the lake. Northerly wind send tidal water into the outer channel which increases the salinity. In the body of the lake, a fairly small range of low salinity exists, ranging from 2-4 ppt (northern sector), 4-6 ppt (central sector) and 6-7.5 ppt (southern sector).

During the summer, evaporation and prevailing southern winds cause greater salinity all over the lake. The northern and central sectors turn mesohaline with salinity ranging from 5.5–33 ppt and 7.5–27 ppt respectively. The southern sector shows a slight increase in salinity (10-11 ppt) (Ramanadhan *et al.*, 1964; Banerjee and Roychoudhury, 1966; Mohanty, 1975). The outer channel is practically like seawater at this point, with salinity at 34-35 ppt. (Sarkar, 1977).

Due to increased silt load, narrowing of the lake mouth and other factors, the overall average salinity for the whole lake has dropped from 22.31 ppt in 1957-1958 to 13.2 ppt in 1960-1961, and to 9.14-11.83 ppt in 1961-1964, but appears to have stabilized after the rapid drop. Such a decrease in salinity would obviously have grave implications for the biodiversity as well as fisheries of Chilika. Recently a new mouth was created opposite to village Sipakuda along the outer channel and a lead channel of 200 metre wide near Magarmukh over a length of 3.2 km. After the desalination of the outer channel as well as the opening of the new mouth increase in salinity flux by 40% and tidal flux by 45% into the lagoon was observed.

2. BIODIVERSITY OF CHILIKA LAKE

Chilika is one of the hot spots of biodiversity in India. It's partly fresh and partly saline water character and presence of a variety of habitats in and around the lake encourage a large number of plant and animal species in the lake.

2.1 Flora

2.1.1 Aquatic vegetation

Emergent macrophytes

Emergent plants such as *Scripus littoralis*, *Eleocharis plantaginea* and *Cyperus* sp. grow almost throughout the year on the north and northwest margins of the lake (Patnaik, 1973a). The most conspicuous marsh plant is *Phragmites karka* which grows mainly in the north-western margins.

Floating macrophytes

The free-floating plants *Eichhornia crassipes* (water hyacinth), *Pistia stratiotes* and *Azolla pinnata* come along with floodwaters in the monsoon. They remain in small patches in the eastern part of the northern sector and die out in the February-March as salinity increases (Patnaik, 1973a).

Other floating plants of seasonal occurrence, seen on the north-western margins during November-March are *Hydrilla verticillata, Vallisneria spiralis, Ceratophyllum demersum* and *Polygonum* sp., *Ipomeo* sp., *Jussieua* sp. and *Marselia* sp.

Submerged macrophytes

Two submerged plants dominate Chilika lake, namely, *Potamogeton pectinatus* and *Najas indica. Potamogeton pectinatus* is locally known as Charidal. This plant tolerates a wide range of salinity. Charidal-made up 78% of the total weed cover of the lake in 1973. From a position of virtual absence in 1932, an area of 20 sq. km was covered by *Potamogeton pectinatus* in 1973 (Patnaik, 1973a). This had increased to about 60 sq. km by 1977, 100 sq. km in 1980, up to 180 sq. km in 1985 (Fig. 3) and still further by 2002. *Potamogeton* dominated areas made up about 125 sq. km (Chatrath, 1992). Although weeds play an important food sources for certain species and support higher concentration of benthic fauna, this unchecked spread of weeds can be disastrous for Chilika.

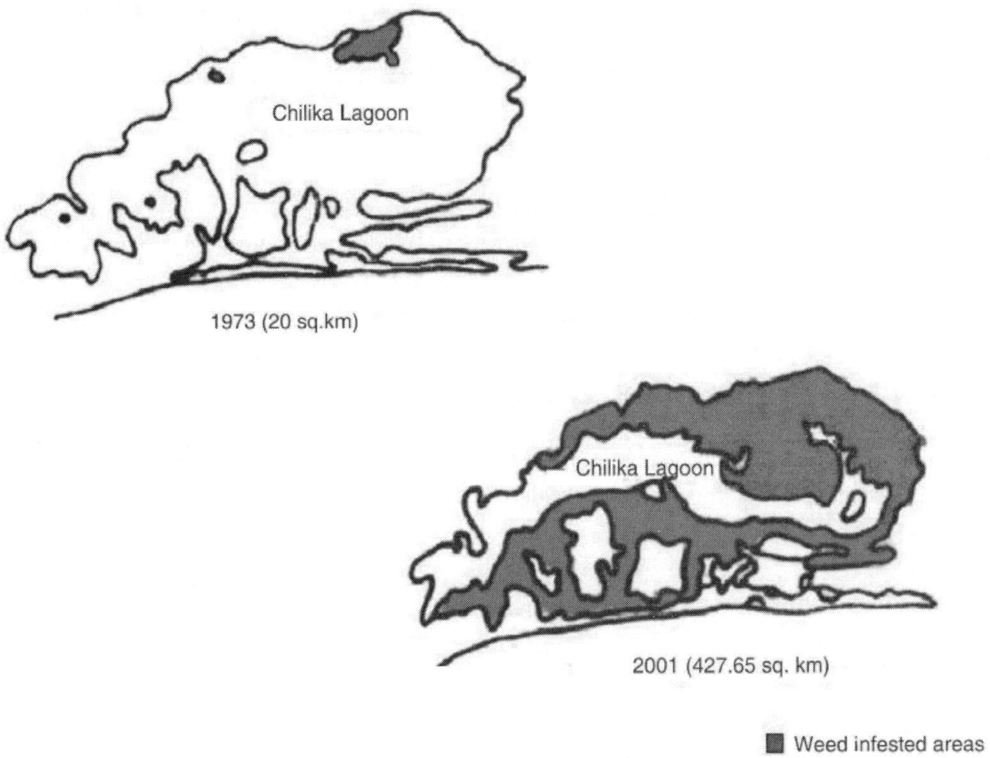

Fig. 3. Map showing weed infested areas of Chilika Lake during 1973 and 2001.

The other dominant submerged plant is *Najas indica,* which was estimated to form 14% of the weed cover in 1973 (Patnaik, 1973a). Locally known as Chinguria Dala the plant grows in patches mixed with *P. pectinatus*. Other submerged plants seen in Chilika include *Halophila ovata* (7% of weed cover, locally known as Kamari), *H. beccarii* and *Ruppia maritime.*

While macrophytes dominate large parts of the northern and southern sectors, the outer channel and deeper parts of the central sector are relatively free of them. According to satellite image weed coverage areas in 1973, 1977, 1985 and 1993 were 20, 60, 200 and 368 km^2. respectively.

2.1.2 Non-aquatic vegetation

Depending on the habitat conditions, the non-aquatic vegetation of Chilika can be described under two categories—littoral scrub jungles and the vegetations on the sand dunes.

Littoral scrub jungles

The mangroves which were once plentiful towards the margin of the lake and its uninhabited islands like Badakuda, Sanakuda, Cherriakuda and Ghantasila, etc. are no longer found due to ruthless cutting. The only vegetation now exist are the littoral scrub jungles along the rocky faces, representing the remnants of one time dominant tidal swamp forests. The vegetation is completely devoid of arboreous species of mangroves through some of their associates like *Aegiceras corniculatum, Salvadora persica, Pongamia pinnata, Cassipourea ceylanica*, etc. could be observed in denuded condition giving the appearance of low shrubs. Other important species constituting this type of vegetation are *Azima tetracantha, Colubrina asiatica, Carmona retusa, Carissa spinarum, Opuntia dillenii*, etc. Climbers/twiners are fairly common like *Cissus quadrangularis, Pentatropis capensis, Capparis roxburghii, Capparis zeylanica, Derris trifoliate* are quite abundant. *Crateva adansoni, Lepisanthes tetraphylla,* species of *Ficus* and *Streblus asper* are the other tree species found.

The ground flora composed of a large number of herbaceous members like *Cressa cretica, Indigofera asplathoides, Heliotropium curassavicum* forms the vegetation in such habitats. *Solanum trilobatum* is fairly common among bushes and scrub jungles (Panda and Patnaik, 1988).

Vegetation on sand dunes

The scorching sun, shifting sands and salt-laden winds on the beaches, sandy islands and the spit support a variety of plants including broad-leaved runners, herbs, shrubs, trees and reeds and sedges along the water's edge. *Hydrophylax maritime, Ipomoea pescaprae* are some of the broad-leaved runners on the dunes. Other dominant herbaceous elements constitute *Cyperus arenarium, Bulbostylis barbata, Hedyotis graminifolia, Becopa monnieri, Crotalaria nana, Gisekia pharnaceoides*, etc. *Pandanus odoratissimum* attains very large dimension due to congenial edaphic factor and is found in a continuous belt along the fringes of the lake. A stretch of land lying in between the Bay of Bengal and Chilika Lake has been planted with *Casuarina equisetifolia* which exhibit luxuriant growth and adequate natural regeneration. Most of these plants found are salt-tolerant. The edges of the island have several marshy plants. Extensive flowering occurs at different times of the year, and the island appears to change colour with the flower (Panigrahi, 1988; Jhingran, 1965; Patnaik, 1973b).

2.1.3 Algal flora

The algal flora of Chilika Lake and its distribution play a very important role in the brackish water environment. Through the number of forms recorded were few, their distribution in space and time was greatly influenced by the peculiar hydrographic conditions of the lake. The algae found in Chilika include benthic algae and those associated with aquatic weeds. Twelve species of Cyanophyta, twenty-three species of Chlorophyta and four species of Rhodophyta have been reported to constitute the benthic algal population (Rath and Adhikary, 2002). *Chaetomorpha linum* is the most dominant and widely distributed algae in the lake followed by *Enteromorpha intestinalis* and *Gracilaria verrucosa.*

Northern sector is the largest and shallowest and showed great fluctuation of salinity (0.31 to 12.24%) because of river discharge, and that's why some red algae like *Polysiphonia, Ceramium* and *Gracilaria* were not growing in this sector but found in rocky islands of southern sector. However, forms like *Chara, Nitella, Anabaena, Spirulina,* etc. have been recorded from comparatively freshwater regions of northern sectors. In the central sector the rocks and hillocks which remain submerged had good growth of algae like *Gracilaria verrucosa, Ceramium* sp., *Enteromorpha intestinali, E. compressa* and *Chaetomorpha linum.* The Nalabana Island of this sector which gets submerged during monsoon develop good quantity of algae like *Spirogyra, Enteromorpha* and *Chaetomorpha.*

The southern sector was comparatively small and deep and showed the least salinity variation. The transparency was comparatively high and highest algal density was obtained during January to March. In this sector *Lyngbya aestuarii* was dominant throughout except during October to December when *Chaetomorpha linum* was the dominant. The rocky islands like Badakuda, Sanakuda, Bird's Island, Somola, etc. of this sector had luxuriant growth of *Enteromorpha, Ceramium, Polysiphonia* and *Gracilaria* on the submerged rocks.

The outer channel sector being the link between the sea and the main lake, tidal effect was felt and during monsoon period floodwater flushed out to the sea through this channel. The algal flora in general is comparatively poor because of constant to and fro flow of water and fluctuation in salinity. However, in the sheltered areas of islands, there was good growth of *Enteromorpha, Chaetomorpha, Gracilaria* and *Lyngbya* species.

Phytoplankton of Chilika Lake consisted of fresh water, brackish water and marine forms which appear and disappear on the salinity conditions. The phytoplankton population in Chilika Lake was constituted mainly by diatoms, blue-green algae, green algae and dinoflagellates. The phytoplankton in the lake as a whole showed two peak seasons. The principal one being from March to June (summer). The second peak season of short duration in November (winter) was possibly due to rapid vegetative multiplication consequent on the rise of salinity and richness of nutrient in the water after monsoon. Out of the total 89 species of phytoplankton found in the lake, 47 species frequently occurred. These belong to the division Cyanophyta, Chlorophyta, Bacillariophyta and Dinophyta, and the numbers of species under these groups were 5, 5, 34 and 3 species respectively. *Cocconis pediculus, Gyrosigma acuminata, Mastigloia minuta, Pinnularia alpina, Pinnularia nobilis* and *Pleurosigma normanii* all belongs to Bacillariophyta and *Lyngbya aesturii* under Cyanophyta group were found occurring in all the sectors of the lake throughout the year. Sector specific distribution of the plankton species was also observed only in the outer channel sector of the lake. These species are *Bacterastrum furcatum, Chaetoceros ebenii* and *Chaetoceros paradoxus* these belong to Bacillariophytes occuring in the Outer channel during summer and winter seasons, however, were absent in the rainy season showing their preference to higher salinity levels. *Bacterastrum furcatum* was especially occurring in higher number in the outer channel during summer season when the salinity exceeds 30 ppt, however, three other diatoms *Chaetoceros affinis, Coscinodiscus centralis* and *Odentila mobilensis* occurred in maximum number in the outer channel of the lake during summer season. Their biomass in terms of number of cells/litre was 1421, 1372 and 1256, respectively, however, in the other two seasons another Bacillariophyte member *Coscinodiscus centralis* was the most dominant plankton in the sector with its number being 1170 and 520 no/litre during winter and rainy season respectively. Distribution of phytoplankton and their relative occurrence in different sectors showed a wide range of variations in physiochemical features and the habitat preferences of these organisms. In the southern sector *Coscinodiscus centralis* was also dominant in the summer season, but in the rainy season *Anabaena flos-aquae* and in the winter *Pleurosigma normaniii* were the most dominant species showing the number 314, 221 225 no/litre respectively. In the central sector also higher number of *Coscinodiscus centralis* was found in the water column in the summer season (285 no/litre) and in the rainy season number of *Anabaena flos-aquae* rose to 1081 no/litre. However, in the winter another diatom species *Thalasionema nitzchiodes* (176 no/litre) was dominant. In the northern sector *Anabaena flos-aquae* was the dominant plankton in the rainy season (1684 no/litre). In the summer and winter season the diatom *Cylindrothica clostratum* was the most abundant organism.

Regarding the total biomass of phytoplankton in the Chilika Lake it was highest during summer (17,748 no/litre) followed by in winter season (15,005 no/litre) and least in rainy season (14,051 no/litre). With respect to quantity of biomass of phytoplankton in different sectors, their number was highest in outer channel sector (20,556 no/litre) followed by in the northern sector (10,194 no/litre), central sector (8300 no/litre) and lowest in southern sector of the lake 7754 no/litre. In southern and outer channel sectors with increase in salinity biomass of phytoplankton increases but in the northern and central sectors with increase in salinity phytoplankton biomass decreases which indicates that in outer channel and southern sectors marine phytoplankton were dominant and in contrast in Central and Northern sector freshwater phytoplankton were dominant. Though there are few exceptions of dominance of a few particular species in the respective seasons, only in the Outer channel sector the number of phytoplankton was highest during the summer season. The phytoplanktons which occur in the northern sector were mostly dominated by a blue-green alga *Anabaena flos-aquae* and the number of diatom in this sector was quite low. The heterogeneity in the composition of phytoplankton in different sectors during the same period is explained by the fact that most of the phytoplankton populations in Chilika Lake are allochthonous in origin.

2.2 Fauna

The pioneering work on the fauna of the lagoon was carried out by Annandale and Kemp (1915a, 1915,b 1916) and Annandale (1922, 1926). They have recorded their findings in 13 volumes published in the form of memories of Indian museum. This is the first comprehensive report on the faunal resources of Chilika lagoon, which has not been further followed up in systematic manner. Several organizations and individuals have subsequently worked on different groups of plants and animal species, prominent among those include Central Inland Fisheries Research Institute, Zoological Survey of India and Academic and Research Institutes in Orissa and Andhra Pradesh. These investigations have brought out information on various taxonomic groups of plants and animals in and around Chilika lagoon. Choudhury (1916 a,b,c; 1917,

1923) and Hora (1923) reported 118 species of fishes. The survey of Zoological Survey of India (1995) recorded 217 species of fishes, and in addition, 61 species of Protozoa, 29 species of platyhelminthes, 37 species of parasitic nematodes, 31 species of polychaetes, 3 species of stomatopoda, 28 species of *Brachyara crustacea*, 30 species of decapoda, 136 species of mollusca, 3 species of echicurans, one species of protochordata, 37 species of amphibians and reptiles, 156 species of birds and 118 species of mammals. The diverse fauna of this lagoon can be broadly described as pelagic, benthic and phytal based on their taxonomy, place of occurrence and ecological status. They can be grouped as zooplankton, zoo-benthos, phytal fauna, icthyofauna, avifauna and mammals. The most recent account of fauna of Chilika has been published by Zoological Survey of India (1995) documenting 800 species. Starting from Sarkar (1977), several investigations have been made dealing with the species composition and seasonal distribution of zooplankton in Chilika (Devasundaram and Roy 1954; Rajan, 1964; Jhingran and Natarajan, 1966; Patnaik, 1986). In total 26 faunal groups comprising 170 species have been identified from this lagoon. Copepods constitute the most dominant group having 43-70% of the total zooplankton communities.

2.2.1 *Benthic fauna*

Much work has been done on the systematics and ecology of zoo benthos of this lagoon. In total, 117 species of benthos have been identified (Sarma *et al.,* 1988; Subba Rao *et al.,* 1988). The prominent groups include foraminifers, nematodes, polychaetes, copepods, isopods, amphipods and gastropods.

2.2.2 *Phytal fauna*

The macrophytes and seaweeds of the lagoon provide conducive habitats for the growth of many phytal fauna. A total of 77 invertebrates occurring on macro algae and other macrophytes have been identified from the lagoon.

2.2.3 *Avifauna*

Chilika is known worldwide for its resident as well as migratory birds. It falls on the major fly route of the migratory birds. 205 species of birds have been recorded from this lagoon by Bombay Natural History Society (2002) of which 97 are intercontinental migrants coming from Caspian Sea, Baikal, Aral Sea, remote parts of Russia, Kirghiz Steppes of Mangolia, Central and South-East Asia, Ladakh and the great Himalayas. Some rare birds from Pakistan, Iran and Rann of Kutch also constitute a part of its avian fauna. The more important bird species of the lagoon are *Anser clypeata, Aytha ferina, Anas querquedula, Anas penelope, Anas fuligula, Anser strepera, Limnodromus semipalmatus* and *Eurynorhynchus pygmeus.* The other interesting and threatened species are Spoonbilled Sandpiper *Eurynorhynchus pygmeus* and the Asian dowitcher *Limnodromus semipalmatus,* Spotbilled Pelican *Pelecanus philippensis*, Dalmatian Pelican *Pelecanus crispus* and Pallas Fishing Eagle *Haliaeetus leucoryphus.* Each year more than one million migratory shore birds, ducks, terns congregate in large numbers at Chilika for feeding and resting.

2.2.4 Mammalian fauna

The mammalian fauna of the lagoon include aquatic species like Dungong (*Dungong dungong*), Irrawady dolphin (*Orcaella brevirostris)* and many terrestrial ones such as *Antelope cervicapra, Axis axis, Hyena hyena* and *Canius aureus*. Considerable decline has occurred in the population density of *Antelope cervicapra, Axis axis,* and *Dungong dungong* and not spotted for quite some time. The Irrawaddy dolphin is a somewhat elusive species, found in various large rivers, bays and estuaries in South and South-East Asia. Its IUCN Red List status is 'undetermined', but suspected to be rare, vulnerable or endangered and it is listed in CITES Appendix II. The status of Irrawady dolphin in Chilika is not yet studied.

2.2.5 Fish fauna

Fisheries of Chilika have got an immense significance from the socio-economic point of view. Over 1.5 lakh fisher folk living in 122 villages in and around the lagoon earn their livelihood primarily from the fish and shellfish harvest. As per the survey by the ZSI-1995, a total of 217 species of fish from 147 genera, 71 families under 15 orders have so far been reported from the lagoon. The faunal composition of finfish has undergone many changes. Jhingran and Natrajan (1966), had identified 158 species, but Subha Rao *et al.* (1988) enlisted 159 species belonging to 83 genera and 48 families. Biswas (1995), on the other hand, reported 166 species of which 115 were brackish water, 24 limnetic and the rest 27 marine. Out of 166 species, 80 species have been listed as commercially important. The more important species are *Mugil cephalus, Mugil macrolepis, Liza parsia, Lates calcaarifer, Hilsa ilisha, Nematolosa nasusa, Pseudosiena coiber, Eleuthorenima tetradactylus, Mystus gulio, Silago sihama, Anabus testudineus, Chanos chanos, Tachysurus caelatus, Therapon jarbua, Liza troschelli* and *Gerres setifer.* A total of 30 species of prawns have been recorded from the lagoon of which, 6 species, viz. *Penaeus monodon, P. semisulcatus, P. indicus, Metapenaeus monoceros, M. affinis* and *M. dobsoni* are considered very important. Of late, occurrence of *Macrobrachium malcolmsonii* has also been reported. The crab population is represented by four species, viz. *Scylla serrata, Neptunus pelagicus, Varuna littoralis, Paratelphus* sp. Existence of edible oysters, viz. *Ostrea talpus, Meretrix* sp. and *Arca* sp. were also reported but have been wiped out due to the drastic fall in salinity.

3. PRIMARY PRODUCTION RATE

Primary production is measured by finding out the total amount of plant matter produced every year. However, phytoplankton primary production has not been measured at Chilika. Preliminary studies on the primary production of the lake indicate that the northern sector is much more productive than the rest of the lake (Asthana, 1979). Based on the observation at Rambha and Balugaon Misra *et al.* (1988) reported a bimodal pattern of production with a peak in May and the other in September. The central sector has comparatively higher benthic algal biomass than the northern sector probably due to lesser variation in salinity, higher transparency and abundance of sheltered areas.

4. THREAT TO THE LAGOON ECOSYSTEM

The lagoon is at present facing the problems like siltation, shrinkage of water spread area, chocking of the inlet as well as the outer channel connecting the sea, decrease in salinity, weed infestation, decrease in fishery resources and an overall loss of biodiversity.

4.1 Siltation

Siltation is attributed as most detrimental to the lagoon ecosystem. As per the monitoring done by the Chilika Development Authority recently 3.65 million cubic metres of silt is pumped in the lagoon through the 52 number of rivers and the rivulets. The catchment of the lagoon (excluding the Mahanadi drainage) is estimated to be 3, 21,189 ha. out of which 37% is arable land, 32% is fallow land, 26% is forest land and 5% is wasteland.

4.2 Shifting of the Mouth and Choking of the Inlet and Outer Channel

Various studies indicate that due to the littoral drift the mouth of the lagoon is shifting in north-eastern direction. The cross section of the outer channel is also significantly reducing due to shoal formation along the outer channel which leads to considerable hydraulic head loss and poor flushing action. The recent bathymetry of the Magarmukh (the gateway between lake and outer channel) indicates that the summer depth is alarmingly low, i.e., only 0.3 metre. This is preventing the discharge of fresh water and sediment during monsoon to sea through outer channel, thereby causing waterlogging in the adjoining villages in the northern sector. This is also reducing the prospect of auto-recruitment of the fish and other animals into the lagoon.

4.3 Shrinkage of Water-Spread Area

The water-spread area as calculated from Survey of India Toposheet (1972-73) was 824 sq. km. - (summer). The Land Sat-5 data of 25th May 1986 indicates the water-spread area to be 790 sq. km. The shrinkage of water-spread from 1973 to 1986 is from 824 to 790 sq. km. From the latest satellite imagery the water-spread area taken during May 2000 was computed to be 845 km^2.

4.4 Weed Infestation

Analysis of satellite data reveals that during the years 1973, 1977, 1985 and 1993 the weed covered area in the lagoon were 20, 60, 200 and 398 sq. km respectively. The weeds are emergent, floating and submerged type. The predominant weeds are *Potomogeton pectinatus* mostly found in central and fringes of southern sector. *Shoenoplectus littoralis* is predominant in northern sector.

4.5 Decrease in Fish-Landing

The species composition of fish in Chilika lagoon is complex in the sense that it includes freshwater, brackish, catadramous and anadromous fish species. A large percentage of the commercial fish species are migratory (either catodromous or anadromous). Jhingran (1963) estimated that 63-75% of the annual fish production was contributed by migratory species. This percentage is reduced to 30% due to relative increase in freshwater species (1995). A large percentage of the fish species feed off plant material and detritus. Because of this trophic charcteristic the Chilika lagoon is highly productive as estimated in terms of grams of carbon fixed on an aerial basis. As per calculation of Biswas, the annual fish production from Chilika average about 90 kg/ha during the 1980's (Range: 65-122 kg/ha).

The total fish landing on an average was 6,000 MTS but it is gradually declining. The fish landing of 1999-2000 was 1737 MTS. (Source: Fishery and Animal Resource Development Dept.) This decline can be attributed to composite factors like over-fishing beyond carrying capacity, and obstruction of migratory route of economic species (due to chocking of mouth and outer channel as well as at Magarmukh). The Orissa Marine Fisheries Regulation Act prevents operation of offshore vessels within 5 km of coast as the fish recruitment into Chilika Lake is mainly by spawners which concentrate within 5 km of restricted fishing zone. There is also provision for restriction of fishing in Palur Canal and outer channel. The decline is also attributed to decrease in salinity level of lagoon.

5. NOTES ON THE CHANGES OF BIOTA IN THE LAKE

As an estuarine lagoon, Chilika is destined to be converted into a marsh and slowly to dry land over geological time. However, a vastly accelerated rate of eutrophication is being seen in Chilika today caused by human activity. The major environmental changes in Chilika are:

(i) Shifting of the mouth and sand casting
(ii) Silting of the lake
(iii) Changes in overall salinity concentration of the lake
(iv) Aquatic weed infestation
(v) Downward trend of fish population
(vi) Changes in species composition of avifauna
(vii) Pollution of the lake

Several reptiles whose presence had been reported earlier (Annandale and Kemp, 1915a, 1922) were not seen in the ZSI survey from 1985-1987. These include crocodiles, sea turtles, gharials, etc. which causes concern. Most of these reptiles were threatened or endangered in India. Mangroves which were once plentiful found along the margin of the lake and on uninhabited islands, are not found now due to ruthless cutting. The other associated mangroves which are now threatened are *Cassipourea ceylanica, Derris indica, Aegiceras curniculatum*, etc. The agar producing red algae *Gracilaria verrucosa* which occurred in the southern and central sectors of the lake until two decades ago is now restricted to the Pathara region of the central sector and *Gracilaria lichenoides* was not observed in any part of the lake during 1980 (Mohanty, 1980). The total absence of *Gracilaria* sp. in the northern sector over years was probably because of siltation, comparatively high turbidity and decrease in salinity of the lake.

6. INSTITUTIONAL MEASURES FOR UPGRADING CHILIKA LAKE ENVIRONMENT

The Ministry of Environment and Forest is actively involved in conserving and improving Chilika and its environment. They have provided funds for soil conservation efforts in the drainage basin, habitat improvement of Nalaban Island, preparation of environmental education and publicity material on Chilika, and setting up an information database on the lagoon. In addition, the ministry is spending research on various aspects of Chilika under the Man and Biosphere and other programme. The ministry suspended the Chilika Aquatic Farms Ltd .located at Balugaon. Due to concerns over environmental impacts it is trying to get the lagoon notified under the Environmental Protection Act so that all major developmental activities would require environmental clearance. Chilika Development Authority was created in 1992 by the Ministry of Environment and Forest to look over all related issues on the lagoon. Recently an Interpretation Centre at Satapada and a Wetland Research Centre has also been established at Balugaon for long- term research on different aspects of Chilika Lake.

REFERENCES

Annandale, N. (1922). Introduction to the Study of the fauna of an island in Chilika lake. *Mem. Asiatic Soc. Bengal.* **7:** 257-287.

Annandale, N. (1924). Fauna of the Chilika lake: Mollusca Gastropoda (revision) Mem. Indian Museums, **5:** 853-873.

Annandale, N. and Kemp, S. (1915a). Fauna of Chilika lake: Mem. Indian Museums, **5:** 1-118.

Annandale, N. and Kemp, S. (1915b). Fauna of Chilika lake: Aquatic insects other than Coleoptrea with notes on some marginal species. *Mem Indian Museums*, **5:** 175-189.

Annandale, N. and Kemp, S. (1916). Fauna of Chilika lake: Mollusca, Gastropoda and Lamellibran. *Mem. Indian Museums.* **5:** 327-366.

Asthana, V. (1979). Limnological studies of lake Chilika, Orissa. Final Report. Indian Programme on Man and Biosphere (MAB). Project No. 112. Department of Science and Technology, New Delhi.

Banerjee, A.C. and Roychaudhuri. N.C. (1966). Observations on some physico-chemical features of the Chilika lake. *Indian J. Fisheries*, **13:** 359-429.

Biswas, K.P. (1995). Ecological and Fisheries development in wetlands: A Study of Chilika lagoon, India. Daya Publishing House, New Delhi, pp. 205.

Chatrath, K.J.S. (1992). The Challenge of Chilika. In: K.J.S. Chatrath (Ed.). Wetlands of India. Ashish Publishing House, New Delhi, 1992 pp. 135-169.

Chatterjee, A. and Goswami, S. (1966). Puri-Chilika coastal area. A geographic study. Geographical Review of India, pp. 18-33.

Choudhury, B.L. (1916a). Description of two new fishes from Chilika lake. *Rec. Indian Museums*, **12:** 105-109.

Choudhury, B.L. (1916b). Fauna of Chilika lake: Fish part I Mem. Indian Museums, **5:** 403-439.

Choudhury, B.L. (1916c). Fauna of Chilika lake: Fish part II Mem. Indian Museums, **5:** 441-458.

Choudhury, B.L. (1917). Fauna of Chilika lake: Fish part III Mem. Indian Museums, **5:** 491-508.

Choudhury, B.L. (1923). Fauna of Chilika lake: Fish part IV Mem. Indian Museums **5:** 711-736.

Das, N.K. and Samal, R.C. (1988). Environmental survey of Chilika. In: S.N. Patro (Ed.). Chilika, the Pride of our Wetland Heritage. Orissa Environmental Society, Bhubaneswar, pp. 96-103.

Devasundaram, M.P. and Roy, J.C. (1954). A preliminary study of the plankton of Chilika lake for the year 1950 and 1951. IPFC proceeding cymp brackishwater plankton in Indo-pacific Bangkok 25 and 26 January, pp. 48-54.

Jhingran, V.G. (1963). Report on fisheries of Chilika lake 1957-1960. Bulletin, Central Inland Fisheries Research Institute, Barrakpore.

Jhingran, V.G. and Natarajan, A.V. (1966). Final report on the fisheries of the Chilika lake (1957-65). Bulletin no. 8, Central Inland Fisheries Research Institute, Barrakpore.

Mishra, P.M., Sahu, B.K. and Panigrahi, R.C. (1988). Studies in the hydrobiological features of Chilika lake, Orissa. Final Report. Department of Environment, Gonernment of Orissa. Bhubaneswar.

Mohanty, P.R. (1980). Seaweed culture with particular reference to *Gracilaria verrucosa* (Huds) Papenfu in Chilika lake 1-11. In: Summer Institute on Brackish water capture and culture fisheries: 1-9 Central Inland Fisheries Research Institute (ICAR), Barrackpore West Bengal.

Mohanty, S.K. (1975). Some observations on the physicochemical features of the outer channel of the Chilika lake during 1971-73. Bulletin of the Department of Marine Sciences, University of Cochin, **VII**: 543-559.

ORSAC, (1988). Interim report on study of Chilika lake resources and environment. Orissa Remote Sensing Application Center. Bhubaneswar.

Panda, P.C. and Patnaik S.N. (1988). A contribution to the flora and vegetation of Chilika Wildlife Sanctuary, Orissa. In: Chilika, the pride of our wetland heritage, S.N. Patro (Ed.). Orissa Environmental Society, Bhubaneswar, pp. 63-80.

Panigrahi, G. (1988). Vegetation and flora of Chilika lagoon. In: Chilika, the pride of our wetland heritage, S.N. Patro (Ed.). Orissa Environmental Society, Bhubaneswar, pp. 63-80.

Patnaik, S. (1971). Seasonal abundance and distribution of bottom fauna of the Chilika lake. *J. Inland Fisheries Soc. India.* **8:** 38-48.

Patnaik, S. (1973a). A study of the aquatic plants of Chilika lake. *Proc. Nat. Acad. Sci. India.* **43B:** 53-65.

Patnaik, S. (1973b). Observation on the seasonal fluctuation of plankton in the Chilika lake. *Indian J. Fisheries* **20:** 43-56.

Patnaik, P.K. (1986). On the spatial distribution of zooplankton in Chilika lake (Brackish water lagoon) Geobios News Reports, **5:** 31-34.

Rajan, S. (1964). Environmental studies of the Chilika lake. Feeding spectrum of fishes. *Indian J. Fisheries Section A,* **2:** 521-532.

Ramanadhan, R., Reddy, M.P.M and Murty, A.V.S. (1964). Limnology of the Chilika lake. *J. Marine Biological Association of India* **6:** 183-201.

Rath, J. and Adhikary, S.P. (2002). Growth and spectral characteristics of four algae and agar yield of *Gracilaria verrucosa* at different salinity gradients of Chilika lake. *Seaweed Research and Utilization* **24:** 41-46.

Sahu, B.N. (1988). Geomorphology, evolution and recession of Chilika lake. Abstract in Seminar on Conservation and Management of Chilika. Department of Science, Technology and Environment, Government of Orissa. Bhubaneswar.

Sarkar, S.K. (1977). Further studies on seasonal and spatial variations of salinity in Chilika lake. *J. Inland Fisheries Soc. India* **2:** 1-9.

Sarma, A.L.N., Rao, D.G, Satapathy, S. and Pattanaik, P.K. (1988). Invertebrate community ecology in the context of conservation and development of Chilika lagoon (Bay of Bengal). Abstract, National Seminar on Conservation and Management of Chilika. Department of Science, Technology and Environment, Government of Orissa, Bhubaneswar.

Subha Rao, M.V., Sekhar, P.S.R., Prasada Rao, P.V.S.V. and Rao, K.K. (1988). Ichthyofauna of Chilika lake. Abstract. National Seminar on Conservation and Management of Chilika. Department of Science, Technology and Environment, Government of Orissa, Bhubaneswar.

Venkatratnam, K. (1966). Formation of the spit and the development of the Chilika lake, a coastal lagoon. Int Indian Ocean Exp WEWSI, New Delhi.

Venkatratnam, K. (1970). Formation of the barrier spit and other sand ridges near Chilika lake on the east coast of India. *Marine Geology* **9:** 101-116.

Diversity of Lower Plants (2014): pp. 190-195
Editors: Rajan Kumar Gupta & Mukesh Kumar
IK International Publishing House Pvt. Ltd.

14

Strategies for Sustainable Biodiversity Conservation

SUSHMA GUPTA AND RAJAN KUMAR GUPTA

Department of Botany, Govt. P.G. College, Rishikesh (Uttarakhand)

ABSTRACT

Biodiversity represents the very foundation of human existence. Biodiversity impoverishment is a common phenomenon all over the world. International communities and government has taken various measures to promote the sustainable biodiversity conservation. Besides national, regional and global conventions and programmes such as conventions concerning the protection of the world culture and national heritage sites, conventions on International trade in endangered species of wild flora and fauna (CITES), etc. National Parks, biosphere reserves, *ex situ* conservation of species in gene banks, curbing contamination on biosphere are some of the measures for sustainable conservation of biodiversity. Biodiversity conservation network is an innovative US-AID funded programme working for Asia and Pacific regions for community based conservation enterprises. The world conservation strategy launched by IUCN, UNEP and WWF emphasizes the global importance of conservation as well as social and economic requirements to achieve sustainable conservation.

The paper emphasizes the various strategies that monitor the biological resource base management and sustainable biodiversity conservation programme.

Keywords: Biodiversity, Protected areas, Biosphere reserves, Gene Bank.

1. INTRODUCTION

Biodiversity represents the very foundation of human existence. It is the vast array of all the species of plants, animals, insects and the microorganisms inhabiting the earth either in the aquatic or the terrestrial habitat. The estimated number of species on earth is supposed to be 5 to 80 million or may be more. Out of this only 1.4 million living species are described, where 750,000 are insects, 41,000 are vertebrates and 250,000 are plants, remainders are invertebrates, fungi, algae and other microorganisms (Wilson, 1988; Mc Neel *et al.,* 1990). The species richness increases from the pole to the equator. Tropical regions have highest richness of mammal species per unit area and vascular plant species diversity is much richer at lower altitude (Reid & Miller 1989). Tropical forests as well as Mediterranean climate regions are highly diverse ecosystems. Wetlands are among the most biologically productive ecosystems in the world. They help to regulate water flows and provide breeding habitats for many species of flora and fauna. In the tropics, countries such as Bangladesh, India, Thailand and Vietnam have lost over 80% of their wetlands (WRI 1990). According to Haven (1988) a quarter of the earth's total biodiversity is in serious danger of extinction. The situation is more alarming in the richest centre of biodiversity, i.e., Asia, Africa and South America, due to severity of threats and lack of authentic information. At present some 3956 species are endangered, 3647 are vulnerable and 7240 are considered rare.

The plant biodiversity is a finite world resource having immense economic and ethical value to mankind. Indian subcontinent is blessed with vast genetic diversity of plants comprising 11.9% of the world flora and designated as one of the eight-vavilovian centres of origin. Approximately, 49,219 higher plant species belonging to 141 genera exist here, of which about one-third are endemic and mostly confined to the Western Ghats, north-eastern regions, north-western Himalayas and Andaman and Nicobar Islands. Biodiversity of tropical regions is under greater risk of endangerment and extinction due to immense biotic and abiotic pressures. About 17 m ha of tropical forests are being cleared annually. At this rate roughly 5-10% of tropical forest species may face extinction within next 30 years. According to IUCN report (1999) as high as 7.7% of the plants in India are under threat of extinction.

Deforestation, the destruction of wetlands and other forms of habitat loss are threatening the stability of local and regional environment and wasting valuable resources. Regional, national and international actions, within agreed global targets, have been initiated since last fifteen years to halt and reverse the pattern of deforestation and habitat loss in all regions. The loss of biological diversity, expressed in the current rapid extinction of species and reduction of genetic variability, is an unnecessary waste of irreplaceable resources needed for sustainable development. Understanding the importance and the measures required for the conservation of biological diversity through educational and public awareness programmes by national and international organizations should be promoted and encouraged. Biodiversity conservation is an attitude towards the proper use and management of bioresources for human sustenance amidst plentiful of other planetary life forms. Linked with the community development and promotion of livelihood opportunities, the biodiversity conservation ensures involvement of local communities. The loss of biodiversity will restrict all socio-economic and environmental benefits and in long run, will compromise the ability of future generations to meet their needs and could cripple the genetic base required for the continued improvement and maintenance of currently utilized species.

Few major international moves are afoot to conserve biodiversity. The World Bank, the United Nations Development Programme (UNDP), and the United Nations Environment Programme (UNEP), established the Global Environmental Facility (GEF) in 1990. Under UNEP's auspices, more then 150 nations worked to negotiate a 'Convention on Biological Diversity" at "Earth Summit", in Rio de Janeiro, Brazil in 1992. The "Earth Summit" produced "Agenda 21", a plan for action on a number of issues, including biodiversity conservation in the 21st century.

2. RESPONSES

Following kinds of actions have been taken by the international community and by the government to promote the conservation and sustainable use of biological diversity.

2.1 Measures to Protect Particular Habitat as National Parks, Biosphere Reserves or Other Protected Areas

Protected areas provide a mechanism for conserving wild biodiversity and most countries today have established at least some protected areas. The protected areas in the world covered 4.9% of the earth's land surface area in 1990. Nationally protected areas have increased from 1478 sites to 6930 in 1990, i.e., about fivefold. The total area of these sites was 164 million ha in 1990. There are a total of 12 biosphere reserves established in the country for protection of certain ecosystems. Yellowstone Park in USA was first to be created in 1872 as a National Park for biodiversity conservation.

With the formation of World Conservation Union (IUCN) emphasis started to be laid on protection of "endangered species" threatened with extinction and those that are "endemic", i.e., confined to particular region of earth. The 4th world congress on National Parks and Protected Areas held in Venezuela in 1992 called to double the world's protected areas. About one hundred particularly valuable and biologically rich and sensitive areas are internationally recognized by UNESCO as "World Heritage Sites". Some of them are Redwood National Park, USA; Yellowstone National Park, USA; Sinharaja Forest Reserve, Sri Lanka; Nanda Devi National Park, India; Keoladeo National Park, India; Kaziranga National Park, India; Sunderbans National Park, India; Chitwan National Park, Nepal; Mount Taishan, China. Some important Indian Biosphere Reserves are Nanda Devi (Uttarakhand), Gulf of Manner (Tamil Nadu), Manas (Assam of India and Bhutan), Sunderbans (West Bengal), Thar Desert (Rajasthan), Rann of Kutch (Gujarat), Kajiranga (Assam), Nilgiri (Karnataka, Kerala and Tamil Nadu).

2.2 Measures to Protect Particular Species or Groups of Species from Over-Exploitation

The most appropriate way is to maintain the wild plants and animal species where they are in the wild habitat like forests, lakes and nature reserves such as biosphere reserves, national parks and wildlife sanctuaries or by introducing them back into agricultural, horticultural and animal husbandry practices so that they are cultivated and used by the farmers at large. With a view to further protect these areas, environmentally sound and sustainable development in the adjacent areas is promoted. Rehabilitation and restoration of degraded ecosystem and to promote the recovery of threatened species "inter alia", through the development and implementation of plans and other management strategies. Means are established to regulate, manage or control the risks associated with the use and release of living modified organisms resulting from biotechnology, which are likely to have adverse environmental impacts that could affect the conservation and sustainable use of biodiversity. The introduction of or eradication of those alien or exotic species, which threatens ecosystems, habitats or species are to be prevented.

Conventionally, *in situ* strategies are being followed for conservation and management of biodiversity. Enormous genetic diversity of crop plants has been maintained since ancient times by saving seeds or propagules for the next planting season by a wide variety of indigenously developed practices.

2.3 Measures to Promote *Ex situ* Conservation of Species in Botanical Gardens or in Gene Banks

Measures for the recovery and rehabilitation of threatened species and their reintroduction into their natural habitats under appropriate conditions and top reserve them in the form of seeds in seed bank (gene bank) or some other suitable materials like tissue cultures are promoted. Several *ex situ* conservation methods have been used depending on propagation method and storage behaviours of the species. Major global emphasis has been to conserve germplasm diversity *ex situ* in medium and long-term gene banks. The different approaches adopted for this include botanical gardens, herbal gardens, field gene banks, seed gene banks and cryogene banks.

Efforts to conserve plant species by establishing gene banks dates back to 1920s when Nikolai I. Vavilov established the first gene bank, known as Russian Vavilov Institute (VIR) in St. Petersburg followed by establishment of IBPGR in 1974, now known as International Plant Genetic Resources Institute (IPGRI). IBPGR has played a catalytic role in developing effective national and international crop genetic resource conservation efforts. Presently, more than a hundred gene banks around the world have come up. Gene bank is a particular type of genetic resource bank that preserves and distributes the DNA samples and provides associated information. The concept of DNA as a genetic resource has emerged out of the revolution in genomic information brought about by the analysis of DNA extracted from numerous plant species in laboratories across the world. The stored DNA not only safeguards the species threatened with extinction but are also actively utilized by the plant scientists and breeders to develop novel varieties.

Biotechnological tools have vast applications in biodiversity conservation and related activities of characterization and evaluation. These tools are used for *in vitro* mass-clonal propagation, *in vitro* conservation and cryopreservation of germplasm. This technology is extensively used for conservation on short and medium-term basis. *In vitro* technology has been applied to more than 1000 species, as a complimentary strategy for conservation of germplasm of different vegetatively propagated and recalcitrant seeded species. Sixty-three countries have tissue culture conservation facilities and the total number of accessions stored *in vitro* worldwide is approximately 38,000 (FAO, 1996). In India, *in vitro* conservation of vegetatively propagated, medicinal and aromatic plants is being undertaken at the National Gene Bank of National Bureau of Plant Genetic Resources (NBPGR). Presently, more than 1900 accessions are being maintained under slow growth conditions in the *in vitro* repository.

An area of current active research for long-term preservation of germplasm is cryopreservation of whole seeds or seed embryos. Cryopreservation reduces cell and tissue deterioration in storage by virtually halting metabolism and hence is an important tool for non-lethal storage of biological materials. It ensures infinite periods of storage (long-term conservation) and causes no change in viability, vigour and genetic make-up of the conserved materials. Cryobanking of orthodox seeds, vegetative tissues, pollen grains and biotech products is being undertaken in several countries. In India cryogene bank at NBPGR holds many thousands accessions of diverse crop germplasms.

3. INTERNATIONAL CONVENTION AND TREATIES

Several national, regional and global conventions and programmes have been formulated to implement the above measures. The Atlantic Treaty (Washington, 1959); the Convention on International Trade in Endangered Species of Wild Flora and Fauna (CITES, Washington, 1973); Convention on Biological Diversity (1992); Convention on the Wetlands of International Importance Especially as Waterfowl Habitats (Ramsar, Iran 1971); Convention on the Conservation of Migratory Species of Wild Animals (Germany 1979) ; The Convention on the Conservation of Antarctic Marine Living Resources (Tasmania 1980); The FAO International Undertaking on Plant Genetic Resources (Italy 1983), etc. are some important conventions. The World Conservation Strategy launched by IUCN, UNEP and WWF reinforces the global objectives of conservation and emphasizes the importance of social and economic requirements that can be met to achieve sustainable development. "Caring for the Earth" emphasizes that biodiversity must be conserved as a matter of principle, as a matter of survival and as a matter of economic benefits. UNEP in cooperation with FAO, UNESCO and IUCN is assisting an Intergovernmental Negotiating Committee established by UNEP's Governing Council to elaborate an international convention on the conservation and rational use of biological diversity.

The Biodiversity Conservation Network (BCN), a component of the Biodiversity Support Programme (BSP), is an innovative USAID-funded program working in Asia Pacific region to provide grants for community based enterprises that directly depend on biodiversity conservation. BCN is testing the hypothesis that if the local communities receive sufficient benefits from a biodiversity linked enterprises; they will act to conserve the resources it depends on. BCN funded projects integrate enterprise development with community organization and policy efforts needed to achieve conservation. BCN and World Wildlife Fund US (WWLUS) had completed a poster describing the conservation and development activities that it was supporting in and around National Royal Chitwan National Park, Nepal.

Several laws were enacted in India for the conservation of flora and fauna and their habitats. Some of them are: the Wildlife (Protection) Act, the Forest (Conservation) Act, Wildlife (Protection) Amendment Act, Conservation of Forest and National Ecosystem, etc.

4. RECOMMENDATIONS

Simple techniques should be developed for monitoring biodiversity, enterprise profitability and social structure of participation. The network needs to persuade local people that they should be helping to develop and implement system of monitoring. The indigenous and local people should be participating in all aspects of projects activities. The BCN needs to have direct contact with these peoples. Plans should be started for a larger enterprise impact by increasing the competitiveness of the private sector in non-timber forest products. The conservation network projects should be implemented in the areas where the infrastructure is poor and they attract relatively few private operators.

International support for preserving the biological resource base in developing countries is massive. Foreign assistance is pouring vigorously in this direction, it will thus help indigenous and local people conceptualize conservation projects, improve their capacity to articulate their needs and applying the foreign resources to meet those needs. Thus, the conservation network enables local people to raise their incomes by monitoring the sustainable, economic utilization of biological natural resources. It mobilizes local people to protect the resource base in their own interest, and provides technical assistance to the monitoring operations so essential to the sustainable use of biological resources for the benefit of overall development.

REFERENCES

FAO. (1996). Report on the state oh the world's plant genetic resources, Rome, Italy.

IUCN /UNEP/WWF. (1991). Caring for the Earth; a strategy for sustainable living, IUCN Gland.

McNeely, J.A. (1990). Conserving the world's biological diversity, IUCN, Gland.

Reid, W.V. and Miller, K.R. (1989). "Keeping options alive. The scientific basis for conserving Biodiversity", World Resource Institute (WRI), Washigton, DC.

UNEP. (1992). Saving our Planet, Challenges and Hopes, Nairobi, 1992.

Vavilov, N.I. (1951). The Origin, Variation, Immunity and Breeding of Cultivated Plants, Chronica Botanica, Vol. 13 pp. 1-364.

Vincente, M.C. and Anderson, M.S. (Eds.). (2006). DNA banks-providing novel options for gene banks topical reviews in Agriculture biodiversity, International plant genetic resources institute, Rome, Italy.

Wilson, E.O. (1988). Biodiversity, National Academy Press, Washington, DC.

WRI. (1990). World Resources 1991 Oxford University Press, New York.

Diversity of Lower Plants (2014): pp. 196-211
Editors: Rajan Kumar Gupta & Mukesh Kumar
IK International Publishing House Pvt. Ltd.

15

Phytocystatins: The New Age Healers

Garima Sharma*, Sandeep Sharma** and M.K. Sharma***

**Department of Botany, Hindu College, Moradabad, Bareilly (UP)*
***Department of Biochemistry, Aligarh Muslim University, Aligarh (UP),*
****Department of Botany, Sahu Jain College, Moradabad, MJP Rohilkhand University, Bareilly (UP)*

ABSTRACT

Plants contain varying amounts of protein proteinase inhibitors, since their discovery these inhibitors have been extensively studied particularly with respect to their significance, physiochemical properties as well as to explore their mechanism of action and inhibition of proteinases.

Cystatins are multifunctional glycoprotein thiol proteinase inhibitors, found both in plants and mammalian systems. These proteins specifically inhibit thiol proteinases like papain, bromelain, ficin of the plant system and cathepsin of the mammalian system. The name cystatin was given to the inhibitor by Barett, because of its distinct inhibitory activity against cysteine proteinases. Structurally, cystatins comprise an unusually diverse superfamily of proteins, ranging from small (type-I) cystatins without disulphide bonds, through the slightly large disulphide bonded (type-II) cystatins to the much more complex type III cystatins containing three divergent copies of type-I cystatin. This class of cystatins are found only in mammalian plasma. Cystatins of the plant system resemble with the properties of type I and type II cystatins of the mammalian system.

In comparison to much available information about the cystatins of the mammalian system, little information is available about thiol proteinase inhibitors of the plant system, which are called phytocystatins. This overview will consider phytocystatins with respect to their evolution, classification, distribution, properties, nutritional and physiological significance, mechanism of action and inhibition as well as importance of cystatins in health and disease.

Keywords: Phytocystatins, Thiol protease inhibitors, Plant cystatins, Cystatins and health, Cystatins in diseases.

*Corresponding author

1. INTRODUCTION, EVOLUTION AND CLASSIFICATION

Proteolytic enzymes or proteinases are of universal occurrence in the living systems. These enzymes catalyse the hydrolysis of susceptible peptide bonds in proteins. Discovery of proteolytic activity in some parts of the plant body was observed around two centuries back, when the presence of a meat tenderizing substance in the latex of papaya plant was noted (Monti *et al.*, 2004).

Proteases are customarily classified as exopeptidases when they hydrolyse only the N-terminal or C-terminal bonds of proteins and endopeptidases when they hydrolyse internal peptide bonds. They are also classified according to the presence of specific amino acid residues at their active sites and their mechanism of action. Proteinases of all the four mechanistic classes (serine, aspartic, cysteine and metalloproteinases) have been found in different tissues. The most studied plant proteinases belong to the cysteine class which is partly due to the long known enzymes papain present in the latex of the papaya plant. Cysteine proteinases are apparently more important then serine proteinases in the mobilization of proteins during germination of seeds and other processes. Control of the activity of these enzymes is certainly performed, at least in part, through the action of cysteine protease inhibitors of the papain which are found in seeds. Cysteine proteinase inhibitors of papain show low concentrations (0.3-6%) in the several seeds examined. This is in sharp contrast with the very wide variation in concentrations found for seed trypsin inhibitors. Furthermore, as compared to serine and trypsin protease inhibitors, papain inhibitors are apparently universally present in seeds (Rele *et al.,* 1980). It is also interesting to note that the levels of cysteine proteinase detected in the seeds seem to be comparable to the levels of their inhibitors in contrast with the different levels of serine proteinases and their inhibitors (Urwin *et al.*, 1995).

In cysteine proteinases, a cysteine residue, activated by the presence of histidine residue at the active site plays the role of nucleophile that attacks the peptide bond. An ideal example of these proteins is papain. The cysteine based active site arose independently at least twice during the course of evolution. The term protease in general includes both endopeptidases and exopeptidases whereas the term proteinase is used to describe only endopeptidases (Ryan & Clarence, 1990).

Some well-known proteinases in plants are papain, bromelain and ficin. These proteinases have been extensively studied particularly in germinating seeds and their activities have been associated with the events of seeding, growth and development. A number of proteolytic enzymes are present in leaves, flowers, fruits and their study is important because of their possible roles in turnover and senescence in the plant tissues. Proteinase and proteinase inhibitors when present in the same organism present a paradoxical situation. Yet it's the natures own way of controlling unwanted, unchecked and often distructive proteolysis with the help of protease inhibitors. Cysteine proteinase inhibitors (cystatins) is one such class of proteins that regulates the activities of cysteine proteinases.

The cystatin superfamily includes three families. Type-I cystatins (the stefins), type-II cystatins (the cystatins) and type-III cystatins (the kininogens). Type-I cystatin represent the simplest form of cystatins. These are single chain proteins and consist of about 110 amino acid residues (Mr.11000) with no disulphide bonds or carbohydrate content. Their presence has been reported in potatoes, barley and mammalian system (Monti *et al.,* 2004).

Type-II cystatins contain single chain proteins with two disulphide bonds towards the carboxyl terminus. These are made of upto 115 amino acid residues (Mr. 13000) and represent an intermediate class of cystatins present both in plant and the mammalian system. Several studies have shown that specific thiol protease inhibitors, cystatins, inhibit proteinases in a variety of nutritionally important foods such as rice and maize (Abe *et al.,* 1992).

The third cystatin family, the kininogens, are large molecules with three type-II like domains, bound carbohydrates and disulphide bonds, these are present only in mammalian plasma, placenta, etc. (Rashid *et al.,* 2005).

During the last decade, a new family (family-IV) belonging to cystatin superfamily has emerged named as phytocystatins. These are the cysteine proteinase inhibitors of the plant origin. Phytocystatins can be classified into two groups, class-I which possess single domain, comprised the majority of phytocystatins (Pernas *et al.,* 1998), and class-II possessing multiple domains, for example, multicystatin isolated from potato tubers and tomato leaves. Cystatins of family-IV include almost all plant cysteine proteinases inhibitors. This family of proteins are known to have characteristics that are found commonly in most family-I and family-II cystatins of the mammalian system. Homology searches show that some plant cystatins resemble family-II cystatins and some resemble family-I cystatins of the mammalian origin. Sunflower cystatins are 59-73% identical in sequence to other animal cystatins. This reinforces the common ancestory but divergent evolution of phytocystatins.

The classification of the cystatin superfamily mainly involves the analysis of cystatins from vertebrate sources, and it is based on the presence of one, two or three divergent copies of the cystatin like segments and presence or absence of disulphide loops. This classification is based on the evolutionary studies, utilizing the sequence comparison of a number of cystains from different sources. These studies have shown that cystatins evolved around 1000 years back as a precursor containing a disulphide loop. Then onwards the evolution preceded with a duplication of segments to give rise to more complex molecules. Since cystatins are a super family of cysteine proteases, several genes including CRES (cystatin related epididymal spermatogene), testatin and cystatin T have been identified that are related to family-II cystatins but lack critical consensus sites important for cysteine protease inhibitors.

2. DISTRIBUTION

Significant amounts of phytocystatins are widespread in plant tissues and their storage organs like seeds and latex (Nallamsetty *et al.,* 2003).

The first known plant thiol proteinase inhibitor was that from *Oryza sativa* (rice) which was discovered in 1987, subsequently, it was detected in other species. Majority of the plant inhibitors studied in the plant kingdom originated from three main families, namely, leguminosae, graminae and solanaceae. Cysteine protease inhibitors have been detected in seeds of plants of all botanical families so far examined from gymnosperms to angiosperms (Rele *et al.,* 1980). Papain inhibitors have been isolated from cereal seeds like maize and most recently from the seeds of a commonly used Indian legume *Phaseolus mungo* (Sharma *et al.,* 2005), rice, wheat, rye, barley and millet. Following are some of the plants from which cystatins of thiol proteases have been isolated and purified: sugarcane (Soares-Costa *et al.,* 2002), avocado (Kimura *et al.,* 1995) sunflower, pineapple (Rowan *et al.,* 1990), seedlings from sorghum, leaves of *Dianthus caryophyllus* and apple fruit, several cystatins isolated from the plant sources cannot be categorized in just one class of cystatin superfamily like oryzacystatins from rice and cystatins from corn have no disulphide bonds or carbohydrate content like type-I cystatin but have many properties similar to type-II cystatin, similar kind of cystatins are found in sunflower seeds, avocado fruit and Chinese cabbage flower bud. Phytocystatins have also been identified and studied in plants such as cowpea, potato, and carrot (Fernande *et al.,* 1993, Ojima *et al.,* 1997). However, limited efforts have been devoted to discover cystatins from other plant sources. Recently, a papain inhibitor has also been purified from the seeds of xerophytic tree Algaroba (*Prosopis juliflora*) (Oliveira *et al.,* 2002). Most recently it has been suggested that *Vigna unguiculata* as well as other leguminous seeds like, *Phaseolus lunatus* and *Adenanthara pavonina* contain at least three size classes of papain inhibitors: a low molecular weight class (5-12 kDa) and a high molecular weight class (60-80 kDa) and an intermediate size class (20-30 kDa). Seeds of mung bean (*Phaseolus aureus*) contain small molecular weight (2-12 kDa) inhibitors active towards a cysteine protease isolated from the same seeds (Baumgartener & Chrispeels, 1976). The non-specific cysteine protease inhibitors isolated from the seeds of a *Bauhinia* species tree with 24 kDa is an e.g., of the above suggested intermediate class. An acidic papain inhibitor of 92 kDa isolated from maize endosperms seems to be related to the rice cystatins. Oryzacystatins, soya bean cystatin as well as a papain inhibitor from *Vigna unguiculata* seeds were shown to inhibit cysteine proteases from the midgut of seed eating bruchids. A cysteine protease inhibitor of 6 kDa was isolated from the seeds of pumpkin (a cucurbit) and was shown to be constituted by three isoforms of pi's 5.9, 7.5 and 10.3. A much larger phytocystatin (Mr = 87 kDa) was isolated from potato tuber, (Walsh & Strickland 1993). Potato cystatin contains no carbohydrate attachments and is comprised of eight repeated cystatin domains with 53-89% identity of residues in the domains. The amino acid sequence of each potato cystatin domain contains 43-50% homology to oryzacystatins. Trypsin treatment of potato multicystatin hydrolyzes the protein into 132 kDa and 510 kDa polypeptides. Potato multicystatins are stored in the form of crystals in the peridermal cells of potato tubers. Unlike potato multicystatin the tomato multicystatin ocurrs only in the leaves of tomato plants treated with methyl jasmonate. Table 1 and Table 2 show the amino acid composition of different cystatins. Table 3 illustrates examples of some of the phytocystatins investigated and their characteristics.

3. PROPERTIES OF PHYTOCYSTATINS

Phytocystatins are quite stable proteins as they remain active in the pH range of 4-11 and are also stable to temperature changes in the range of 30-60°C.They are mostly low molecular weight acidic proteins and contain five conserved residues: Gln-Val-Val-Ala-Gly which are involved in complex formation with the enzymes. In addition to conserved sequences or points in the cystatin superfamily (e.g., Q-V-V-A-G), phytocystatins contain a new conserved sequence identified as: Leu-Ala-Arg [Phe or Tyr] Ala-[Val or Ile] x-x-Asn at a part of the N-terminal α-1 helix (22). Seeds of Wisteria floribunda contain a protease inhibitor which has a sequence Q-V-V-A-G which is also found to be conserved in oryzacystatin. This sequence could be used to identify phytocystatin on protein databanks and to differentiate them from other members of cystatin superfamily. Mammalian cystatins are commonly tight binding competitive inhibitors,

Table 1: Amino acid composition of tomato and potato cystatins

Amino acid	Tomato cystatin (% mole)	Potato cystatin (% mole)
Asx	11.2	12.6
Thr	4.0	3.3
Ser	12.2	1.9
Glx	13.6	12.6
Pro	4.3	3.0
Gly	12.3	5.9
Ala	8.8	7.9
Val	6.9	10.6
Met	1.0	1.3
Ile	4.1	5.3
Tyr	5.0	6.9
Phe	2.8	4.6
His	3.6	6.8
Lys	1.6	1.1
Arg	6.3	11.9
Trp	1.8	2.0
Cys	0.6	2.9

Amino acid composition of potato cystatin from Rodis and Hoff (1984).

Table 2: Amino acid composition of cystatins

Type-I				Type-II			
AA	Human cyst	Human cyst	Bovine stefin	Cyst	Cyst	Chicken cyst	Cyst
	A	B	C	C	S		**F**
Cys	–	1	–	4	4	4	6
Ser	2	6	5	9	5	12	6
Thr	7	7	5	7	5	5	9
Pro	5	5	4	8	5	5	5
Ala	5	8	7	10	7	7	3
Gly	8	3	4	8	5	5	4
An	5	5	5	5	4	4	7
Asp	6	4	7	7	8	6	5
Glu	9	6	8	5	12	9	6
Gln	6	7	7	7	8	7	6
His	1	5	5	5	2	1	4
Arg	1	2	3	8	8	8	7
Lys	12	9	9	7	4	7	10
Met	2	2	2	3	1	2	2
Ile	4	2	4	2	6	6	4
Leu	8	6	6	8	8	10	12
Vol	9	9	11	10	7	9	11
Phe	2	7	7	5	6	3	6
Tyr	6	4	3	4	6	5	3
Trp	–	–	1	–	2	1	2
Total	**98**	**98**	**101**	**122**	**113**	**116**	**118**

Data are calculated from published amino acid sequence for Type-I and II

Table 3: Phytocystatins investigated and their characteristics

Source and Reference	N	N. aa.	M.m. (kDa)	pI
Helianthus annus	Sca	83	~9	5.6
Kouzuma et al., 1996	Scb	101	~11	9.6
Persea americana Kimura et al., 1995	(–)	100	~11	(–)
Daucus carrot Ojima et al., 1997	EIP18	133	18	(–)
Oryza sativa Abe et al., 1987 a,b	OC-I	102	~12	5.3
Vigna unguiculata Fernandes et al., 1993	(–)	97	10.7	(–)
Solanum tubersoum Waldron et al., 1993	PMC	757	~87	(=)
(pineapple stem) Lenarcic et al., 1992	(–)	52	~5	(–)
Glycine max Misaka et al., 1996	(–)	245	26	(–)
Corn seeds Abe et al., 1992	(–)	135	18	(–)
Vigna unguiculata	(–)	97	10.7	(–)
Castanea sativa Pernas et al., 1998	CsC	(–)	11.2	6.9
Wisteria floribunda Hirashiki et al., 1990	WCPI-3	(–)	~15.7	5.7
Prosopis julfilora Oliveira, 2000	PjC	(–)	~20.5	(–)

N, nomination; N.a.a., number of amino acids; M.m., molecular mass; (–), date non-found; (=) domain-dependent, The molecular mass calculated by SDS-PAGE and/or filtration gel.

however, it has been found that phytocystatins are inhibitory proteins that show non-competitive type inhibition kinetics towards papain. These are corn cystatin-I and oryzacystatin-I soya bean, chestnut seed cystatin and cystatins from *Prosopis juliflora* seeds.

Studies have indicated a profound role for glycine at the N-terminal region of many cystatins. It is found to be conserved in many of the plant and mammalian cystatins. This role of glycine could be attributed to its small side chain which allows flexibility in the structure. Corn cystatin (CC), a phytocystatin, shows a wide inhibitory spectrum against various cysteine proteinases, transgenic rice plants were produced by using CC cDNA as a first step to obtain a rice plant with insecticidal activity. This attempt was based on the observation that many insect pests like oryzacystatin show a narrow inhibition spectrum and is present in ordinary rice seeds in insufficient amounts to inhibit the cysteine proteinases of rice insect pests. The transgenic rice plants generated contained high levels of CC mRNA and CC protein in both seeds and leaves, the CC protein content of the seed reaching 2% of the total heat soluble protein. CC activity was also recovered from seeds and it is found that the CC fraction efficiently inhibited both papain and cathepsin H, whereas the corresponding fraction from non-transformed rice seeds showed much lower or undetectable inhibitory activities against these cysteine proteinases. Furthermore, CC prepared from transgenic rice plants showed potent inhibitory activity against proteinases that occur in the gut of the insect pest, *Sitophilus zeamais* (Irie *et al.,* 1996).

A cysteine proteinase inhibitor purified and characterized from chestnut seeds (*Castanea sativa*) was compared for its similarity with its full-length cDNA clone, isolated from an immature chestnut cotyledon library. Identity of both seed and recombinant cystatin was confirmed by matrix-assisted laser desorption/ionization mass spectrometry analysis, two-dimensional electrophoresis and N-terminal sequence indicated all the motifs that are thought to be essential for inhibitory activity, and shows significant identity to other phytocystatins, especially that of cowpea (70%). Recombinant CsC inhibited papain (Ki 29 nM), ficin (Ki 65 nM), chymopapain (Ki 366 nM), and cathepsin B (Ki 473 nM). By contrast with most cystatins, it is also effective towards trypsin (Ki 34.89 nM). CsC is active against digesting proteases in the insect *Tribolium castaneum* and the mite *Dermatophagoides farinae*, these are the two important agricultural pests. Its effects on the cysteine proteinase activity of two closely related mite species revealed the high specificity of the chestnut cystatin.

Despite several studies, the importance of the N-terminal region of cystatins, and of a conserved glycine within this region, remains unclear. It is found that deletion of the N-terminal 21 amino acids abolishes the inhibitory capacity of oryzacystatin-I. The importance of a conserved glycine residue (Gly10) has also been examined by replacing it with 11 other amino acids. Three other glycine residues (Gly-5, -6 and -11) in this N-terminal region of oryzacystatin-I were found to be insignificant. Only those variants in which Gly10 was substituted show any significant change in inhibitory capacity compared with wild-type oryzacystatin-I. The inhibitory characteristics of hybrid cystatin molecules comprising regions of chicken egg white cystatin and oryzacystatin suggested that in common with animal cystatins, the N-terminal region of the plant cystatin, oryzacystatin-I, and in particular the highly conserved Gly10 residue are important for effective inhibition of papain.

4. HOMOLOGY BETWEEN PLANT AND ANIMAL CYSTATINS/ PROPERTIES

Contrary to serine proteinase inhibitors the plant cysteine proteinase inhibitors were relatively recently investigated. Several cystatins have been isolated and sequenced, out of these some showed homology with serine proteinase inhibitor differing also in their effectiveness of inhibition towards various cysteine proteinases. The potato tuber cystatin represented a distinct group of proteins that could belong to Kunitz type of soya bean trypsin inhibitor. They did not possess the conserved region (Gln-X-Val-Gly) that usually characterizes the cystatin superfamily.

Oryzacystatin represents the best example of homology between the two groups (animal and plant cystatins). The alignment of the amino acid sequence with those of other cystatins reveals that plant cystatin like oryzacystatin is highly homologous to members of the cystatin superfamilies of the mammalian origin and both have evolved from a cognate ancestrol gene. Oryzacystatins isolated from rice seeds in 1987 is one of the first thiol protease inhibitors found in plants and some of the most important properties of oryzacystatin-I and II are described in details.

Rice grain contains cysteine proteinase inhibitors, termed oryzacystatins-I and II (OC-I and II). OC-I is a polypeptide with a molecular size of about 11.5 kDa (determined by SDS-PAGE and amino acid analysis) and 12 kDa (determined by gel filtration using Sephadex G-75). OC-I occurs in a number of rice cultivars at an average of 2-3 mg/kg of seed. OC-I is heat stable inhibitor, can withstand temperature as high as 100°C or higher. The amino acid sequence of OC-I has a significant similarity to those of representative cystatins of animal origin, which led to the suggestion that this inhibitor belongs to the cystatin superfamily. OC-I bears no disulphide bonds, therefore, it could be classified as family-I cystatin. However, in terms of the sequence homology, OC-I resembles more closely to family-II cystatins. OC-I has 102 amino acids, and no cysteine residues. The predominant amino acids include Asp, Glu, Val, Ala, and Ly. OC-I, like animal cystatins, has the conserved central pentapeptide Gln-Val-Val-Ala-Gly as a probable target enzyme-binding site, but lacks disulphide bonds. OC-I inhibits papain in a stoichiometric manner, with a K_i value of 10^{-8}M. However, OC-I inhibits cathepsin H with a K_i value of 10^{-6}. OC-I also inhibits ficin, but to a lesser extent. However, it failed to inhibit bromelain by only inhibiting up to 4% or less. OC-I does not inhibit serine proteinases (trypsin, chymotrypsin, and subtilisin), aspartic proteinases, and carboxyl proteinases.

OC-II shows 59% identity with OC-I in terms of nucleotide sequences. This OC has a molecular size of 12 kDa, as shown by SDS-PAGE (64). OC-II is composed of 107 amino acid residues. Like OC-I, it does not contain disulphide bonds, and can be classified as belonging to family-I of the cystatin superfamily. However, with respect to amino acid sequence, OC-I, like OC-II is more closely related to family-II cystatins. Again, the commonly conserved sequence Gln-Val-Val-Ala-Gly in the central part of the molecule of most cystatin superfamily members including OC-I is also present in OC-II. OC-II gives a K_i value of 10^{-6} against papain and a K_i value as small as 10^{-8} against cathepsin H. OC-I and OC-II inhibit papain and cathepsin H non-covalently. Neither OC-I nor OC-II showed any effective inhibition against cathepsin B or L.

5. MECHANISM OF INHIBITION OF PHYTOCYSTATINS

Thiol proteinases play an important role in the intracellular proteolysis such as catabolism of peptides, processing of prohormones and precursor proteins. These proteins contain two principal catalytic amino acid residues, cysteine and histidine at their active site which are involved in the cleavage of peptide bonds connecting the amino acid residues of a protein or peptide substrate. Papain is the most extensively studied plant thiol protease. It is a monomeric polypeptide with a molecular weight of 23,000 consisting of 212 amino acid residues with three disulphide bridges (Cys-22-Cys-63, Cys-56-Cys-95, Cys-153-Cys-200). Its sequence and three-dimensional structure is well known. X-ray structural analysis has also been performed on papain-inhibitor complexes with the inhibitors like E-64. Papain is isolated in an inactive form, in which the cysteine residue at the active site is blocked by a disulphide bond with Cys-22 and Cys-63 having a free thiol group. Activation is possible by intramolecular disulphide exchange with thiol reagents or reducing agents. Papain is formed from a classical zymogen (Propapain). This proregion has been shown to be essential for correct folding and can act as a high affinity inhibitor of the enzyme. Other plant cysteine proteases whose known fragment sequences indicate a relationship to papain, e.g., the chymopapains A and B isolated from *Carica papaya*, Ficin (EC.3.4.22.3) from the latex of various ficus species (Moraceae), the bromelains from the latex of pineapple tree stem (Bromeliaceae) and other recently characterized plant enzymes which also belong to the papain family are aleurain, phytolacain R and a ginger protease. These legumains are asparaginyl endopeptidases and constitute a unique group of cysteine proteases distinct from the papain family. Papain from papaya, Bleomycin hydrolase as well as Actinicin from kiwi fruit, *Actinidia chinesis* and *caricain* are the only cysteine proteases for which the complete amino acid sequences as well as the tertiary structures are known.

A major mode of control of proteases is through the use of protease inhibitors. Biochemists have profitably studied plant protease inhibitors as model systems to explore the mechanism of action and inhibition of proteases. Several interaction models between cystatins and their target cysteine proteinases were proposed that suggested three regions of the interactions: Gly residue near the amino terminal, Gly-Val-Val-Ala-Gly region is conserved, and second hairpin loop near the carboxyl terminal region was also found to be conserved. The data results of the X-ray crystallography did not show the involvement of the disulphide bonds in the interaction (Machleidt *et al.,* 1989). In order to determine which regions were responsible for the papain inhibitory activity, Abe and Arai evaluated the effects of the wild oryzacystatin and oryzacystatin truncated forms against papain activity. They observed that the primary region of interaction of the oryzacystatin responsible for the papain inhibitory activity was the Gly-Val-Val-Ala-Gly sequence, Pro-Tyr dipeptide region could also act as a cofactor in the interaction, but the possibility that some other regions also contributed to inhibitory activity could not be ruled out (Table 4). As far as the mechanism of action of phytocystatins is concerned, it has been established that no disulphide bond is formed between the active site of cysteine and the inhibitor because the complexes dissociate when denatured without reduction and carboxymethylation of enzyme and inhibitor does not prevent complex formation, showing that the complex formation is reversible.

Table 4: Effects of the wild oryzacystatin and several oryzacystatin truncated forms against papain activity.

Wild oryzacystatin-I (WOC-I)	PPIA[a]
Mutant-1: Lacking 21 amino-terminal and 11 carboxyl-terminal amino acids	PPIA[a]
Mutant-2: Lacking 38 amino terminal and 35 carboxyl-terminal amino acids	ACI[b]
Mutant-3: Substitution of Gln-Val-Val-Ala-Gly by Pro-Val-Val-Ala-Gly or Leu-Val-Val-Ala-Gly	DR-PPIA[c]
Mutant 4: Substitution of Gln-Val-Val-Ala-Gly by Gln-Val-Asp-Ala-Gly	MPIA[d]
Mutant-5: Substitution of Gln-Val-Val-Ala-Gly by Gln-Val-Gly-Ala-Gly	E-PPI[e]

Wild oryzacystatin (WOC-I); PPIA, a: potent papain inhibitory activity; b: ACI, almost completely inactive; c: DR-, reduced drastically; d: MPIA, moderate papain inhibitory activity; e: E-PPIA, potent papain inhibitory activity equivalent to that of WOC-I. In boldface refer amino acid substituted by mutation. Data were obtained on references Abe et al., 1988; Arai et al., 1991.

In fact, most mammalian cystatins are reversible, tight binding competitive inhibitors of cysteine proteinases, which form equimolar complexes with their target enzymes. Specifically, the mechanism of interaction of mammalian cystatins with papain shows that three parts of the inhibitor are in close contact with the active site cleft of papain, the amino terminus with conserved Gly-9, a first hairpin loop containing the prototype sequence, QVVAG and a second hairpin loop containing a conserved Trp-104. This is also observed that the major contribution is from the first hairpin loop containing the QVVAG sequence. According to the proposed model, the N-terminal segment of cystatin which is more flexible bridges over the active site Cys-25 residue of papains without completely burying it and additionally the side chain of Leu-8 binds to the S-2 subsite of papain which determines the substrate specificity of papain. The side chain groups of QLVSG segment makes the most intimate contacts with putative S-1 subsite of papain, Fig. 1A shows the schematic diagram of enzyme inhibitor complex of E-64 and papain and Fig. 1B gives the terminology of protease specificity. This explains the presence of Gly-9 in the close vicinity of papain but not in direct contact of Cys-25 of papain. Removal of Leu-7 and Leu-8 leads to an approximately 5000-folds lower affinity for papain. However, Nycander and Berk emphasized the role of Trp-104 in the inhibition of cysteine proteinases. According to their model Trp-104 of cystatin interacts with two tryptophan side chains in the active site cleft of papain, Trp-177 and Trp-181, interact in such a manner that the indole ring of Trp-104 stacks on the side chain of Trp-177 and lies on the indole ring of Trp-181.

(A)

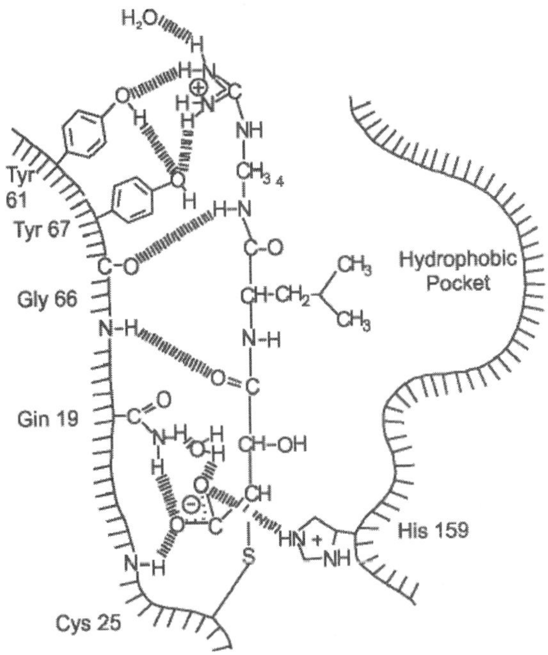

Fig. 1A: Schematic diagram of the enzyme inhibitor complex of E-64 and Papain (American Chemical Society).

(B)

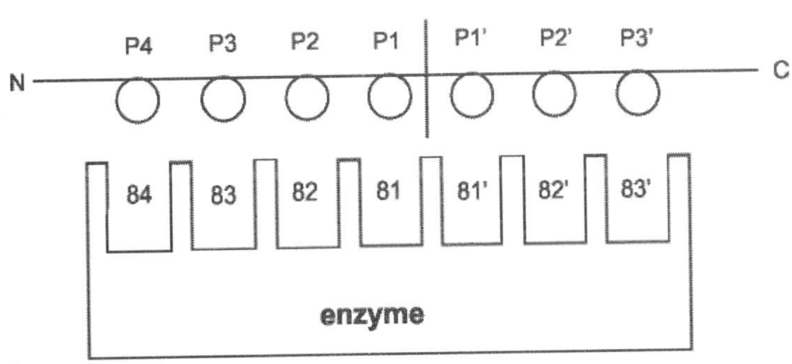

Fig. 1B: Terminology of protease specificity (According to Berger and Schechter).

According to Nycander, initial weak interactions in which N-terminal of the inhibitor binds to the proteinase is followed by a conformational change. Subsequently, the occluding loop of the proteinase that partially obscures the active site is displaced by the inhibitor bringing about another conformational change in the proteinase. Furthermore, the inhibitors contain active sites for the inhibition of proteolytic enzymes that apparently endow them with their specificity.

The three-dimensional structure of cystatins and other evidence suggest that a flexible N-terminal region of these inhibitors may bind to target proteinases, independent of the two rigid hairpin loops forming the remainder of the inhibitory surface. According to Estrada (Estrada *et al.,* 2000), the N-terminal region of the cystatin interacts with the proteinase subsequent to the interaction of the hairpin loops. Researchers believe that such two-step binding of cystatin to target proteinases may be the characteristic of most cystatin protease reactions. Although the detailed mechanism of the mode of inhibition by plant cystatins is not known, Oryzacystatin-I and II are reported to show a significant degree of homology with egg white cystatin and other animal cystatins.

6. PHYTOCYSTATINS IN HEALTH AND DISEASE

Many viruses require proteolytic cleavage to become infectious. Cystatins possess activity against a variety of viruses such as poliovirus, rhinovirus, cornonavirus, and herpes simplex virus. Although most likely the antiviral activity of cystatins is due to their proteinase inhibitory properties, the viral target enzymes have not been identified yet.

Cystatins have antibacterial activity as well since they play a role in the inhibition of bacterial cysteine proteinases. Cystatins have been reported to inhibit the growth of *Porphyromonas gingivalis* and *Staphylococuss aureus*.

Schlep and Pongpaew and Troll suggested that proteinase inhibitors present in cereals like rice and maize can prevent certain types of cancer. Various scientists have reported anti-tumor activity of plant cysteine proteinase inhibitors (Saito *et al.,* 1980). Furthermore, various cysteine proteinase have been found to activate proinflammatory mediators. Periodontitis and rheumatoid arthritis are inflammatory diseases catalysed by cysteine proteinases of the mammalian system.

7. NUTRITIONAL SIGNIFICANCE AND FUNCTIONS IN PLANTS

Phytocystatins are rich in cysteine and lysine contributing to a better and enhanced nutritional quality. The most attractive role of these inhibitors is that they provide powerful regulatory systems to the plants as they are important agents for controlling endogenous proteinases and prevention of unwanted proteolysis. They have a storage function which is supported by their presence in large quantities in seeds and tubers. They constitute about 6% of the soya bean protein and upto 10% of the soluble proteins of barley grains and potato tubers.

Cysteine proteinases are probably present in seeds of all plants where they seem to be involved in the mobilization of protein reserves. These proteinases are also associated with the processing of proteins and in plant senescence in which case they are probably associated with ubiquitin dependent system requiring ATP.

Phytocystatins are important in a variety of ways as they are participants in the mechanisms of programmed cell death. Apoptosis have been implicated in several processes such as xylogenesis, some forms of senescence, in the pathogen attack response and stress response (Lamb & Dixon 1997). They are expressed in response to biotic and abiotic stress and in various plant defence processes including insect and pest attacks. Cysteine protease inhibitors of which papain inhibitors are the most well known have been recently associated with the control of protein mobilization.

Oryzacystatins-I and II have been well studied for their functions. Rice phytocystatins, namely, OC-I and OC-II occur in ripened rice seeds and possibly play a role in regulating the actions of cysteine proteinases present in rice. Three molecular species of cysteine proteinases are present in rice, named oryzains α, β and γ (OZ-α, OZ-β, and OZ-γ). These are the probable endogenous target enzymes of OC-I and OC-II, since their proteolytic activities are well regulated by these cystatins *in vitro*. OZ-β is involved in proteolysis of seed storage proteins. OZ-α is involved in proteolysis of some functional proteins in the aleurone, and OZγ is involved in intracellular protein turnover. OZ-α and OZ-β are very homologous to papain, whereas OZ-γ shows a very high homology to cathepsin H. Therefore, it is highly probable that in rice seeds OZ-β as well as OZ-α functions as proteinases of the papain type and OZ-γ functions as a proteinase of the cathepsin H-type. The function of these three proteinases of the two types may contribute to either or both of the processing for maturation of storage proteins such as glutenin during ripening of rice seeds and the proteolysis of the storage proteins during germination.

8. ROLE OF PHYTOCYSTATINS IN PROTECTION FROM INSECTS

Cystatins have been well established to play a potent defensive role against predators and pathogens, although, diverse endogenous functions for these proteins have been proposed, evidence for many of these roles is either partial or confined to isolated examples. On the other hand, many protease inhibitors have been shown to act as defensive compounds against pests by expression in transgenic crop plants and a body of evidence for their role in plants has been accumulated consistently. The possible role of protease inhibitors in plant protection was investigated as early as 1947 by Mickel and Standish, following these early studies, there have been many examples of protease inhibitors active against certain insect gut proteinases. It has been reported that many proteinase facilitate the invasion of plant tissue by hydrolysing proteins in and between cell walls. Proteinase inhibitors function in retarding the proteolysis of cell wall and membrane proteins reducing the disruption of cell organization, thus proving to be an asset to the plant by arresting the extracellular proteinases of the pathogens. Cystatins are also involved in inhibiting the proteinases of the insects, which, in turn cut off the supply of proteins to the insect and eventually helps in the repulsion of the invader, an example is the insects of the order *Coleoptera* which are repulsed by oryzacystatins, *in vivo* deleterious effects of E-64 and phytocystatins on growth and development in insects of the order coleopteran have been reported by various authors. E-64, when ingested by the larvae of the *Z. dicemlineata, A. obtectus Say* showed adverse physiological effects on these organisms. Plant cysteine proteinase inhibitors are encoded by gene families, but little is known about regulation of these genes. Expression of the proteinase inhibitor genes is usually limited to specific organs or to particular phases during plant growth: germination (Huang *et al.*, 2001), and early leaf senescence (Cowgill *et al.*, 2002). The cytosolic localization of inhibitors suggests that they are involved in plant defence against insects. Transgenic potato expressing two cystatin genes and transgenic rape plants expressing rice cystatin I became resistant to nematode (potato) and aphids (rape). Insect induced damage of the plant body causes the induction of PIIF (proteinase inhibitor inducing factor) within 48 hours of the attack, this leads to the accumulation of large amounts of proteinase inhibitors.

9. PROTECTION FROM VIRUSES AND PESTS

As the processing mechanism of all known potyviruses involves the activity of cysteine proteases, tobacco cell lines expressing the foreign gene for cystatin at varying levels were examined for resistance against Tobacco Etch Virus (TEV) and showed direct correlation between the level of cystatin message, inhibition of papain and resistance to TEV and PVY, in all lines tested. Thus, phytocystatins could be used against potyviruses and potentially against other viruses, whose replication involves cysteine proteinase activity.

There are a variety of human viruses including picornaviruses that enter the intestine per se. These viruses in particular, when entering animal cells, synthesize a high-molecular weight polyprotein, which is subsequently divided into a number of viral components. The cleavage of this polyprotein is carried out by virus encoded cysteine proteinases. Therefore, the inhibition of this viral cysteine proteinase by cystatin would lead to stopping the proliferation of virus particles in the infected cells. Arai used, poliovirus as an example of picornavirus. The virus yields in the presence of each OC-I and OC-II remained much lower than that obtained in its absence. OC-I also showed a strong inhibitory effect on the proliferation of herpes simplex virus. It is inferred that phytocystatins like OC-I and OC-II can be seen as useful factors for molecular breeding of virus resistant crop cultivars, as well as of functional food materials with an antiviral effect. Oryzacystatin-I and II has been reported to show antiviral action against polio virus and herpes simplex virus (type I), the *in vivo* antiherpetic effect of oryzacystatin-I and II has made possible the topical treatment of HSV-I (herpes simplex virus) infection in humans and this could be a potential biotechnological application of phytocystatins. Though nothing is known experimentally concerning the mechanism involved in these processes, it is speculated that OC-I molecules enter the infected cells by internalization to inhibit the molecular event in which the polyprotein is processed into functional products.

Plant proteinase inhibitors inactivate proteinases in the digestive tract of insects or pests. Parasites from cowpea and soya beans were fed with diets containing OC-I and II. Results showed that OC-I retarded growth, or even killed those insects at higher concentrations. Another pest from rice seeds, namely, *Sitophilus zeamais*. Both OC-I and II retarded the growth of the rice pest and even killed it at larval stage. This suggests that cystatins, which do not interfere with protein digestion in humans, may target specific cysteine proteinase inhibitors of insect pests.

An active area of research nowadays is the development of transgenic plants which express protease inhibitors with improved features. The techniques of DNA shuffling and phage display are being extensively utilized in this regard.

Bruchids are considered to be the most serious pests attacking food legume seeds during storage. Substantial amounts of stored beans such as common bean (*Phaseolus vulgaris*), cowpea (*Vigna unguiculata*) and common bean (*Phaseolus aureus*) are lost due to damage caused by three main storage pests, the bean weevil *(Acanthoscelides obtectus*), mexican bean weevil (*Zabrotes subfasciatus*) and cowpea weevil (*Calosobruchus maculates*), (Lawrence & Koundal 2002). These bruchids use mainly cysteine proteinase to degrade proteins. All these findings seem to firmly point to a metabolic role for cysteine protease inhibitors in seeds as contrasted with a defensive role attributed to other protease inhibitors (e.g., serine protease inhibitors). It is well known that in developing countries the problem of competition from insect pests is complicated with a rapid annual increase in the human population (2.5-3.0%) in comparison to 1% increase in the food production. In order to feed the ever expanding population, crop protection plays a vital and integral role in the modern-day agricultural production to minimize yield losses. Currently, the crop protection practice in such agricultural systems relies on the use of agrochemicals, although a few specific cases do exist where inherent varietals, resistance and biological control have been successfully employed. The production of transgenic crops has seen rapid advances during the last decade with the commercial introduction of Bt transgenics, but the major concern with these crops has been the development of resistance by pest and public acceptability. Hence, there has been a need to discover new effective plant genes which would offer resistance/protection against pests. Thus, protease inhibitors are one of the prime candidates with highly proven inhibitory activity against insect pests and are also known to improve the nutritional quality of food.

10. PROTECTION FROM FUNGI

The presence of phytocystatins in plants is also significant in view of their antifungal activity. Purified sugarcane cystatin is found to inhibit the growth of filamentous fungus *Trichoderma reesi* (Williamson & Hussey 1996), suggesting that it could be employed to inhibit the growth of pathogenic sugarcane fungi. Phytocystatins are also reported to exhibit a very broad spectrum of activity including suppression of pathogenic nematodes like *Globodera tabacum*. Cysteine protease inhibitors from pearl millet is found to inhibit the growth of fungi including *Trichoderma reesi* (Pernas *et al.,* 1998). Plant productivity can be affected by plant's susceptibility to fungal diseases that results in production losses.

Hence, a relevant commercial application of cystatins is the development of plants expressing high amounts of phytocystatins that would impart potent antifungal property to the plant. These advantages make phytocystatins an ideal choice to be used in transgenic crops resistant to phytophagous insects and pests which can avoid poor crop yield and finally the production losses.

11. CYSTATIN EXPRESSION IN ABIOTIC AND BIOTIC STRESS

Cystatin from chestnut seeds (*Castanea sativa*) showed induction and accumulation in chestnut seeds after fungal infection as well as from treatment by jasmonic acid (Ryan, 1990). Furthermore, the levels were also raised both in roots and leaves, when chestnut plantlets were subjected to cold and saline shocks, and also in roots by heat stress. This suggests that chestnut cystatin is not only involved in defence responses to pest and pathogen invasion, but also in those related to abiotic stress. Studies on plant proteinase inhibitors are also important due to their involvement in defence mechanisms and in protection of seed reserves from premature hydrolysis.

Thus, in conclusion, it can be said that phytocystatins found particularly in germinating seeds, flowers, fruits, cereals, latex and leaves are valuable systems for studying the role of proteinases in the process of plant development as a whole.

REFERENCES

Abe, M., Abe, K., Kudora, M. and Arai, S. (1992). Corn kernel cysteine proteinase inhibitor as a novel cystatin superfamily member of plant origin *Eur. J. Biochem* **209:** 933-937.

Aoki, H., Akaike, T., Abe, K., Kuroda, M., Arai, S., Kamura, R.O., Negi, A. and Maeda, H. (1995). Antiviral effects of oryzacystatin, a proteinase inhibitor in rice, against herpes simplex virus type-1 *in vitro* and *in vivo*, *Antimicrob Agents Chemother.* **39:** 846-9.

Arai, S., Matsumoto, I. and Abe, K. (1998). Phytocystatins and their target enzymes: from Molecular Biology to Practical Application: A Review, *J. Food Biochem.* **22:** 287-299.

Baumgartener, B. and Chrispeels, M.J. (1976). Partial characterization of a Protease inhibitor which inhibits the major endopeptidase present in the cotyledons of Mung Beans, *Plant Physiology* **58:** 1-6.

Cowgill, S.E., Wright, C. and Atkinson, H. (2002). Transgenic potatoes with enhanced levels of nematode resistance do not have altered susceptibility to non target aphids, *J. Mol. Ecol.* **11:** 821-7.

Estrada, S., Olson, S.T., Raub-Segall, E. and Bjork, I. (2000). The N-terminal region of cystatin A (Stefin A) binds to papain subsequent to the two hairpin loops of the inhibitor. Demonstration of two step binding by rapid kinetic studies of cystatin A labeled at the N-terminus with a fluorescent reporter group, *Protein Sci.* **9:** 2218-24.

Fernandes, K.V.S., Sabelli, P.A., Barrett, A.J., Richardson, M., Xavier-Filho, J. and Shewry, R. (1993). The resistance of cowpea seeds to bruchid beetles is not related to levels of cysteine proteinase inhibitors, *Plant Mol. Biol.* **23:** 215.

Huang, Y.J., To, K.Y., Yap, M.N., Chiang, W.J., Suen, D.F. and Chen, S.C.G. (2001). Multifunctional role of plant cysteine proteinases, *Physiol Plant.* **113:** 384-91.

Irie, K., Hossoyama, H., Takeuchi, T., Iwabuchi, K., Watanabe, H., Abe, M., Abe, K. and Arai, S. (1996). Transgenic rice established to express corn cystatin exhibits strong inhibitory activity against insect gut proteinases, *Plant Mol. Biol.* **30:** 149-57.

Kimura, M., Ikeda, T., Fukumoto, D., Yamasaki, N. and Yonekura, M. (1995). Primary structure of a cysteine proteinase inhibitor from the fruit of Avocado (*Persea Americana* Mill), *Biosci. Biotechnol. Biochem.* **59:** 2328-2329.

Korant, B.D., Towatari, T., Ivanoff, L., Petteway, S. Jr., Brzin, J., Lenarcic, B. and Turk, V. (1986). Viral therapy: Prospects for protease inhibitors, *J. Cell Biochem.* **32:** 91-5.

Lamb, C. and Dixon, R.A. (1997). The oxidative burst in Plant disease resistance, *Annu. Ver. Plant Physiol, Plant Mol. Biol.* **48:** 251-275.

Lawrence, P.K. and Koundal, K.R. (2002). Plant proteinase inhibitors in control of phytophagous insects, *Electronic Journal of Biotechnology.* **5:** 93-109.

Machleidt, W., Thiele, U., Laber, B., Arsfald-Machleidt, I., Esterl, A., Wiegand, G., Kos, J. Turk, and Bode, W. (1989). Mechanism of inhibition of papain by egg white cystatin: Inhibition constants of N-terminally truncated forms and cyanogen bromide fragments of the inhibitor, *FEBS Letters.* **243:** 234-238.

Monti, R., Contiero, J. and Goulart, A.J. (2004). Isolation of natural inhibitors of papain obtained from *Carica papaya*, Brazilian Archives of Biology and Technology **47:** 747-754.

Nallamsetty, S., Kundu, S. and Jagganadham, M.V. (2003). Purification and biochemical characterization of a highly active cysteine proteinase Ervatamin A from the latex of *Ervatamia coronaria*, *J. Protein Chem.* **22:** 1-12.

Ojima, A., Shiota, H., Higashi, K., Kamada, H., Shimma, Y.I., Wadamasata, P. and Satoh, S. (1997). Inhibition of cathepsin B by cystatin C, due to inhibitor displacing the occluding loop of the enzyme, *Plant Mol. Bio.* **34:** 99.

Oliveira, S., Adelinana Pereira, A., Railene Lima Liziane, M., Morais Ana, H.A., Melofrancislete, R., Franco Octavio, L., Bloch Carlos, Jr., Grossi-de-Sa Maria, F. and Sales Mauricio, P. (2002). Activity toward Bruchid pest of a Kunitz-type inhibitor from the seeds of Algaroba tree (*Prosopis juliflora* D.C.), *Pesticide Biochemsitry and Physiology* **72:** 122-132.

Pernas, M., Sanchez-Monge, R., Gomez, L. and Salcedo, G. (1998). Antifungal activity of a plant cystatin, *Plant. Mol. Biol.* **38:** 1235-1242.

Rashid, F., Sharma, S. and Bano, B. (2005). Comparison of guanidine hydrochloride (GdnHCl) and urea denaturation on inactivation and unfolding of human placental cystatin (HPC), *The Protein Journal* 2005 Jul; **24(5):** 283-92.

Rele, M.V., Vartak, H.G. and Jagannathan, V. (1980). Proteinase inhibitors from Vigna *unguiculata* subsp. Cylindrical: III properties and kinetics of inhibitors of papain, subtilisin and trypsin, *Archives of Biochemistry and Biophysics* **202:** 117-128.

Rowan, A.D., Buttle, D.J. and Barrett, A.J. (1990). The cysteine proteinases of the Pineapple plant, *Biochem. J.* **266:** 869-75.

Ryan, A. and Clarence, (1990). Proteinase inhibitors in plants: Genes for improving defense against insects and pathogens, *Annual Review of Phytopathology* **28:** 425-449.

Saito, D., Sawamura, M., Umezawa, K., Kanai, Y., Furihata, C., Matsushima, T. and Sugimura, T. (1980). Inhibition of experimental blood borne lung metastasis by protease inhibitors, *Cancer Res.* **40:** 2539-42.

Sharma, S., Rashid, F. and Bano, B. (2005). Biochemical and biophysical changes induced by fungicide sodium diethyl dithiocarbamate (SDD), in phytocystatin purified from *Phaseolus mungo*: A commonly used Indian Legume, *J. Agri. Food Chemistry.* ACS 2005 Jul 27; **53(15):** 6027-34.

Soares-Costa, A., Beltramini, L.M., Thiemann, O.H. and Henrique Silva, F. (2002). A sugar cane cystatin: recombinant expression, purification and antifungal activity, *Biochem. Biophys. Res. Commun.* **296:** 1194-9.

Urwin P.E., Atkinson, H.J. and McPherson, M.J. (1995). Involvement of the NH2 terminal region of oryzacystatin in cysteine proteinase inhibition, *Protein Eng.* **12:** 1303-7.

Walsh, T.A. and Strickland, J.A. (1993). Proteolysis of the 85 kilodalton crystalline cysteine proteinase inhibitor from potato releases functional cysteine domains, *Plant Physiol.* **103:** 1227-1234.

Williamson, V.M. and Hussey, R.S. (1996). Nematode pathogenesis and resistance in plants, *Plant Cell.* **8:** 1735-1745.

Diversity of Lower Plants (2014): pp. 212-221
Editors: Rajan Kumar Gupta & Mukesh Kumar
IK International Publishing House Pvt. Ltd.

16

Aquatic Biodiversity: Nature and Characterization

Manisha Kaushal, G.K. Sharma*, Desh Deepak and Mukesh Kumar**

Department of Botany, Hindu College, Moradabad–244 001 (U.P.)
**Department of Botany, Sahu Jain College, Najibabad–246 763 (U.P.)*

ABSTRACT

Aquatic biodiversity has enormous economic and aesthetic value and is largely responsible for maintaining and supporting overall environmental health. Humans have long depended on aquatic resources for food, medicines, and materials as well as for recreational and commercial purpose such as fishing and tourism. Aquatic organisms also rely upon the great diversity of aquatic habitats and resources for food, materials, and breeding grounds. In recognition of the anthropogenic effect on water bodies, Government of India enacted a legislation to prevent and control water pollution and conservation of aquatic biodiversity way back in 1974. These efforts have promoted studies on water quality and the effect of water pollution on aquatic biodiversity throughout India.

Keywords: Biodiversity, Aquatic biodiversity, Riverine vegetation conservation.

1. INTRODUCTION

Aquatic biodiversity can be defined as the variety of life and the ecosystems that make up the freshwater, tidal, and marine regions of the world and their interactions. Aquatic biodiversity encompasses freshwater ecosystems, including lakes, ponds, and reservoirs, rivers and streams, groundwater, and wetlands. The banks of streams or riprarian areas are also important areas associated with freshwater systems. It also consists of marine ecosystems, including oceans, estuaries, salt marshes, sea-grass beds, coral reefs, kelp beds, and mangrove forests.

Aquatic ecosystems also provide a home to many species including phytoplankton, zooplankton, aquatic plants, insects, fish, birds, mammals, and others. They are organized at many levels, from the smallest building blocks of life to complete ecosystems, encompassing

**Corresponding author

communities, populations, species, and genetic levels. In summary, aquatic biodiversity includes all unique species and habitats, and the interaction between them.

Aquatic biodiversity is particularly sensitive to changes in river health, and it has declined in many river systems since European settlement. For example, one-third of river length in Australia has lost 20-100 per cent of the various kinds of aquatic invertebrates that should live there. Conversely, many rivers or river reaches, perhaps the majority, may not be significantly degraded due to a lack of development. Many of them are not suitable for dam construction, or they are located in less populated mountain areas or in the northern tropical regions of Australia, Indeed, there is growing concern that man of these rivers, especially the larger pristine rivers in tropical Australia, will come under increasing pressure as sources of water to support irrigation development are exhausted in southern Australia.

2. LOSS OF AQUATIC DIVERSITY

Factors including overexploitation of species, the introduction of exotic species, pollution from urban, industrial, and agricultural areas, as well as habitat loss and alteration through damming and water diversion all contribute to the declining levels of aquatic biodiversity in both fresh water and environments, valuable aquatic resources are becoming increasingly susceptible to both natural and artificial environmental changes. Thus, conservation strategies to protect and conserve aquatic life are necessary to maintain the balance of nature and support the availability of resources for future generations.

"The health of the lakes and their biological diversity is directly related to the health of each component of the ecosystem. Similarly, the lakes are adversely affected when disturbances occur in one of the systems. For example, alteration in the upper watershed can impact the entire lake ecosystem. When a forest is cleared, not only is the physical structure of the terrestrial ecosystem altered, the tributary streams, coastal areas and the open lake can also be affected. When vegetation is removed near a tributary, precipitation is allowed to run off directly into the river or stream, causing flows in these steams to increase much more quickly (following rainfall). This run-off also picks up more soil than it otherwise would, and the load of sediment in the tributary will be increased. The increased sediments can destroy of habitat required by fish and insect species and can prevent the spawning of anadromous fishes which spend most of their lives in the open lake. These sediments can also accelerate the formation of sand bars and blockages at river mouths, altering nearby coasts."

During the nineteenth and twentieth centuries, major changes to Australia's land and water resource use led to altered flow regimes, increases in sediment and nutrient loads, habitat, destruction, the introduction of exotic species and the consequent loss of aquatic biodiversity. Whether rives health has been getting better or worse or has been stable at a national scale over the last five years is difficult to assess because of a lack of data.

3. MANAGEMENT OF RIVERINE VEGETATION

There are some that the management interventions over the past decade may have stabilized the decline in river health in some regions. In 2005 the Victorian Government provided the first scientifically robust assessment of trends in river health across an entire state. It is concluded that there had been no broad change in Victoria's rivers during 1999-2004, and that the previous deterioration in stream condition had been arrested. The study did report that local changes were likely, with improvement in river health in some places and deterioration in a few other (DSE 2005). While far from scientifically or nationally conclusive, these results do provide some causes for optimism. However, the observations in Victoria make it clear that if there are to improvements in river health, they will be hard-earned and a long time coming. Along the way, some rivers will continue to decline in health while other improve.

Frogs are very sensitive indicators of aquatic ecosystem health and extent. Their populations have decreased markedly in the last decade or so, with 27 species of frog listed as endangered or vulnerable.

Riparian vegetation, a key of riverine ecology, has also declines. Extensive revegetation is required to improve river health, and many areas remain under significant pressure due to the combined impacts of human activity and the drought. For example, in 2003, 80 per cent of remaining River Red Gums on the Murray River floodplain in South Australia were stressed to some degree in 2003, and 20-30 per cent of them were severely stressed; this severe decline has occurred in the previous 12 months. In lowland rivers, 'de-snagging' and loss of water plants has also modified much in-stream habitat. While important locally, the scale and scope of recovery programmes are small compared to the scale of the problem. For example, in a sample of river basins, there was only a small increase of some 3700 kilometres in the extent of riparian vegetation between 1991 and 2004. There have also been some limited projects aimed at brining snags, which are essential habitat, back into some rivers.

The physical river environment has been extensively modified as dams and wiers have been built. Not only do these structures physically restrict the movement of native animals, but they also regulate river flow, so there are fewer floods and fewer dry periods, and they divert water out of the river. In New South Wales alone, there are over 4300 weirs and dams. Only 28 have fishways constructed that are effective for native fish.

The impact of wetlands has been dramatic. As many as 231 nationally important wetlands are under pressure across Australia. Of the 64 Ramsar wetlands, latest assessments indicated that 22 have changed in ecological character or have the potential to change (DEH 2002). In the Macquaries Marshes, for example, there has been a significant long-term decline in river flows as a result of river regulation and subsequent diversions upstream. There are now fewer waterbirds, and fewer species of waterbird. In the Lowbidgee floodplain, at least 76% of the area has been destroyed by dams, diversions and floodplain development, reducing the amount of water reaching the floodplain by at least 60%. Waterbirds number in that area have decreased by 90%. Breeding success has been improved in local experiments in the Macquarie Marshes, by allocating water for bird breeding through environmental flows (Kingsford and Porter 2005.) While much attention is being paid to restoring degraded rivers, there is also growing movement to develop a system of state and national river 'reserves' to protect rivers that remain in a healthy, near-pristine condition.

With so much of Australia's land affected by dryland salinity and with so much more at risk, it is not surprising that a third of catchments assessed by the NLWRA (200lb) showed signs of increased salts loads in rivers and steams. The problem is particularly severe in the southern Murray-Darling Basin, along the south-east coast of Australia, and in catchments in south-west Western Australia. In other areas, such as the Lower Murray River, salt interception schemes are helping to ameliorate in stream salinity levels. Whether river's salinity has improved is not yet certain because, with only 8 to 10 years of data in most places, there has not been enough time for any significant trends to emerge from the effects of climate variability and variations inflow.

There is little evidence the increased salt concentrations have yet had a broad-scale impact on aquatic plants and animals, but some wetland ecosystems have shown signs of salt degradation. Impacts on sensitive species should be expected if the salt concentration in Australian fish, for example, are generally believed to be quite salt tolerant, recent research shows that the larval and juvenile stages of certain fish, such as Murray Cod (*Maccullochella peelii peelii*) are particularly sensitive to salt.

Throughout Australia, soil erosion has significantly increased sediment loads to rivers and estuaries. River water from forested catchments, or from catchments with large areas of relatively unmodified native vegetation, tends to be clearer than river water from agricultural and urban catchments. Extensively cleared catchments, such as large areas of the wheat and sheep zones of eastern and south-western Australia, are worst affected. In many parts of Australia, because much of the eroded material has still not yet worked its way through the bigger river systems, large amounts of sediments are stored in river channels in low-gradient areas (Wasson *et al.*, 1996; Martin and McCulloch, 1999; Prosser *et al.*, 2001). Sand slugs in the lowland reaches have smothered various in stream habitats, which has significant impacts on aquatic plants and animals. Rehabilitation techniques for sand slugs are being explored, including placing railway sleepers across a riverbed to increase the natural scouring process (Bond *et al.*, 2004).

Erosion of sediments is dramatically increased after fire. The large-scale fires that burned across the Victorian and New South Wales high country and the Australian Capital Territory in 2001 and 2003 caused greatly increased erosion in these areas, leading to poor water quality and the deposition of large amounts of soil and other materials in stream beds and lakes downstream of the fires (DSE 2003). In the Australian Capital years' worth of sediments into the water supply dams (Carey *et al.*, 2003). Erosion and sedimentation rates were still high in 2005, especially after rainfall, and the territory has had to upgrade its water treatment facilities to maintain drinking water quality to the city (ACTEW 2005). For all of these areas, the risk of further deterioration will remain until the vegetation recovers.

Excess nutrients were a major water quality issue in about 60% of basins assessed by the NLWRA in 2001. For most rivers, the largest source of increased phosphorus loads is gully and stream bank erosion, rather than farm fertilizers (Martin and McCulloch, 1999; Prosser *et al.*, 2001). In contrast, increases in nitrogen levels are typically from fertiliser use, animal's wastes and sewage discharges (Caraco and Cole 1999). Nearly 19,000 tonnes of total phosphorus and 141,000 tonnes of total nitrogen were estimated to be transported down rivers to the coast each year from areas of intensive agricultural activity.

Table 1: Exceedance of water quality guidelines for Australia (number of river basins)

	Major exceedances	Significant exceedances	Number of basins assessed
Nutrient: total nitrogen	19	19	50
Nutrient : total phosphorus	40	20	75
Salinity : electrical conductivity	24	18	74
Turbidity	41	10	67
pH	7	6	43

Blue-green algae are more widespread as changes to in-stream habitat and catchment land use have altered the stream metabolism of inland water. All the changes seem to favour increased toxic algal blooms, and more frequent periods of oxygen depletion in the bottom waters of pools, weirs and dams. The current cost of algal blooms is in the range of $180 million to $240 million per year (Atech, 2000). Between 2001 and 2005, SA water recorded the occurrence of 19 algal blooms in 12 freshwater bodies in South Australia. Most blooms occurred during the summer-autumn months and some lasted several months (Baker 2006).

4. INTRODUCTION OF THE SPECIES

Introduced species appear to be favoured by the large-scale landscape and waterscape management practices found across large areas of the continent. Introduction of pests—such as carp *(Cyprinus carpio)*, mosquito fish *(Gambhusia holbooki)* and oriental weatherloach *(Misgurnus anguillicandatus)*, and weeds – have continued to expand their range. In the Lower Murray-Darling catchments, for example, exotic species make up 56% of the total biomass of fish (Gilligan, 2005). There has been a corresponding decline of native species, with little evidence of effective control of invasive species. Species such as carp not only compete with native species but also alter habitats, and their management is not straightforward. Native fish are more successful in breeding in wet years, and the reintroduction of environmental flows and over bank flows is needed for ongoing improvement.

There are some possible solutions being developed, such as a research programme aimed at producing 'daughterless carp' to help reduce carp number and distribution. Another technique uses the ability of carp (unlike most native fish) to 'jump' over in stream into some floodplain lakes and wetlands.

Sharma *et al.* (2007) studied the aquatic flora of different water bodies in district Bikaner (Rajasthan). Bikaner district covers an area of 27,338 km^2. It lies below latitude 27°11 to 29°3′ north and longitude 71°52′ to 76°15′ east. It is bounded on north by Sri Ganganagar district on west part by Jaisalmer, partly by Bahawalpur district (Pakistan), on south-east by Jodhpur. Aquatic flora of different water bodies, viz. drains, ponds, ditches, pits, puddles, marshes and Indira Gandhi Canal (IGC) covered under the area studies. 27 species belonging to 26 different genera to four different group of aquatic vegetation, viz. algae (2), bryophytes (1), pteridophytes (2) and angiosperms (22) have been recorded. The economic importance of some species has also been described in this chapter.

The species which normally stand in water and grow at least a part of their life cycle in either completely submerged or emerged condition are called aquatic plants (Muencher, 1994). The aquatic plants are practically unaffected by climate and seasonal variation in comparison to the land plants. The work on aquatic plants of India is well documented (Subhramanyan, 1962). Kumar & Singh (1976) and Malhotra (1985) have worked on freshwater alga and limnology. M.M. Bhandri (1978) worked on flora of Indian desert, Shetty and Singh (1988) worked on Flora of Rajasthan, Sharma and Tyagi (1979) worked on flora of the north-east Rajasthan. Parmer *et al.* (1985) studied on the vegetation of Bikaner district (Rajasthan).

Bikaner has many water reservoirs such as drains, ponds, ditches, pits, puddles, marshers, etc. rich in aquatic flora. Some aquatic plants are introduced from Punjab to Bikaner by Indira

Gandhi Canal. A thorough survey of aquatic vegetation was made during 2002-04. The criteria for the selection of plants as specimen was that the plant must either be floating or submerged in lentic or lotic condition. The aquatic flora is concerned with different water bodies. The specimens were collected in the polythene bags and brought to the laboratory for the preparation of herbarium material.

The identification of the species is based on the monographs and Floras of Biswas & Colder (1936). Subhramanyam (1962), Sharma & Khan (1980), Sharma & Tyagi (1979), Shetty & Singh (1988), Bhandari (1978). Some standard publication of Arber (1920), Gangulee, Das & Dutta (1972), Gupta (2001), Joseph & Joseph (1986), Kumar & Singh (1996) were also consulted the herbarium specimens with the vaucher specimens kept in the BSI, Jodhpur & Herbarium of Botany Department, Dungar College, Bikaner (Raj.) The system of classification followed is based on Benthan & Hooker.

Table 2: Taxonomic enumeration of the species in Rajasthan (After Sharma *et al.,* 1985).

Kingdom Frequency	Name of Species/Family	Common Name (Hindi/English) Occurrence	Characteristics including Habitat/Economic importance, if any.	%age
ALGAE	*Chara zeylanica* Wild (Characeae)	Stone wort Ponds	Submerged, dirty green, well branched, upto 30 cm in height, in deep water	14.81%
ALGAE	*Nitella furcata* (Roxb.) Ag. (Characeae)	Stone Wort or Little Star, Ponds	Submerged, golden green, well branched, height upto 20 cm in deep water.	11.31%
BRYOPHYTES	*Funaria hygrometrica* Hedw.(Funariaceae)	Moss Ponds	A marshy herb, on moist soil near bank of ponds, length upto 2 cm, leaves with a distinct midrib, divided into capsule, stem and leaves.	8.45%
PTERIDOPHYTES	*Marsilea minuta* Linn. (Marsileaceae)	Pepper Wort Ponds, Ditches	An amphibious herb, usually with 2 sporocarp, leaves with long petiole upto 15 cm.	25.55%
PTERIDOPHYTES	*Adiantum caudatum* Forsk (Pteridaceae)	Walking Fern Ponds, Lakes	An evergreen stone species, pinnae oblongobtuse, being wedged-shaped at the base	12.25%
ANGIOSPERM	*Eichornia crassipes* (Mart) Solms. (Ponterideriaceae Kunth)	Jalkumbhi	A free floating herb, in polluted water, light violet flowers round the year, bulbous and spongy petiole and used as food for horded and pigs.	24.25%
ANGIOSPERM	*Hydrilla verticillata* (Linn.f.)Royle, (Hydrocharitaceae Juss)	Jhangi Ponds	A submerged and rooted herb, slender and elongated stem, leaves in whorls no. 5-6, flowers during Sep.-Dec.	51.89%

Table 2: Contd.

Kingdom Frequency	Name of Species/Family	Common Name (Hindi/English) Occurrence	Characteristics including Habitat/Economic importance, if any.	%age
ANGIOSPERM	*Celosia argentea* Linn. (Amaranthaceae)	Imarti, Kukari Small Ditches, Canal	A very variable plant, assuming different appearance according to habitat, common weed is cultivated.	24.55%
ANGIOSPERM	*Ipomoea pescapre* (Linn.) Swet (Convolvulaceae Juss.)	Do Patti lata Ponds, Ditches	A perennial glabrous trailing and rooting at nodes or twinning, purple colour flower during Aug.-March.	21.54%
ANGIOSPERM	*Jussiaea repens* Linn. (=*Ludwigia repens* Forst.) (Onagraceae Juss.)	Water Priomrose Canals, Ponds.	A herb, creeping and floating by white spongy respiratory roots, yellowish while flowers during Feb.-March.	21.89%
ANGIOSPERM	*Wolffia globosa* (Roxb.)Hartog & Van der Plas (Lemnaceae S.F. Gray).	Water Meal Ponds, Ditches	A pinhead size thallus, forming green colour on water suface of ponds.	45.50%
ANGIOSPERM	*Nymphoides cristatum* (Roxb.) O. Ketz (Gentianiceae Juss.)	Water Fringe Ponds	A small floated and rooted herb, smooth leaves, white flowers during Jun-Aug.	24.00%
ANGIOSPERM	*Polygonum plebium* R. Br. (Polygonaceae Juss.)	Water Pepper Ponds, Ditches.	A rooted amphibious herb, fully open flowers are bright, pinkish or red, but the colour fed after.	35.66%
ANGIOSPERM	*Aponogeton natans* (L.)Engl. & Kraus. (Aponogetonaceae).	———- Ponds, tanks.	A submerged, aquatic herb, long petiolated leaves, leaves, flowers during Sept.-Oct.	42.88%
ANGIOSPERM	*Rumex dentatus* Linnj. (Polygonaceae Juss.)	Pani Paloak Ponds, Ditches, Drains.	A marshy herb, near drains, ditches, etc., flowering during Feb.-April, red dye obtained from roots.	77.7%
ANGIOSPERM	*Ranunculus sceleratus* Linn. (Ranunculaceae Juss.)	Jal Dhania Ponds, Ditches.	An erect, submerged and root here, 8-60 cm tall, fistular stem, dissected leave, lobe, ovate, yellow flowers.	57.55%

Table 2: Contd.

Kingdom Frequency	Name of Species/Family	Common Name (Hindi/English) Occurrence	Characteristics including Habitat/Economic importance, if any.	%age
ANGIOSPERM	*Trapa bispinosa* Roxb. (Trapaceae Dumort.)	Water chestnut Ponds.	A floating and rooted herb, spongy and swollen petiole, white flowers during Sept.-Oct. Fruit edible.	25.08%
ANGIOSPERM	*Veronica anagallis-aquatica* Linn. (Secrophulariaceae Juss.)	Water Speed Weed Ditches, Drains Marxhes.	A marshy, submerged and rooted herb, opposite decussate leaves, white flowers during Mar.-April.	12.22%
ANGIOSPERM	*Ceratophyllum, demersum* Linn. (Ceratophyllaceae S.F. Gray).	Hornh Wort Ponds	A submerged free loating herb, root absent, leaves, finely dissected, flowers during Sept-Dec.	58.02%
ANGIOSPERM	*Potamogeton crispus* Linn. (Potamogetonaceae Dumort,. =Najadaceae Juss.).	Crisped Leaf Pond Weed Ponds.	A submerged, floating and rooted herb, long spongy, stem, linear serrated leaves, pedicellated yellow	39.07%
ANGIOSPERM	*Potamogeton nodosus* poir. (Potamogetonaceae Dumort.)	Crisped Leaf Pond Weed Ponds.	A submerged herb, rooting at nodes, long. Spongy stem flowering during May.-Sept.	24.45%
ANGIOSPERM	*Typha elephantina* Roxb. (Typhaceae Juss.)	Pater Ponds, Canal, Ditches.	An amphibious herb, in waterlogged soil, leaves linear, semi-cylindrical, blade above sheath, flowering during April-Dec., used in making carpets, hand fans in pain relief and to increase milk in mothers.	45.33%
ANGIOSPERM	*Cyprus rotundus* Linn. (Cypraceae Juss.)	Nagarmotha Canals, Ponds.	A marshy herb of rice crop, upto 1 ft. in height, leaves used in making handfans and also used in various herbal preparations	78.45%
ANGIOSPERM	*Eclipta alba* L., (Asteraceae Dumort.).	Bhringraj Ditches, Drains.	A marshy herb, near drains, ditches, etc., serrated and ovate leaves, juice used in spleen enlargement and oil used as a hair tonic.	56%

Table 2: Contd.

Kingdom Frequency	Name of Species/Family	Common Name (Hindi/English) Occurrence	Characteristics including Habitat/Economic importance, if any.	%age
ANGIOSPERM	*Saccharum bengalensis* Retz. (=*S. munja* Roxb.) (Poaceae Barnh.).	Mnja Ponds, Ditches	A submerged and rooted herb, near the bank in ponds, flowers during Jan.–March, used in thatching.	84.48%
ANGIOSPERM	*Naja graminea* Del. (Najadaceae).	——- Tanks, Ponds	A submerged grass-like herbs, muricated stem, rooting at lower nodes, leaves linear, marginal spinules minute, numerous, sheathing at the base.	25%
ANGIOSPERM	*Bergia suffruticosa* (delile)Frenzl. (Elatinaceae)	Kharbuja, Rowan Tanks, Ponds.	A shrubby plant, leaves, long, glandular hair on both sides, commonly found in cultivated fields.	35.50%

Alam and Khan (1996) worked on dynamics of plankton communities in four freshwater lentic ecosystems in relation to varying dominant biota. Barush *et al.* (1996) worked on heavy metal distribution in river Jhangi in India. Das & Sinha (1994) studied on pollution status of river Ganga at Patna (Bihar). Chapman (1996) completed studies on water quality assessment a guide to the use of biota, sediments and water in environmental monitoring. Jamil (1993) has worked on the role of macrophytes in aquatic ecosystems. Jabanesan & Selvanayagam (1994) studied on population dynamics of aquatic hemipterans in river Cooum, Madras. Kataria & Iqbal (1995) analysed the trace elements in lower lake water Bhopal. Kaur *et al.,* (1996) worked on abiotic and biotic components of a freshwater pond of Patiala. Kumaresan & Bagavathiraj (1996) studied on the physico-chemical and microbiological aspects of courtallam water. Munawar (1974) the limnological studies on the freshwater ponds of Hyderabad, India. Nag & Das (1994) studied on quality of drinking water in the Birbhu district of West Bengal. Parvateesam & Sudha (1994) worked on physico-chemical characteristics of a lake receiving effluents from textile mills in Rajasthan. Ramachandra & Ahalya (2001) studied the monograph on essential in limnology and GIS. Borse (1988) worked on frequency of occurrence of marine fungi from Maharashtra. Chaudhari (1988) studied on biotic communities in endangered aquatic and mangrove environment. Das *et al.* (1994) worked on the karyotype diversity in three species of Heritiera a common mangrove tree on the Orissa coast. Mani (1992) worked on natural phytoplankton communities in Pitchavaram mangroves in India.

In recognition of the anthropogenic effect on water bodies, Government of India enacted a legislation to prevent and control water pollution and conservation of aquatic biodiversity way back in 1974. These efforts have promoted studies on water quality and the effect of water pollution on aquatic biodiversity throughout India.

REFERENCES

Alam, Aftab, and Khan, A.A. (1996). Dynamics of plankton communities if four freshwater lentic ecosystems in relation to varying dominant biota. *Polln. Res.* **15(3):** 287-291.

Arber, A. (1920). Water plants: A Study of Aquatic Angiosperms, Cambridge Univ Press, London.

Baurah, N.K. (1996). Heavy metal distribution in river Jhangi, *Indian J. Environ. Port*, **16(4):** 290-293.

Baurah, N.K., Kotoky, P., Bhattacharya, K.G. and Borah, G.C. (1996). Heavy metal distribution in river Jhangi, *Ind. J. Environ. Port.* **16(4):** 290-293.

Bentham, G. and Hooker, J.D. (1862-1883). Genera Plantarum, etc. 3 Vols, London.

Bhandari, M.M. (1978) Flora of The Indian Desert, Scientific Publishes, Jodhpur.

Biswas, K. and Calder, C.C. (1936). Handbook of Common Water & Marsh Plants of India & Burma, Bishen Singh & Mahendra Pal Singh Publ., Dehradun.

Borse, B.D. (1988). Frequency of occurrence of marine fungi from Maharashtra coast, India. *Indian J. Mar. Sci.* **17(2):** 165-167.

Gangulee, H.C., Das K.S. and Dutta, C. (1972). College Botany Vol I. The Central Book Agency, Kolkata.

Chaudhari, A.B. (1988). Biotic communities in endangered aquatic and mangrove environment. *Sci. Cult.* **54(6):** 182-187.

Das, A.B, Basak, U.C. and Das, P. (1994). Karyotype diversity in three species of Heritiera, a common mangrove tree on the Orissa coast. *Cytobios* **80:** 71-78.

Das, N.K. and Sinha, R.K. (1994) Pollution status of river Ganga at Patna (Bihar). *Ind. J. Freshwater Bio.* **6(2):** 159-161.

Gupta, O.P. (2001). weedy Aquatic Plants : Their Utility Mence & Management, Agrobios (India), Jodhpur.

Jamil, K. (1993). The role of macrophyte in aquatic ecosystems, *J. Freshwater Biology* **5(2):** 29-36.

Joseph, J. and Joseph, K.M. (1986). Insectivorous Plants of Khasi and Jaintia Hills, Meghalaya, BSI, Kolkata.

Kataria, H.C. and Iqbal, S.A. (1995). Analysis of trace elements in lower lake water Bhopal. *Oriental. J. chem.* **11(3):** 288-289.

Kaur, H., Dhillon, S.S., Bath, K.S. and Mander, G. (1996). Abiotic and biotic components of a freshwater pond if Patiala (Panjab), *Polln. Res.* **15(3):** 253-256.

Kumar, H.D. and Singh, H.N. (1976). A Text Book of Algae, East-West Press, New Delhi.

Malhotra, M.B. (1985). Work Book on Limnology, Pratibha Publ, Sagar.

Muencher, W.C. (1994). Aquatic Plants of the United States, Comstock Publ Co., New York.

Nag, J.K. and Das, A.K. (1994). Quality of drinking water in the Birbhu district of West Bengal, *Ind. J. Environ. Prot.* **14(7):** 516-519.

Parmer, P.J., Pandey, R.P. and Roy, G.P. (1985) Studies on the Vegetation of Bikaner District, Rajasthan, *Ibid.*

Ramachandra, T.V. and Ahalya, N. (2001). Monograph on "Essentials in Limnology and GIS", Publication of Karnataka Environment Research Foundation.

Sharma, Shiva, and Tyagi, B. (1979). Flora of the North-East Rajasthan, Kalyani Publ., New Delhi.

Sarma, Y.S.R.K. and Khan, M. (1980). Algal Taxonomy in India Bot. Reco. & Monograph (2), Today & Tomorrow's Print and Publ., New Delhi.

Shetty, B.V. and Singh, V. (1988). Flora of Rajasthan Vol. I BSI, Kolkata.

Singh, U., Wadhwani, A.M. and Johri, B.M. (1996). Dictionary of Econamic Plants in India, ICAR, New Delhi.

Subhramanyam, K. (1962). Aquatic Angiosperms, Botanical Monograph (3) CSIR, New Delhi.

Diversity of Lower Plants (2014): pp. 222-234
Editors: Rajan Kumar Gupta & Mukesh Kumar
IK International Publishing House Pvt. Ltd.

17

Threats to Phytodiversity

SUMEET GAIROLA, K.N. BHARDWAJ* AND C.S. RANA

Department of Botany and Microbiology, HNB Garhwal University,
Srinagar Garhwal–246 174, Uttarakhand
e-mail: kailash0011@rediffmail.com

ABSTRACT

The basic aim of biological conservation is to prevent individual species, or sometimes whole communities, from becoming extinct either regionally or globally. But without proper knowledge of threats that particular species or community is facing for its survival, no conservation strategy can be fruitful. In this review paper, we have tried to summarize the threats that phytodiversity is facing today. In this study habitat loss, overexploitation and harvesting, pollution, exotic species invasion and climate change are identified as the major threats to the phytodiversity. More research is needed in this matter for proper understanding of the mechanism and extent of these threats. Authors suggest that for proper implementation of any conservation programme planners should also take these threats into consideration. As far as threats endangering microbial diversity are concerned, there is still much research to be done.

Keywords: Habitat loss, Forest clearance, Anthropogenic disturbances, Exotic species.

1. INTRODUCTION

Biodiversity is essential for human survival and economic well-being and for the ecosystem function and stability (Singh, 2002). The current biological diversity is a product of millions of years of evolution (Wilson, 1992). Many kinds of environmental changes influence or determine processes that can both augment and erode diversity (Sheil, 1999). Various ecosystem functions and services may be negatively affected by the loss of biodiversity (Schulze & Mooney, 1994). In the long term, declining species diversity may lead to a decrease in the ecosystem's ability to buffer against perturbation (Naeem & Shibin, 1997). Extinction is a principle of evolution and well over 99% of the species that have ever lived have become extinct. Nevertheless, we seem

*Communicating Author

at present to be living in a period of mass extinction, the 6th of its kind, in which human activity such as that which causes habitat fragmentation seems to be the driving force (Weinbauer & Rassoulzadegan, 2007). Large number of plant species have already been lost and summary of the plant taxonomic classes which are listed as extinct, extinct in the wild and globally threatened with extinction have been given in Table 1. Political and scientific concerns have been raised as we are experiencing an increase in species extinction rates caused by anthropogenic activity (Ehrlich & Wilson, 1991). Habitat destruction, overexploitation, pollution and species introduction are the major causes of biodiversity loss (UNEP 2001). Conservation biology relies on an understanding of the threats faced by individual species and by whole community. For proper management and conservation of biodiversity, there is grave need of understanding the threats it is facing. Keeping aforesaid facts in view, in this review article we have tried to summarize major threats biodiversity is facing across the globe, which will be discussed in the following sections.

Table 1: Summary of the numbers of the species of the plant taxonomic classes which are listed as extinct, extinct in the wild (the species has been extirpated from its natural habitat), or globally threatened with extinction

	Extinct	Extinct in the wild	Threatened
Bryopsida	2	0	36
Anthocerotopsida	0	0	2
Marchantiopsida	1	0	42
Coniferopsida	0	1	140
Ginkgoopsida	0	0	1
Magnoliopsida	69	14	5099
Liliopsida	1	2	291
Total	73	17	5611

Adapted from Hilton-Taylor, 2000

1.1 Habitat Loss and Fragmentation

Although, biodiversity is essential to humankind, brings innumerable benefits, and has other important values, humans have had strong negative impacts on biodiversity. The primary threat to biodiversity is habitat loss and fragmentation. A conceptual model illustrating humanity's direct and indirect effects on the earth system and biodiversity is given in Fig. 1. Forest fragmentation replaces large areas of native forest by other ecosystems leaving isolated forest patches, with deleterious consequences for most of the native forest biota (Murica 1995). Habitats may be adversely affected by following three main ways.

1.1.1 Destruction

The destruction of habitats is the primary reason for the loss of biodiversity in terrestrial and coastal ecosystem (Sala *et al.*, 2000). In it a proportion of habitat available to a particular species may simply be destroyed, for urban and industrial development or for production of food and other natural resources such as timber (Townsend *et al.*, 2000). Some of the reasons for destruction of habitats are as the following:

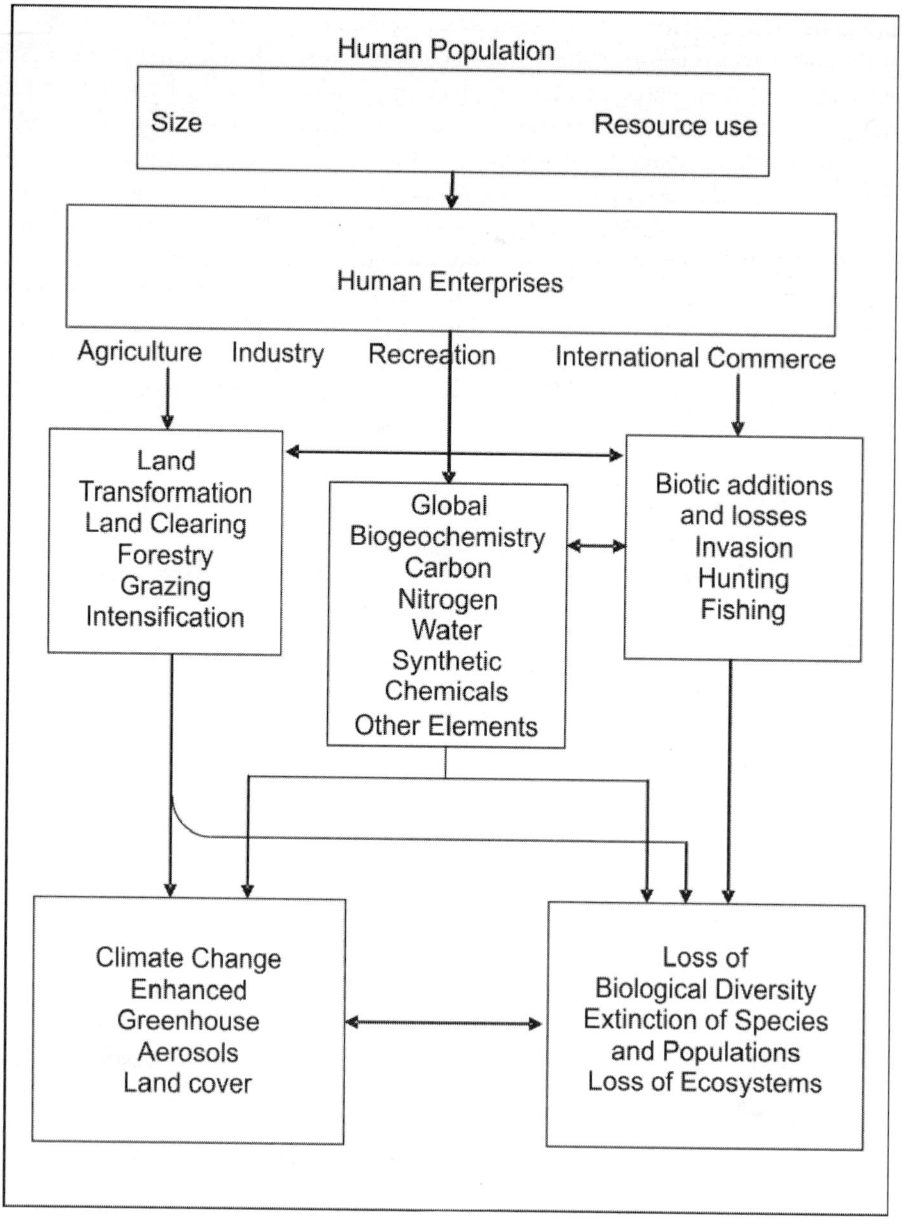

Fig. 1. A conceptual model illustrating humanity's direct and indirect effects on the earth system (From Vitousek *et al.*, 1997b).

1.1.2 Forest clearance

One of the pronounced effects of human activities on the global biosphere is the dramatic change in ecosystem productivity due to deforestation and the resulting loss of biological diversity (Kimmins, 1997). Deforestation is a primary driver of biotic extinctions in the tropics (Sala *et al.*, 2000; Sodhi & Brook, 2006). Forests covered about 50% of the earth's land area 8000 years ago, as opposed to 30% today (Ball, 2001). Today, the forests are being cleared at faster pace than ever before. With this speed most of the forests of the world will be lost leading to large-scale loss of biodiversity. For example, with the current level of deforestation, by 2100 only about 10% of the land area of the Indian Himalaya will be covered by dense forest (> 40% canopy cover)—a scenario in which almost a quarter of the endemic species could be wiped out, including 366 endemic vascular plant taxa and 35 endemic vertebrate taxa (Pandit *et al.,* 2007).

1.1.3 Land use change (LUC)

LUC is one of the major reasons behind destruction of the habitat. Large portions of forests are being encroached for fulfilling requirement of ever-increasing human population. Land conversion from natural ecosystems to simplified agro-ecosystems result in a substantial reduction in biodiversity and a decrease in habitat for displaced wildlife and plant communities (Cassman *et al.*, 2003).

2. RIVER VALLEY PROJECTS

Other cause of habitat destruction is construction of large hydroelectric projects, which have several benefits such as power, irrigation, tourism and industrial development, etc. But from the biodiversity point of view, such projects are unethical as they impede the biological web of the environment, as a large hydel project leads to a complete alteration of a terrestrial habitat into an aquatic ecosystem (Gaur, 2007). Modifications in normal flow of river systems result in several complexities to people, forests, fauna, flora, microorganisms and land use pattern (Sharma, 2006). In general, loss of biodiversity cannot be compensated with the economic growth, and such projects aimed to overall development of populace, ever remain questionable as to their real value (Gaur, 2007).

3. DEGRADATION

Due to unplanned use of natural resources and indiscriminate use of chemical fertilizers, pesticide and various other chemicals to increase agricultural productivity, huge pressure has mounted on environment (Gairola *et al.*, 2009). The application of strong dosage of pesticides has resulted in the killing of several useful microorganisms, friendly parasites, birds, beneficiary predators and helpful worms and nematodes. By clearing all the weeds from the fields using deadly chemicals (herbicides), the food crop plants have become more susceptible to pests/diseases. The modern scientific strides of man have contributed to faster and ruthless extinction of large number of farmer friendly species and crop varieties (Narayana, 2006). Acid rain also degrades various ecosystems as diverse as forest, agriculture, pond, river, etc., causing large-scale harm to biodiversity. Aquatic environments are particularly vulnerable to pollution. Land use changes, waste disposal, and water impoundment and abstraction can profoundly affect their patterns of water flow and quality (Allan & Flecker, 1993). In many countries of the developing world slash and burn (Jhum/shifting cultivation or swidden agriculture) greatly affects forest structure and species composition by creating a mosaic of forest patches of different ages. Intensive forestry, agriculture and eutrophication are some of the other causes of habitat degradation.

4. DISTURBANCE

Disturbance of an ecosystem has implications for the maintenance and restoration of biodiversity at all hierarchical levels (Pickett & Parker, 1994). Disruption of forest structure by natural and anthropogenic disturbance alters species richness and other ecosystem properties (Kumar & Ram, 2005). Intensive and prolonged human disturbances may lead to a decline in species diversity, change in species composition and forest degradation (Sagar *et al.*, 2003). Each disturbance affects forests differently. Some cause large-scale tree mortality, whereas others affect community structure and organization without causing massive mortality (e.g., ground fires) (Dale *et al.*, 2001). Forest disturbances can be classified into two categories, viz., natural and human-induced or anthropogenic. Disturbances, both human-induced and natural, shape forest systems by influencing their composition, structure, and functional processes (Dale *et al.*, 2001).

Natural disturbances: These are disturbances which occur naturally without any influence of humans. These disturbances include landslide, wild fire, lightening, windfall, thunderstorms, cyclones, earthquakes, hurricane, tornado, hailstorms, tsunami, landslide, floods, droughts, etc. There is likelihood that coupling of these disturbances with climate change may result in devastating consequences. Thus, if climate change can alter the distribution, extent, frequency, or intensity of any of these disturbances, large impacts (such as loss of species regeneration) could be expected (Dale *et al.*, 2001).

5. ANTHROPOGENIC DISTURBANCES

These disturbances can be directly or indirectly attributed to the human activities. Some of the major anthropogenic disturbances are human induced fires in forests, forest clearance for timber extraction, fuelwood collection, fodder extraction, extraction of non-timber forest products, land use changes, road making, etc. Fuel wood gathered from forested commons is an important source of domestic energy in rural areas of many poor countries (Heltberg *et al.*, 2000). It has been estimated that more than 2.4 billion people rely directly on traditional biomass fuels for their cooking and heating, and in poor countries biomass use represents over half of residential energy consumption (IEA, 2006). A demand for fuelwood by subsistence agricultural households is the leading cause of world deforestation (Amacher *et al.,* 1996), which in turn causes loss of biodiversity (Heltberg *et al.*, 2000). Uncontrolled grazing and overstocking of livestock prevent regeneration of the tree cover (Kumar & Shahabuddin, 2005), which may negatively affect biodiversity and ecosystem function in fragile ecosystems.

6. EXOTIC SPECIES INVASION

As compared to native species, which are indigenous and found naturally in an environment, exotic species are alien or non-native species, which are introduced from other countries and areas. Invasive alien species are those subsets of alien species whose establishment and spread threatens ecosystems, habitats or species with economic or environmental significance (McNeelay *et al.*, 2001). After habitat destruction, the exotic species are now considered the second greatest threat to biological diversity (Simberloff, 1995). Invasive species may lead to loss of species diversity, increase in erosion and affect the physical structure of the habitat (Singh *et al.*, 2006). At the continental and global scale, the invasive species diminish the regional distinctiveness of flora and fauna (Vitousek *et al.*, 1997a). Impact of invasion on community structure and ecosystem process in different parts of the world is shown in Table 2. All the exotic species are not harmful, some introductions have enriched human existence and most invaders have minor consequences. For example, about 18% of the Indian flora constitutes adventive aliens, of which 55% are American, 10% Asian, 20% Asian and Malaysian and 15% European and Central Asian species (Nayar, 1977). Williamson (1996) suggests that as a useful rule of thumb, 10% of introduced invaders become established, and 10% of those established become pests. Introduced species have most frequently caused species extinctions through predation/parasitism. Perhaps some of the best-documented examples have concerned the introduction of exotic predators to lakes and islands and the consequent extinction of plants and animals that had evolved no defences against them (Gaston & Spicer, 2004). List of the some of world's worst invasive alien species of plants is given in Table 3. *Chromolaena odorata* (*Eupatorium odoratum*), *E. adenophorum*, *Lantana camara*, *Mikania micrantha*, *Mimosa invisa* and *Parthenium hysterophorus* among terrestrial exotics, and *Eichhornia crassipes* and *Pistia stratiotes* and aquatics, have posed serious threat to the native flora of India (Singh *et al.*, 2001).

Table 2: World's worst invasive alien species of plants. These were chosen according to their adverse effects on biodiversity and/or human activities and their illustration of important issues surrounding biological invasion

Aquatic Plants	
Caulerpa seaweed (*Caulerpa taxifolia*)	Common cordgrass (*Spartina anglica*)
Wakame seaweed (*Undaria pinnatifida*)	Water hyacinth (*Eichhornia crassipes*)
Land Plants	
African tulip tree (*Spathodea campanulata*)	Black wattle (*Acacia mearnsii*)
Brazilian pepper tree (*Schinus terebinthifolius*)	Chromolaena (*Chromolaena odorata*)
Cluster pine (*Pinus pinaster*)	Cogon grass (*Imperata cylindrica*)
Fire tree (*Myrica faya*)	Giant reed (*Arundo donax*)
Gorse (*Ulex europaeus*)	Hiptage (*Hiptage benghalensis*)
Japenese knotweed (*Polygonum cuspidatum*)	Kahili ginger (*Hedychium gardnerianum*)
Koster's curse (*Clidemia hirta*)	Kudzu (*Pueraria montana*)
Lantana (*Lantana camara*)	Leafy spurge (*Euphorbia esula*)
Leucaena (*Leucaena leucocephala*)	Melaleuca (*Melaleuca quinquenervia*)
Mesquite (*Prosopis glandulosa*)	Miconia (*Miconia calvescens*)
Mile-a-minute weed (*Mikania micrantha*)	Mimosa (giant sensitive plant) (*Mimosa pigra*)
Prickley pear cactus (*Opuntia stricta*)	Privet (*Ligustrum robustum*)
Pumpwood (*Cecropia peltata*)	Purple loosestrife (*Lythrum salicaria*)
Quinine (*Cinchona pubescens*)	Shoebutton ardisia (*Ardisia elliptica*)
Strawberry guava (*Psidium cattleianum*)	Tamarisk (Saltcedar) (*Tamarix ramosissima*)
Wedetia (Singapore daisy) (*Wedelia trilobata*)	Yellow Himalayan raspberry (*Rubus ellipticus*)

Adapted from Lowe *et al.*, 2000.

Table 3: Impact of invasion on plant community structure and ecosystem processes

Plant community structure
• Reduced species diversity in area invaded by *Heracleum mantegazzianum* compared with un-invaded region in Czech Republic. • Lower richness of seedling and sapling in area invaded by Norway maples. • *Orbea variegata* reduced the diversity of annual plants and performance of chenopods in Australian shrubland (*Orbea* impact was mediated via reduced water availability). • Impact of *Tatarian* honeysuckle on the diversity and cover of native understorey in New England forest due to light competition. • A succulent perennial *Carpobrotus edulis* reduces soil water availability to native shrub which reduces the growth and reproduction in coastal chaparral. • *Bromus tectorum* invasion reduces the amount of soil water available for other plants in Nevada grasslands. • Reduction in competitive performance of *Fynbos* shrub due to increased nitrogen availability caused by *Acacia* invasion.
Nutrient cycling
• *Myrica* and *Acacia* tend to increase available nitrogen in systems they invade in Hawaii. • *Bromus tectorum* reduces nitrogen mineralization rates by having greater carbon-nitrogen and lignin-nitrogen ratios than native species, while similar litter quality effects did not explain nitrogen mineralization under invasive *Hieracium* in New Zealand grassland. • *Hyparrhenia rufa* lowers rate of nitrogen cycling in Costa Rica.
Hydrology
• *Tamarix* in south-western North America increases evapotranspiration by 300-450 mm/yr due to high leaf area. • Increase in water use upto 105-120 mm/yr by *Centaurea solstitialis* in annual grasslands of western North America due to its active growth period in summer. • Decrease in community water use due to water loss by invasion of *Bromus tectorum* in New Zealand. • Exotic annual grasses in California displace competitively deep-rooting native perennials. • Excessive water use by *Acacia* and *Hakea* sp. in South Africa has led to major water loss (estimated at 3×10^9 m^3/yr) and for this reason many rivers do not flow at all or flow only infrequently.
Fire regimes
• Introduced grasses have increased fire frequency more than threefold in seasonally dry shrubland and woodland, effecting biodiversity in Hawaii. • Invasion of annual grasses from Mediterranean grasslands in desert shrubland, converting shrubland to grasslands thereby affecting biodiversity.

Adapted from Singh *et al.*, 2006

7. HARVESTING AND OVEREXPLOITATION

The use of forest resources by local people for day-to-day living is common in much of the tropics. The groups of such people are hunter-gatherers, shifting farmers, forest-in-migrants, and permanent field farmers living near forests. People in Himalayan region, as an example collect firewood, fodder (from woody plants, shrubs and herbs), and litter from forest floor to prepare organic manure. Apart from these, people in tropics collect medicinal plants, mushrooms and morels, berries, seeds, flower, honey, lichens, dyes, resins, canes, fodder, firewood, fibres and other non-timber forest products (NTFPs). In forests extraction of biomass by local people in little quantities continuously is a major problem. Although at one day small extraction from forest may seem negligible but over the period of time it may gradually reduce over all biomass of the forest. These disturbances occur in chronic form involving removal of a small amount of biomass at any given time, but persisting all the year, without any respite for recovery (Singh, 1998). Chronic disturbances like harvesting and overexploitation do not provide time for ecosystem recovery and arrest the regeneration. Species of several trophic levels are tied to tree species and therefore clear-cut logging has a large impact on biodiversity. For example, approximately 300 species of insects and several bird species may depend on a single tree species. Thus, tree species extinctions are conspicuously disastrous (Singh *et al.*, 2006).

8. CLIMATE CHANGE

Climate change is a natural phenomenon and has been occurring since the earth came into being. It has always occurred naturally but at a pace to which the earth has adapted well. United Nations Framework Convention on Climate Change (UNFCCC) defines climate change as "A change in climate which is attributed directly or indirectly to human activity that alters the composition of the global atmosphere and which is in addition to natural climate variability observed over comparable time periods". Global climate change is a complex and multifaceted issue (Sharma & Gairola, 2007), and its impact on phytodiversity needs immediate attention. The unsustainable human activities have resulted in increased concentration of greenhouse gases (GHGs) in the atmosphere. The average atmospheric CO_2 concentration has increased from pre-industrial concentration of 280 μ mol mol^{-1} to 364 μ mol mol^{-1} in 1994, and is currently increasing at a rate of about 1.5 mol mol^{-1} (Kelling & Whorf, 1998). The increase in GHGs are projected to lead to 1-3.5°C in global mean surface temperature and associated increase in sea level by 15-95 cm by 2100. The earth's climate has warmed by approximately 0.6°C over the past 100 years (IPCC, 2001).

Possible consequences of global warming on phytodiversity are summarized in Table 4. International Rice Research Institute, Manila has indicated that with 1°C increase in atmospheric temperature production of rice could reduce by 10%. Temperature decreases by latitude at the rate of 5-10°C/km across various mountains of the world. With increasing warming species and ecosystems are likely to shift from lower latitudes and altitudes to higher ones. A sustained increase of as little as 1°C in mean annual temperature is sufficient to cause significant changes in the growth and regeneration capacity of many tree species. It has been estimated that each 1°C rise in temperature will displace the limits of tolerance of land species some 125 km towards the poles or 150 m vertically on mountains. Climatic zones i.e. the ecosystems and agricultural zones could shift towards the poles by 50-550 km in the mid-latitude regions. Climate change may directly alter plant fitness (Galen & Stanton, 1993), as well as alter the reproductive success of plants and their interactions via impacts on flowering phenology

(Hughes, 2000). Khanduri *et al.* (2008) explored literature on effects of climate change on plant phenology of more than 650 species, and have noticed that an average advancement of 1.9 days per decade in spring events, an average delay of 1.4 days per decade in autumnal phonological events, and an extension of the average length of the growing season by 3.3 days per decade over a time period ranging from 12 to 44 years have occurred.

Table 4: Possible consequences of climate change on phytodiversity

- Increased mean global temperature can disrupt physiological processes of the plant species having narrow tolerance range.
- Increased mean precipitation (5-10% in Northern Hemisphere) causing more case of floods.
- Ice melting in polar and high mountain areas leading to global mean sea level rise and flooding of coastal cities leading to destruction of coastal biodiversity.
- Widespread retreat of ice glaciers making perennial rivers seasonal especially Himalayan rivers.
- Shift of plant range in pole ward or up in elevation leading to loss of habitat.
- Vegetation zones could shift to higher elevations, resulting in the loss of some species and ecosystems. Species and ecosystems with limited climatic ranges could disappear.
- Increased droughts leading to failure of traditional agricultural system.
- Greater occurrence of insect and disease outbreaks in existing plantations as trees comes under increasing stress.
- Changes in phenology of plants, e.g. lengthening of growing season and earlier flowering of trees.
- Increase in cases of exotic species invasion causing threat to native species.
- Increase in natural disturbance, viz. cyclone, wildfire, tornado, ice storms, landslides, torrential rainfalls, etc. leading to destruction of habitat.

9. CONCLUSION

The preservation of genetic diversity is important first because of the long-term evolutionary potential it provides. Small populations tend to have less variation and hence lower evolutionary potential. Each and every species is important and unique in itself. In many cases, the extinction of one species may lead to the extinction of others. Indeed, this is inevitable where this species provides critical resources for others such as specialist herbivores, parasites or predators, or perhaps itself acts as a specialist pollinator of dispersal agent (Gaston & Spicer, 2004). Species with very narrow geographical range, small population size, low reproduction rate, little genetic variability, specialized niche requirements are more vulnerable to these threats. All of these threatening factors operate at different scales but over the period of time they may work in synchrony to reduce the effective population size of the species. Extinction vortices may progressively lower population sizes, leading inexorably to extinction (Fig. 2). In case of higher plants much of the research work is done and assessment of threats is being worked out but in case of microbial diversity lots of work is still needed to be done (see Box 1).

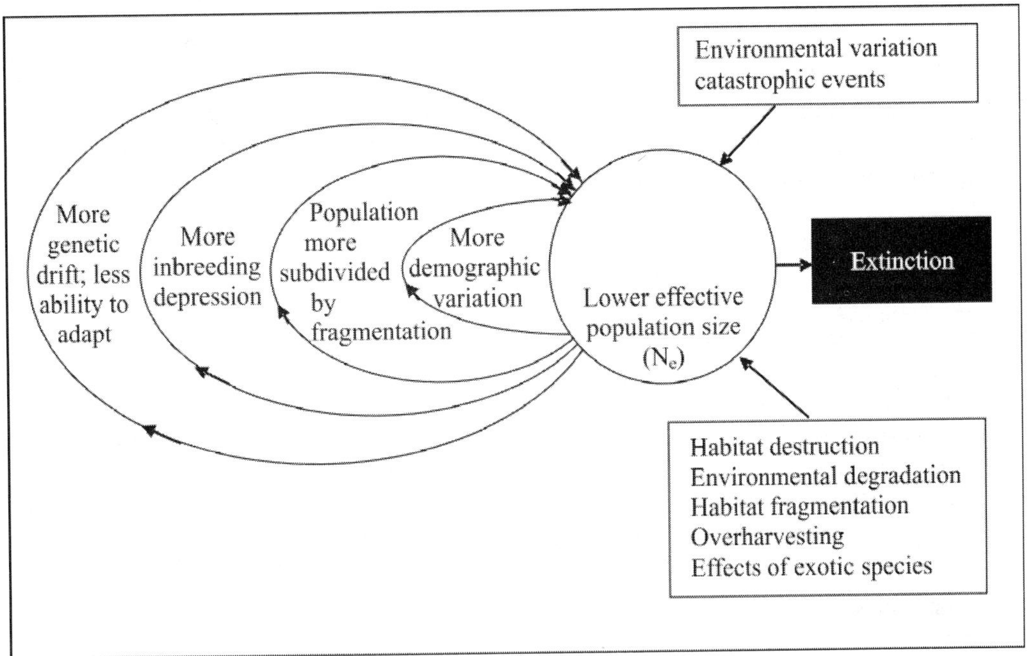

Fig. 2. Extinction vortices may progressively lower population sizes, leading inexorably to extinction (From Primack, 1993).

Box 1

Threats to Microbial Diversity: A Treatise

Various threats to biodiversity influence direction of the evolution. Various biological activities in accordance with the biological interactions accompanied by natural physical and chemical processes render a self-regulating mechanism of the earth (Gaia Hypothesis given by James Lovelock, 1988). It is, however, the fact that anthropogenic activities have affected this self-regulating system to the verge of its critical state. These anthropogenic activities lead to greater ecological dysfunction, which in turn leads to potential threat to biodiversity and microorganisms are not exceptions.

Microbes constitute a great genetic diversity on the planet earth but with a setback of culturability (Torsvik *et al.*, 2002), only of 1% (rest 99% are still unculturable). Due to their diverse metabolic functions, wide and efficient dispersal, tolerance of wide environmental gradients, adaptive genetic recombination and horizontal gene transfer, presence of resting or dormant life forms and wide distributional pattern they have been considered ubiquitous in nature without defined biogeography. Falkowsky *et al.* (2008) consider these characteristics for having low probability of local extinction. The hypothesis of 'everything is everywhere but environment select' (Cosmopolitan Hypothesis) (de Wit and Bouvier, 2006) is in fact based on these characteristics of superspeciality or 'buffering capacity' of these tiny creatures. This concept might be correct but distracted attention from important question of microbial biogeography. Recent advances in this matter clearly indicate that even free-living microbes too have defined biogeography and therefore might be subjected to at least local extinction. Detection limits of rare microbes are however big lacuna while studying the threats of these microbes.

The concept of biogeography has not been sufficiently addressed for free-living microbes. Some evidence (Whitaker *et al.*, 2003; Papke and Ward, 2004; Fierer and Jackson, 2006; Sogin *et al.*, 2006; Weinbauer, 2004) suggests that endemicity can occur in free-living microbes and even biogeography of the abundance in some taxa (although they might be cosmopolitan). Endemicity of microorganisms can mean potential (local) extinction where the respective environments are endangered (Weinbauer and Rassuoulzadegan, 2007). Cosmopolitan microbial populations, even if they are very rare in some environments, are not endangered; nevertheless, local reduction can have an effect on their genetic diversification and their effect on ecosyste functions. Some more evidence of endemicity and biogeography is available for microorganisms associated with plants (e.g., Mycorrhiza) and animals (Rohwer *et al.*, 2002; Taylor *et al.*, 2004).

Local extinction (or reductions in abundance of microbial species) probably occurs quite frequently due to clearing of forest, agricultural activity, erupting volcanoes, forest fire and degradation of natural habitat by anthropogenic activities or by climate change. Hot-spring biodiversity is also drawing attention of the scientific community worldwide (Bhardwaj and Tiwari, 2008) as their habitat may get lost due to excessive resource utilization. Most free-living microbes are likely not endangered, although their local distribution might vary considerably and local extinctions (in the sense of population densities below the detection limit) might occur. Associated species under various biological interactions may be threatened in the same way as the multicellular host or its partner organisms. The conditions and consequences of causes for extinction of microbes remain basically unknown and is the matter of debate among biologists. Clearly more research is needed in this field.

REFERENCES

Allan, J.D. and Flecker, (1993). Biodiversity conservation in running waters. *Bioscience* **43:** 32-43.

Amacher, G.S., Hyde, W.F. and Kanel, K.R. (1996). Household Fuelwood demand and supply in Nepal's Tarai and Mid Hills; Choice between cash outlays and labor opportunity. *World Develop* **24(11):** 1725-1736.

Ball, J.B. (2001). Global forest resources: History and dynamics. In: Evans, J. (Ed.). Forests Handbook Volume 1, Blackwell Science, Oxford, 3-22.

Bhardwaj, K.N. and Tiwari, S.C. (2008). Geothermal energy resource utilization: Perspectives of the Uttarakhand Himalaya. *Curr. Sci.* **95(7):** 846-850.

Cassman, K.G., Dobermann, A., Walters, D.T. and Yang, H. (2003). Meeting cereal demand while protecting natural resources and improving environmental quality. *Ann. Rev. Environ. Resour.* **28:** 315-358.

Dale, V.H., Joyce, L.A., McNulty, S., Neilson, R.P., Ayres, M.P., Flannigan, M.D., Hanson, P.J., Irland, L.C., Lugo, A.E., Peterson, C.J., Simberloff, D., Swanson, F.J., Stocks, B.J. and Wotton, B.M. (2001). Climate change and forest disturbances. *BioScience* **51(9):** 723-734.

de Wit, R. and Bouvier, T. (2006). 'Everything is everywhere, but, the environment selects'; what did Bass Becking and Beijerinck really say? *Environ. Microbiol.* **8:** 755-758.

Ehrlich, P.R. and Wilson, E.O. (1991) Biodiversity studies: science and policy. *Science* **253:** 758-762.

Falkowski, P.G., Fenchel, T. and Delong, E.F. (2008). The Microbial Engines that Drive Earth's Biogeochemical Cycles. *Science* **320:** 1034-1039.

Fierer, N. and Jackson, R.B. (2006). The diversity and biogeography of soil bacterial communities. *Proc. Natl. Acad. Sci.* USA **103:** 626-631.

Gairola, S., Umar, S. and Suryapani, S. (2009). Nitrate accumulation, growth and leaf quality of Spinach beet (*Beta vulgaris* Linn.) as affected by NPK fertilization with special reference to potassium. *Ind. J. Sci. Tech.* **2(2):** 35-40.

Galen, C. and Stanton, M.L. (1993). Short-term responses of alpine buttercups to experimental manipulations of growing season length. *Ecology* **74:** 1052-1058.

Gaston, K.J. and Spicer, J.I. (2004). Biodiversity: An introduction. Blackwell Publishing Company, Oxford, UK.

Gaur, R.D. (2007) Biodiversity and river valley projects in Uttarakhand. *Proc. Nat. Acad. Sci. India* **77(B)III:** 253-262.

Heltberg, R., Arndt, T.C. and Sekhar, N.U. (2000). Fuelwood consumption and forest degradation: A household model for domestic energy substitution in rural India. *Land Econ.* **76(2):** 213-232.

Hilton-Taylor, C. (Comp.) (2000). IUCN Red List of Threatened Species. IUCN, Gland. (Also available at http://www.redlist.org).

Hughes, L. (2000). Biological consequences of global warming: is the signal already apparent? *Trends Ecol. Evol.* **15:** 56-61.

Intergovernmental Panel on Climate Change (IPCC) (2001). Third assessment report on climate change. *In*: McCarthy, J.J., Canziani, O.F., Leary, N.A., Dokken, D.J., White, K.S. (Eds.). Impacts, adaptation and vulnerability. Cambridge University Press, Cambridge.

International Energy Agency (IEA) (2006). Working Party on Renewable Energy Technologies, Exploiting Alternative Energy Sources. Mobilizing Energy Technology. International Energy Agency (IEA), Head of Publications Service, Paris, France, 32-42.

Kelling, C.D. and Whorf, T.P. (1998). Atmospheric CO_2 records form sites in the SIO air sampling network, in trends: A compendium of data of global change Carbon dioxide. Information Analysis Center, Oak Ridge National Laboratory, Oak Ridge, TN.

Khanduri, V.P., Sharma, C.M. and Singh, S.P. (2008). The effects of climate change on plant phenology. *Environmentalist* **28:** 143-147.

Kimmins, J.P. (1997) Biodiversity and its relationship to ecosystem health and integrity. *For Chron.* **73:** 229-232.

Kumar, A. and Ram, J. (2005). Anthropogenic disturbances and plant biodiversity in forests of Uttaranchal, Central Himalaya. *Biodiver. Conserv.* **14:** 309-331.

Kumar, R. and Shahabuddin, G. (2005). Effects of biomass extraction on vegetation structure, diversity and composition of forest in Sariska Tiger Reserve, India. *Environ. Conserv.* **32:** 248-259.

Lovelock, J.E. (1988). The Ages of Gaia: A Biography of Our Living Earth. Norton, New York.

Lowe, S., Browne, M., Boudjelas, S. and De Poorter, M. (2000). 100 of the World's Worst Invasive Alien Species: A Selection from the Global Invasive Species Database. IUCN/ISSG.

McNeelay, J.A., Mooney, H.A., Neville, L.E., Schei, P. and Waage, J.K. (Eds.). (2001). A Global Strategy on Invasive Alien Species. IUCN Gland, Switzerland, and Cambridge, UK, in collaboration with the Global Invasive Species Programme.

Murica, C. (1995). Edge effects in fragmented forests: Implications for Conservation. *Trends Ecol. Evol.* **10:** 58-62.

Naeem, S. and Shibin, L. (1997). Biodiversity enhances ecosystem reliability. *Nature* **390:** 507-509.

Narayana, K.S. (2006). Biodiversity conservation should be the mantra of the century. *Kurukshetra* **55(2):** 3-11.

Nayar, M.P. (1977). Changing patterns of the Indian flora. *Bull. Bot. Surv. Ind.* **19:** 145-154.

Pandit, M.K., Sodhi, N.S., Koh, L.P., Bhaskar, A. and Brook, B.W. (2007). Unreported yet massive deforestation driving loss of endemic biodiversity in Indian Himalaya. *Biodiver. Conserv.* **16:** 153-163.

Papke, R.T. and Ward, D.M. (2004). The importance of physical isolation to microbial diversification. *FEMS Microbiol. Ecol.* **48:** 293-303.

Pickett, S.T.A. and Parker, V.T. (1994). Avoiding the old pitfalls: opportunities in a new discipline. *Restor. Ecol.* **2:** 75-79.

Primack, R.B. (1993). Essentials of Conservation Biology. Sinauer Associates Sunderland, M.A.

Rohwer, F., Seguritan, V., Azam, F. and Knowlton, N. (2002). Diversity and distribution of coral-associated bacteria. *Mar. Ecol. Prog. Ser.* **243:** 1-10.

Sagar, R., Raghubanshi, A.S. and Singh, J.S. (2003). Tree species composition, dispersion and diversity along a disturbance gradient in a dry tropical forest region of India. *For Ecol. Manage* **186:** 61-71.

Sala, O.E., Chapin, F.S. III, Armesto, J.J., Barlow, E., Bloomfield, J., Dirzo, R., Hubersonwald, E., Huenneke, L.F., Jackson, R.B., Kinzig, A., Leemans, R., Lodge, D.M., Mooney, H.A., Oesterheld, M., Poff, N.L., Sykes, M.T., Walker, B.H., Walker, M. and Wall, D. (2000) Global biodiversity scenarios for the year 2100. *Science* **287:** 1770-1774.

Schulze, E.D. and Mooney, H.A. (Eds.). (1994). Biodiversity and ecosystem function. Springer, New York, 25.

Sharma, C.M. and Gairola, S. (2007). Prospects of Carbon Management in Uttarakhand: An Overview. Samaj Vigyan Shodh Patrika, Special Issue (Uttarakhand-1) 23-34.

Sharma, V. (2006). State of water resources in the country. *In*: Nautiyal, R. and Nautiyal, P. (Eds.). Compendium: Environmental Audit of Hydroelectric Projects for Sustainable Development, Srinagar-Garhwal: Printmedia, 14-21.

Sheil, D. (1999). Tropical forest diversity, environmental change and species augmentation: after intermediate disturbance hypothesis. *J. Veg. Sci.* **10:** 851-860.

Simberloff, D. (1995). Why do introduced species appear to devastate islands more than mainland areas? *Pacific Sci.* **49:** 87-97.

Singh, J.S. (2002). The biodiversity crisis: a multifaceted review. *Curr. Sci.* **82:** 638-647.

Singh, J.S., Singh, S.P. and Gupta, S.R. (2006). Ecology Environment and Resource Conservation. Anamaya Publishers, New Delhi, India.

Singh, S.P. (1998). Chronic disturbance, a principal cause of environmental degradation in developing countries (Editorial). *Environ. Conserv.* **25:** 1-2.

Singh, N.P., Sharma, J.R., Singh, K.P. and Mudgal, V. (2001). Species diversity in angiosperms, *In*: Singh, N.P. and Singh, D.K. (Eds.). Angiosperms Species Diversity and Conservation Strategies in India, *Botanical Survey of India* **4:** 1631-1674.

Sodhi, N.S. and Brook, B.W. (2006). South-east Asian biodiversity in crisis. Cambridge University Press, London, UK.

Sogin, M.L., Morrison, H.G., Huber, J.A., Welch, D.M., Huse, S.M., Neal, P.R., Arrieta, J.M. and Herndl, G.J. (2006). Microbial diversity in the deep sea and the underexplored 'rare biosphere'. *Proc. Natl. Acad. Sci. USA* **103:** 12115-12120.

Taylor, M.W., Schupp, P.J., Dahllof, I., Kjelleberg, S. and Steinberg, P.D. (2004). Host specificity in marine sponge-associated bacteria, and potential implications for marine microbial diversity. *Environ. Microbiol.* **6:** 121-130.

Torsvik, V., Ovreas, L. and Thingstad, T.F. (2002). Prokaryotic diversity–magnitude, dynamics, and controlling factors. *Science* **296:** 1064-1066.

Townsend, C.R., Harper, J.L. and Begon, M. (2000). Essentials of Ecology. Blackwell Science Inc. Oxford, England.

UNEP. (2001). India: State of the Environment-2001. United Nations Environment Programme.

Vitousek, P.M., D'Antanio, C.M., Loope, L.I., Rejmanek, M. and Westbrooks, R. (1997). Introduced species: A significant component of human-caused global change. *New Zealand J. Ecol.* **21:** 1-16.

Vitousek, P.M., Mooney, H.A., Lubchenco, J. and Melillo, J.M. (1997). Human domination of Earth's ecosystems. *Science* **277:** 494-499.

Weinbauer, M.G. (2004). Ecology of prokaryotic viruses. *FEMS Microbiol. Rev.* **28:** 127-181.

Weinbauer, M.G. and Rassoulzadegan, F. (2007). Extinction of microbes: evidence and potential consequences. *Endang. Species Res.* **3:** 205-215.

Whitaker, R.J., Gorgan, D.W. and Taylor, J.W. (2003). Geographic barriers isolate endemic populations of hyperthermophilic archaean. *Science* **301:** 976-978.

Williaamson, M. (19960. Biological Invasions. Chapman & Hall, London.

Wilson, E.O. (1992). The diversity of life. WW Norton & Co., New York, USA.

Diversity of Lower Plants (2014): pp. 235-243
Editors: Rajan Kumar Gupta & Mukesh Kumar
IK International Publishing House Pvt. Ltd.

18

Himalayan Biodiversity with Reference to Uttarakhand and Its Conservation Strategies

KIRAN DIMRI* AND N.K. SHARMA

Department of Botany, Govt. P.G. College, Rishikesh. kdimri0@gmail.com

ABSTRACT

Biodiversity means the variety and variability in living beings. It includes diversity within species, between species and ecosystem. India is one of the megabiodiversity centres of the world. It is considered to possess 2 hot spots extending from Western Ghats on one side, to the Eastern Himalayas on the other side.Uttarakhand comprises 64.81% forest area of total geographical region. There occur about 21 forest types in Himalaya which are the indicators of the presence of vast biodiversity. The main species found in these forests are Chir pine, Deodar, Kail, Fir, Spruce, Oak, Sal, Teak, Sissoo, Eucalyptus, Khair and miscellaneous forest and their associates. Himalayan region supports about 8000 species of angiosperm, 44 species of gymnosperm, 600 species of pteridiophytes, 1737 species of bryophytes, 1159 species of lichens and 6900 species of fungi. Majority of these plants are extracted for drugs from the wild. Among the families, Asteraceae contains the highest number of ethnomedicinal plants, i.e., 129 species. According to survey conducted by Dhar *et al.* (2000), a total of 280 plants species utilized in 316 formulations by herbal drug companies. The socio-economic climate is changing worldwide at a faster rate than global physical climate, especially in relation to use and trade of medicinal plants and is directly affecting the availability and diversity. Due to overexploitation and lack of cultivation practices about 210 species of medicinally valuable plants of Himalayan region are in Red Data Book. Even this, topography and edaphic variations make Himalayan region a rich source of biodiversity.

Keywords: Biodiversity, Conservation strategies, Hot spot, Sustainable conservation.

*Communicating Author

1. INTRODUCTION

Diversity is a concept of the range of variation or differences among entities. Biodiversity is the variety and differences among living organisms from all sources, including terrestrial, marine, and other aquatic ecosystems and the ecological complexes of which they are a part. This includes genetic diversity within and between species and of ecosystem. Thus, in essence, biodiversity represents all life. It is the most significant national asset and constitutes an enduring source for supporting the continued existence of human societies. The diversity of plants can be analyzed from three viewpoints, phytotaxonomy, ecophysiology and chemotaxonomical diversity.

1.1 Hot Spots of Biodiversity

India is one of the megabiodiversity centres of the world. It has 2 hot spots of biodiversity which contribute significantly to this latitudinal biodiversity trends with 2.4% of the world area. Among these two hot spots of biodiversity, one is located in the Western Ghats and another in the eastern Himalaya. The Charak Samihata and Susruta Samita describe Himalayas as the richest sources of biodiversity.

1.2 Indian Himalayan Region

The mountain region in India is spread over Indian Himalayan Regions (IHR) covering a geographical area of 500,000 km^2. Majority of people live in villages and belong to diverse cultures and communities. The large human populace with diverse lifestyles, beliefs, traditions and cultural heritage inhabiting Himalayan has learnt to utilize natural resources and products in various ways. The recently carved mountain state of Uttarakhand is located between latitude 29°5′-31°25′ N and longitude 77°45′-81° E. It comprises 13 districts. The total geographical area of Uttarakhand is 53,483 km^2, out of which 34,661.52 km^2 is under forests, which comprises 64.81% of the total geographical area. Main species found in these forests are Chir pine, Deodar, Kail, Fir, and Spruce, Oak, Sal, Teak, Sissoo, Eucalyptus, Khair and miscellaneous forest and their associates. According to the classification of Chamion & Seth (1968) the main forest types found in Uttarakhand are tropical moist deciduous, riverine and

A view of Indian Himalayan Region

swampy forests, tropical dry deciduous forests, tropical thorn forests, subtropical chir pine forests, Himalayas dry temperate forests, moist alpine scrub forests, dry alpine scrubs and other forest (Uttarakhand Forests Statistics 2001). There occur 21 forest types in Himalayas, which itself is an indicator of the presence of vast biodiversity (Singh & Singh, 1992). Himalayan region supports about 8000 species of angiosperm, 44 species of gymnosperm, 600 species of pteridophytes, 1737 species of bryophytes, 1159 species of lichens, 6900 species of fungi (Singh & Hajra 1996). About 6% of the total population of India lives in Indian Himalaya region. Majority of them live in villages and use plant medicines to cure various ailments. People of Uttarakhand have been using vegetative medicines to cure ailments for thousands of year. Majority of these plants are extracted for drugs from the wild. This has adversely affected the very existence of a number of plants. Due to increase in demand and global interest in natural substances in the health-care system, the natural population of medicinal plants is under tremendous pressure. To meet ever-increasing demand of medicinal plants, it is imperative to go for cultivation of medicinal plants in private lands. It is the demand of time to replenish this demanding area.

1.3 Himalayas: A Region of "Hot Spots" of Biodiversity

The Himalayan region is bestowed with varied landscape features that provide multitude of habitats to a diverse array of faunal communities including several species of wild ungulates (Prater, 1980). The region covers nearly 11% of India's geographical area and ranges from sub-tropical to alpine zones. The region is well known for extensive alpine biome, temperate conifer and broadleaf forests, subtropical (foothill) forests, temperate grassy slopes, and various other habitats. Despite tremendous potential, most of the areas in this region exhibit low abundance of wild ungulates.

Himalayan region is "hot spot" of biodiversity due to topography and edaphic variations. It consists of 21 forests types, which itself is an indicator of the presence of vast biodiversity. The Indian Himalayan region has about 18,440 species of plants (Singh & Hajra, 1996) including 1748 species of medicinal plants and 675 species of wild species (Samant & Dhar, 1997), because of its rich biological and cultural heritage, the east Himalaya has been identified as a hot spot (Myers, 1988). The 50% of ethnomedicinally important species are collected from the Himalayan region (Samant *et al.,* 2000). Among the families, Asteraceae contains the highest number of ethnomedicinal plants, i.e.129 species. (Dhar *et al.,* 2000). According to a survey conducted by Dhar *et al.* (2000), a total of 280 plants species are utilized in 316 formulations by herbal drug companies.

The northern part of India harbours a great diversity of medicinal plants because of the majestic Himalayan range. So far about 8000 species of angiosperms, 44 species of gymnosperms and 600 species of pteridophytes have been reported in the Indian Himalaya (Singh & Hajra, 1996), of these 1748 species are known as medicinal plants (Samant, 1998). The maximum medicinal plants (1717 species) have been reported around the 1800 m elevation range. On the regional scale, the maximum species of medicinal plants have been reported from Uttaranchal (Kala, 2004) followed by Sikkim and North Bengal (Samant, 1998).The trans-Himalaya sustains about 337 species of medicinal plants (Kala, 2002) which is low as compared to other areas of the Himalaya due to the distinct geography and ecological marginal conditions (Kala, 2002).

The socio-economic climate is changing worldwide at a faster rate than global physical climate, especially in relation to use and trade of medicinal plants and is directly affecting availability and diversity. Due to overexploitation, collected from the wild and lack of cultivation practices of some medicinal valuable plants, 210 species of Himalayan region are in Red Data Book (Nayar *et al.,* 1990).The representative biodiversity of Himalayas protected through a Protected Area Network including Biosphere Reserves (Nanda Devi Biosphere Reserves), National Parks and Wildlife sanctuaries and Research centres.

2. NEED TO STUDY BIODIVERSITY

For developing the 'herbal industries', the northern India possesses a rich diversity of medicinal plant species across the various forest types along an altitudinal gradient. Such a high diversity of medicinal plants would be helpful for further scientific research on exploring their medical efficacy, value addition, and use in curing various old and new diseases (Kala, 2004). India has already established a reputation as a low-cost manufacturer of high quality generic drugs in the global market (Mashelkar, 2005). This fact can be used as an important tool for the marketing of herbal products produced in India. It is expected that India's aim to build a golden triangle between traditional medicine, modern medicine, and modern science will be a boon for developing the traditional herbal medicine and the medicinal plants sector (Mashelkar, 2005).

2.1 Consequences of Human Interferences

Several plant species are endemic to the Himalayan region. Out of total known number of higher plants from India, approximately 30% of the endemic species are found in the Himalayas, most of which are collected at high altitudes (Farooquee & Saxena, 1996). Of the total medicinal plant species, sixty-two species of medicinal plants are endemic to the Himalaya and 208 extend their distribution to the adjacent areas, and are therefore classified as near endemic (Chatterjee,

Villagers disturbing biodiversity

1939). Over 200 species of Himalayan medicinal plants are consumed raw, roasted, boiled, fried, cooked, or they are used in the form of oil, spices, jams or pickles (Dhyani, 1994). The indigenous communities use some medicinal plant species as a source of food, fodder, timber as well as various other ethnobotanical purposes. For example, apart from the use of *Myrica esculenta* and *Terminalia bellirica* as medicines, the fruits of these species are edible, the leaves are used for fodder and the wood is used for fuel (Dhayani, 1994). Approximately 81 species of Himalayan medicinal plants are known to be used for the extraction of oil. Of the total 675 species of Himalayan wild edibles, 171 are used for the treatment of diseases (Samant *et al.*, 2001). The crop plant's diversity is also a source of traditional medicine (Arora *et al.*, 1984).

2.2 Habitat Loss

Excessive degradation and fragmentation eventually leads to habitat loss. Systematic studies documenting loss of ungulate habitats in the Himalayan region are lacking. Hence, it is difficult to point out clear cases of this phenomenon. However, in most of the sectors, there are plenty of evidence indicating the shrinkage of wildlife/ungulate habitat. Green (1986) reported that over 70% of potential musk deer habitat has already been lost due to habitat loss and habitat degradation in the southern side of the Greater Himalaya. Based on the spatial time-series analysis of remote sensing data, Awasthi (2001) has pointed out that in upper regions of Bhagirathi Valley and Garhwal Himalaya, there has been a considerable increase in the area of human inhabitation and cultivation during the last 30 years. This study also indicates that much of the sub-Alpine and temperate broadleaf forests have converted into scrub vegetation. Conversion of forested habitat into scrub is in a way loss of habitat especially for the sensitive ungulates such as Himalayan musk deer. Displacement of human populations for the developmental projects, construction of roads along the sensitive habitats, encroachment of forests for agriculture and heavy infestation of exotic weeds are other causes of habitat loss in the region.

3. NEED OF SUSTAINABLE DEVELOPMENT

All biological species of an ecosystem are in state of dynamic equilibrium forms in a close net. Due to increase in world demand and global interest in natural substances in the health-care system, the natural popultion is under tremendous pressure. Today, the goals commonly expressed by environmentalists include reduction and clean-up of man-made pollution, with future goals of zero pollution, reducing societal consumption of non-renewable fuels, development of alternative, green, low carbon or renewable energy sources, conservation and sustainable use

of scarce resources such as water, land and air, protection of ecosystem, preservation and expansion of threatened or endangered species or ecosystem from extinction and protection of biodiversity and ecosystem upon which human and other life on earth depend.

4. SUSTAINABLE CONSERVATION

The sustainable conservation which speaks not of growth zero, it means growth and development with conservation. It requires economic growth to be pursued with environmental sensitivity. Now it's time to make a global effort not only with strong determination but also with governmental support to make the Green Planet habitable. We must follow the obligation of International Treaties of the past and pursue the new global order to save the earth. There are numerous international environmental agreements made to protect the environment in different ways. Many of these are legally binding.

4.1 Major Conservation Issues

Several factors have led to low abundance and poor conservation status of ungulates in the Himalayan region. Firstly, the Himalayan ecosystem is relatively young, fragile and low in primary productivity. With the increase in human population, the area has undergone rapid degradation, fragmentation and loss of wildlife habitat. Poaching and trade of animal parts, competitive exclusion by the domestic livestock, faulty land use practices, and human-wildlife conflicts are other factors affecting wild ungulates and other faunal communities (Sathyakumar 1994; Vinod & Sathyakumar, 1999). Some of the issues related to conservation of wildlife ungulates are discussed below.

4.2 General Strategies for Conservation

Plate : Different Polyhouse for growing plants

- Most of the Protected Areas (PA) in the Himalayan region lack adequate manpower and funds for proper management. In addition, several PAs have ill-defined boundaries leading to conflicts between the local communities and PA management. There is an urgent need to strengthen the management of most of the PAs allocating more well-trained and motivated staff, budget and infrastructure.
- Several PAs and Reserved Forests in the Himalayan region need to be brought under community reserves where local people could be made partners in conservation and management. Through various programmers such as community based ecotourism and participatory management of natural resources, the illegal activities such as trade of animal parts and poaching could be gradually brought down.
- There is an urgent need to increase trans-boundary cooperation between India–Nepal, India–China and India–Myanmar to control the illegal trade of wildlife products. Measures should be taken to register all the arms (licensed guns, etc.) with the Wildlife Warden or the Divisional Forest/Wildlife Officer.

- Currently, there is no organized system of harvesting wild medicinal and aromatic plants and uncontrolled harvest often results in the degradation of habitat. There is a need to evolve policies related to rotational harvest of medicinal plants for the benefit of communities and there is a need to control excessive pressure on the land. As part of eco-development measure cultivation of medicinal plants needs to be promoted in the buffer zones of various PAs where human pressure for these commodities is excessive.

4.3 Traditional Conservation

Many traditional societies all over the world value a large number of plant species from the wild for a variety of reasons for food, fibre and shelter or medicine. The practice of nature conservation is a very ancient tradition. The practice of conservation of plant species by the traditional societies of Garhwal Himalaya has a rich tradition of nature conservation through socio-religious constraints on profligate use of common property resources. The ancient ethics changed by rejection and replacement of traditional practice under the influence of western cultures and by the advent of modern industry. Sacred grooves are one of the first instances of traditional conservation. Increasing threats to biodiversity loss, demands new conservation approaches enabling fair share of the wider values of conservation to the local communities and positive local attitudes towards conservation goals. Nature worship has been a key force of shaping the human attitudes towards conservation and sustainable utilization of natural resources. Garhwal Himalaya is referred to as land of gods as many important religious shrines are located besides the confluence of five tributaries of sacred river Ganges. Although, biological diversity of Himalaya is very rich, however, very little is known about the sacred grooves of this region. These include the river Ganges and tributaries and their confluence amongst the religious shrines Badrinath, Kedarnath, Yamunotri and Gangotri. The sacred mountain peaks are Nanda Devi, Trishul, Chaukhamba, Kailash, Binsar and Syahi Devi (Anthwal *et al.,* 2006). Well-preserved sacred grooves are storehouse of valuable medicinal and other plants having economic value, and serve as a refuge to threatened species.

For sustainable point of view in India, a unified National Act called the Wildlife Protection Act (1972) came into being under which area could be constituted and managed as National Parks, Wildlife Sanctuaries, and Biosphere Reservoir, which contributes significantly to this latitudinal biodiversity trends with mere 2.4% of the world area (Gupta *et al.*, 2006). In India presently, 492 sanctuaries, 14 biosphere reserves and 92 national parks are doing work for sustainable conservation. The representative biodiversity rich areas of the IHR have been protected through a Protected Area Network, including 5 biosphere reserves, 28 national parks and 98 wildlife sanctuaries covering an area of 51,899 km^2 (Mathur *et al.*, 2000; Joshi 2002). These Protected Areas (PAs) include tropical, subtropical, temperate, subalpine and alpine regions, with the respective ecosystems of these zones. In Uttarakhand presently, 6 sanctuaries, 1 biosphere reserve and 6 national parks are doing work for sustainable conservation.

REFERENCES

Anthwal, A., Sharma, R.C. and Sharma, A. (2006). Sacred Groves: Traditional way of conservation plant diversity in Garhwal Himalaya, Uttarakhand. *J. Amer. Sci.* **2:** 35-43.

Arora, R.K. and Nayar, E.R. (1984). Wild relatives of crop plant in India. *NBPGR Science Monograph* **7:** 97.

Awasthi, A. (2001). Ecological Impact of Anthropogenic Pressures on High Altitude Forests Along Bhagirathi Catchment. Ph.D. Thesis. Wildlife Institute of India, Dehradun.

Champion, H.G. and Seth, S.K. (1968). A Revised Survey of forest types of India. Manager of publication, Govt. of India Press, Delhi.

Chatterjee, D. (1939). Studies on the endemic flora of India and Burma. *J. Royal Asiatic Soc. Bengal* **5:** 19-67.

Dhar, U., Majhkhola, S. and Bisht, A.K. (2002). Current Status and Future Strategy for Development of Medicinal Plants Sector in Uttarakhand. *Cur. Sci.* **83:** 25.

Dhyani, P.P. and Dhar, U. (1994). *Myrica esculenta Box myrtle Kaphal: A promising underexploited tree crop of the Himalaya.* A report G.B. Pant Institute of Himalayan Environment and Development, Almora.

Farooquee, N.A., and Saxena, K.G. (1996). Conservation and utilisation of medicinal plants in high hills of the central Himalayas. *Environ. Conser.* **23:** 75-80.

Gupta, N. (2006). Biodiversity of Mothronwala Swamp, Doon Valley, Uttaranchal. *J. Amer. Sci.* pp. 25-31.

Gupta, N. and Anthwal, A. (2006). Biodiversity of Mothronwala Swamp Doon Valley, Uttarakhand. *J. Amer. Sci.* vol 2.

Green, M.J.B. (19860. The distribution, status and conservation of the Himalayan musk deer *Moschus chrysogaster. Biol. Conserv.* **35:** 347-375.

Joshi, H.C. and Samant, S.S. (2005). Plant Diversity and Conservation Status of Nanda Devi National Park of the Indian Himalaya Region. *Intern. J. Biodiv. Sci. Managt.* pp. 65-73.

Joshi, K., Chavan, P., Warude, D. and Patwardhan, B. (2004). Molecular markers in herbal drug technology. *Curr. Sci.* **87:** 159-165.

Kala, C.P. (2004). Revitalizing traditional herbal therapy by exploring medicinal plants: A case study of Uttarakhand State in India, pp. 15-21. *In*: *Indigenous Knowledges: Transforming the Academy, Proceedings of an International Conference.* Pennsylvania State University, Pennsylvania.

Kala, C.P. (2002). *Medicinal Plants of Indian Trans-Himalaya.* Bishen Singh Mahendra Pal Singh, Dehradun.

Kala, C.P. and Mathur, V.B. (2002). Patterns of plant species distribution in the trans-Himalayan region of Ladakh, India. *J. Veget. Sci.* **13:** 751-754.

Mashelkar, R.A. (2005) India's R & D: Reaching for the top. *Science* **307:** 1415-1417.

Mathur, V.B., Kathayat, J.S. and Rath, D.P. (2000). *Envis Bulletin*: Wildlife and Protected Areas Vol. 3(1). Dehradun: Wildlife Insititute of India.

Myers, N. (1991). The world's forests and human population: the environmental interconnections. *Popul. Develop. Rev.* **16:** 1-15.

Prater, S.H. (1980). The book of Indian animals. Bombay Natural History Society, Oxford University Press, pp. 324.

Samant, S.S., Dhar, U., Palni, L.M.S. (1998). *Medicinal Plants of Indian Himalaya: Diversity Distribution Potential Values* G.B. Pant Institute of Himalayan Environment and Development, Almora.

Samant, S.S., Dhar, U. and Rawal, R.S. (2001). Diversity and distribution of wild edible plants of the Indian Himalaya, pp. 421-482. *In*: Pande, P.C. and Samant, S.S. (Eds.). *Plant Diversity of the Himalaya*, Gyanodaya Prakashan, Nainital.

Sharma, A.B. (2004). Global Medicinal Plants Demand May Touch $5 Trillion By 2050. *Indian Express*, New Delhi.

Singh, D.K. and Hajra, P.K. (1996). Floristic diversity, pp. 23-38 *In*: *Biodiversity Status in the Himalaya.* A report British Council, New Delhi.

Sathyakumar, S. (1994). Habitat ecology of ungulates in Kedarnath Musk Deer Sanctuary, Western Himalaya. Ph.D. Thesis, Saurashtra University, Rajkot. pp. 242.

Vinod, T.R. and Sathyakumar, S. (1999). Ecology and conservation of ungulates in Great Himalayan National Park, Himachal Pradesh. *In*: An Ecological Study of the Conservation of Biodiversity and Biotic Pressures in the Great Himalayan National Park Conservation Area – An Ecodevelopment Approach. *Forestry Research Education and Extension Project – Great Himalayan National Park (FREE-GHNP),* Final Project Report.

Diversity of Lower Plants (2014): pp. 244-247
Editors: Rajan Kumar Gupta & Mukesh Kumar
IK International Publishing House Pvt. Ltd.

19

Intellectual Property Rights (IPR): An Effective Tool to Protect Indigenous Technical Knowledge

POONAM PALIWAL* AND M.K. PALIWAL

Department of Botany, I.P. College, Bulandshahr (U.P.)

ABSTRACT

One of the consequences of economic globalization has been the conversion of traditional and local knowledge into global knowledge. Consequently, the issues of protection of traditional knowledge and biopiracy have gained international attention during recent years. As expected, the traditional knowledge available in various fields offers basic information and guidelines for advance research in these areas. Hence, there must be an appropriate law with clear-cut guidelines to safeguard the interest of indigenous people and ultimately the wealth of a country. The existing IPR systems in most of the countries are not capable of fully protecting various categories of technical knowledge. Evidently, to improve the role of IPR in this area, there is an urgent need to pay emphasis on Benefit-sharing Model, documentation of technical knowledge and creation of digital libraries. As such, a proper form of IPR is the only available choice for indigenous people to save their natural wealth and heritage for the present and pass it on to the posterity.

Keywords: IPR, Traditional knowledge, CBD, TRIPs.

1. INTRODUCTION

The expression Indigenous Technical Knowledge (ITK) refers to the unique traditional know-how developed around the specific parameters in a particular geographical area. In the true sense, this is knowledge of tribal people, i.e., the group of individuals who have jealously guarded their perpetual modes of life, customs and systems.

On the one hand, it is stored in people's memories and activities, its expression on the other is in the form of stories, songs, folklore, proverb dances, myths, cultural, beliefs, rituals,

*Communicating Author

community laws, local language, agriculture practices, equipment and materials (Paul & Ramanathan, 2002) has been repeatedly witnessed. However, this knowledge should be advanced and protected by appropriate law since the local communities or individuals do not have the knowledge or means to safeguard their property.

India, being one of the twelve megabiodiversity country with approximately 8% of the world's flora and fauna, is undoubtedly a great storehouse of traditional knowledge for use of the medicinal and aromatic plants, the herbal drugs, as well as the by-products of animal origin. Tribal/indigenous people also provide basic but deep knowledge about their flora and fauna, their habits, habitat as well as the seasonal behaviour. (e.g., during Tsunami, the tribals, moved to deeper forest areas on getting the signals of disaster). This traditional knowledge and technology is socially desirable, economically affordable, sustainable and involves minimum risk. Taking this aspect into consideration the indigenous people deserve due recognition and a partnership in benefit-sharing particularly in respect to commercialization of biodiversity and related knowledge. Further, it is not desirable to borrow knowledge but also add substantially and sincerely values to transform it into meaningful applications or marketable products (Ghat *et al.*, 1999). It is at this point that intellectual property rights came into play which can be broadly defined as the legal rights established over creative or inventive ideas and are divided into two broad categories, outlined below:

1. Industrial property covering IPRs such as patents, trademarks, geographical indications and industrial designs.
2. Copyright and related rights, covering artistic and literary works performances and broadcasts.

Such of the IPRs that do not fit into this classical division are termed ***sui generis***, meaning one of their own kind, e.g., plant breeder's rights. These rights are territorial in nature, and are applicable within the boundaries of a country for a specified duration. On the expiry of the term of protection, the protected technology would enter the public domain or become generic and may be utilized by all for any use. This is true in case of patents, copyrights and designs. However, the trademarks if maintained can be protected indefinitely.

According to WIPO (2001), from the intellectual property perspective, traditional knowledge is not limited to any specific field of technology or arts. Components of traditional knowledge are medicines, healing, biodiversity conservation, environment, food, agriculture, music, dance and 'artisanat' (i.e., designs, textiles, plastic arts, crafts, etc.) of the people.

2. WHO OWNS THE TRADITIONAL KNOWLEDGE?

Whereas much of the products of technical knowledge are in public domain, the related knowledge may be held by a person, persons in a family or community or by the community as a whole. The intangible nature of the knowledge and the lack of documentation make the question of ownership of technical knowledge more complex.

In a vast majority of the national IPR systems, the technical knowledge has not been kept as it is generally being treated as the intellectual property of nobody. Exploiting the loopholes in the national IPR legislation in some countries, people copy the undocumented traditional knowledge from developing countries and receive patents on the inventions claiming that it is new. Evidently, thus the patent claims over biodiversity and indigenous knowledge on the innovations, creativity and genius of indigenous people are acts of biopiracy (Shiva, 2001).

3. EQUITABLE SHARING OF BENEFITS

The biological resources and associated traditional knowledge owned by the indigenous communities are misappropriated by TNCs without proper acknowledgement of the source and equitable sharing of the benefits arising out of the commercial utilization of such knowledge and resources. Neither the TRIPs agreement (Trade Related Aspects of Intellectual Property Rights) nor the national legislation on IPRs in most countries recognize the innovations based on traditional knowledge which are in public domain. Cases of turmeric, basmati rice, neem, vacha, ashwagandha and amla are examples of the lacunae in the present-day IPR regimes.

A number of provisions in the CBD (Convention on Biological Diversity) refer to the equitable sharing of benefits arising from genetic resources. Similarly, Agenda 21 of the Rio declaration also underscores the need for preserving and promoting the traditional knowledge of indigenous communities.

A number of models have been suggested for protecting the technical knowledge and ensuring equitable sharing of benefits arising out of the commercial utilization of such knowledge. Most of the models suggest following provisions for ensuring equitable of sharing of benefits:

1. Disclosure of all places of origin of the material/knowledge used in the application for IPR.
2. Disclosure of all communities/persons of origin.
3. Proof of prior informed consent obtained from community/person of origin.
4. Proof of benefit-sharing arrangements with community or person of origin.
5. Prior public notice in all relevant languages in the places or communities of origin.

A successful example of benefit-sharing agreement between indigenous people and research agency is the arrangement made between Kani tribes on the one hand and the Tropical Botanical Garden and Research Institute (TBGRI) in Kerala. The antifatigue, antistress drug—**jeevani** has been patented by TBGRI which was developed by the disclosure of its benefits and formula by the tribal people to the scientist. The agreement between the TBGRI and the drug manufacturer provides 2% ex-factory sales prices to be paid to the Kani tribe as royalty.

The innovative model developed by TBGRI for benefit-sharing received wide attention in the world summit on sustainable development at Johannesburg in 2002 (Joseph, 2006).

4. OPTIONS FOR INTELLECTUAL PROPERTY PROTECTION OF TRADITIONAL KNOWLEDGE

The present-day IPR regimes recognize only knowledge produced by organized commercial and research-based activities. These often neglect the role played by the indigenous people of developing countries in generating traditional knowledge about plants, animals, and other biological resources. Following steps can enable the achievement of the noble aims, of equitable sharing of benefits and statutory protection:

1. Benefit-sharing contracts should be made between bioprospectors (companies developing new drugs from herbal resources) and indigenous people/tribal groups/Govt./and in case of community by creating a trust.

2. For benefit-sharing contracts the economic valuation of biological resources and traditional knowledge should be carried out by separate agencies.
3. Biosmuggling should be prohibited by implying proper law and order provisions conditions.
4. In spite of the criticism over inappropriateness of IPR mechanisms in technical knowledge domain, protection particularly through copyrights (for folklore or expression of folklore), trademark (to protect traditional names, symbols, marks, words or their combinations), geographical indications (for traditional agricultural products, foodstuffs, handicrafts), patents (indigenous drugs) and *sui generis* system should be appreciated. Attempts are called for to make the existing laws more vigorous, comprehensive, and dynamic.
5. A database on all kinds of traditional knowledge available in a country must be created and before any Patent or Trademark or Design is granted, this database should be searched in order to ascertain that there is no prior use or prior knowledge or art. Documentation of technical knowledge plays a critical role in patent applications, therefore, should be the top priority target for developing countries. The documented technical knowledge could also act as a starting point in the benefit-sharing process (In India, the Ahmedabad based National Innovation Foundation identifies and nurtures local innovations and technical knowledge).
6. The digital libraries carrying traditional knowledge should be incorporated in and linked to every patent office and should be searched while processing patent application. This should be encouraged as in case of Traditional Knowledge Digital Library (TKDL) Project which completed its first phase in October, 2003, a joint initiative of CSIR and Indian Systems of Medicine and Homoeopathy to document technical knowledge available in all systems of indigenous medicine in any form, written or oral.

However, in the absence of legally-binding national and international legislation for the exclusive protection of technical knowledge digitalization alone will not ensure equitable benefit-sharing for the owners of the technical knowledge.

It is to be printed out that although the Indian Govt. has taken initiatives, to protect national IPR by introducing Patent Amendment Bill, Biological Diversity Bill, Plant Variety Protection and Farmer's Rights Act, these are insufficient to protect biopiracy. Hence, we should concentrate on adopting domestic or traditional knowledge-based legal frameworks which can serve our basic needs. This is essentially because ultimately, a nation's prosperity is determined by its ability to convert knowledge into wealth and social goods.

REFERENCES

Anonymous. (2001). Intellectual Property Needs and Expectations of Traditional Knowledge Holders. WIPO, Geneva.

Ghate, U., Gadgil, M. and Seshangiri Rao, P.R. (1999). IPR on Biological resources: Benefiting from biodiversity and peoples knowledge. *Curr. Sci.* **77:** 1418-1425.

Joseph, J. (2006). Protection of Traditional Knowledge: Options and Challenges. pp. 79-100. Proc. National Seminar in: Arvind Kumar & Govind Das. 'Intellectual Property Rights: Plant varieties & Genome Conservation'. Oct 14-15, 2006. University of Allahabad, Allahabad.

Paul, S. and Ramanathan, A. (2002). Conservation of rural biodiversity through indigenous knowledge. *Kurukshetra* **50:** 3-7.

Shiva, Vandana, (2001). Patents: Myths and Reality. Penguin Books. New Delhi.

Diversity of Lower Plants (2014): pp. 248-263
Editors: Rajan Kumar Gupta & Mukesh Kumar
Ik International Publishing House Pvt. Ltd.

20

An Updated List of Pteridophytes of Uttarakhand

N. PUNETHA, B.S. KHOLIA*, RICHA PUNETHA* AND KAMLESH BHAKUNI

Department of Botany, Government Postgraduate College, Pithoragarh–262 502, Uttarakhand
**Botanical Survey of India, Sikkim Himalayan Circle, Gangtok, Sikkim*
Email: punethan_bot@indiatimes.com

ABSTRACT

Based on the collections made by the authors from the eastern part of the state and on the available literature, the ferns and fern allies of Uttarakhand are listed. Efforts have been made to provide correct nomenclature and the place of their occurrence in Uttarakhand. Fern allies are grouped into four families whereas ferns are grouped into 39 families. A total of 341 species of 89 genera are being listed. Thelypteridaceae is the largest family with 13 genera. *Athyrium, Dryopteris and Polystichum* are the largest genera. Presence of many hybrids reported recently and rare occurrence of some species suggest that the endemism is not uncommon.

Keywords: Ferns, Pteidophytes, Distribution, Uttarakhand.

1. INTRODUCTION

Except the foothills and the terai region in the south, most part of the Uttarakhand state is hilly with lofty mountains in the inner Himalayan ranges. While the west part, adjoining to Himanchal Pradesh, is relatively dry, the east part bordering with Nepal is moist where occur many east Himalayan elements. These features contribute significantly to the floristic diversity of the state. Bulk of literature is now available on the pteridophytic flora of Uttarakhand, either in the form of enumerations of certain regions (Pangtey and Punetha, 1987; Punetha and Kaur, 1987; Pande 1990 a, 1990 b; Khullar *et al.*, 1991; Kholia, 1992; Kholia and Punetha, 1995) or in the form

*Communicating Author

of detail taxonomic treatments of the taxa of this group of plants (Khullar, 1994, 2000; Pande and Pande, 2002a, 2002b, 2003; Dixit and Ramesh, 2002). Apart of these publications, papers on new records and new species are also available. Fraser-Jenkins (1997) while making critical studies (as he claims) on the taxonomic status/nomenclature of the Indian subcontinental pteridophytes, did not agree with most of the new species or records reported by Indian workers in recent years and corrected these mistakes by applying correct names following the code. Showing his devotion and honesty towards his studies he, however, corrected some of his mistakes (?) he did earlier (1997) in his very recent publication (2008). Then, one can ask if the pteridophytic flora of Uttarakhand is explored and worked out taxonomically, what is the need of preparing yet another list of the pteridophytes of Uttarakhand.

This list has been prepared for two main reasons: (i) A similar article was prepared in 2002 for the book entitled "Plant Wealth of Uttarakhand" which was to be edited by some people of Department of Botany, University of Delhi, Delhi but for the reasons better known to the editors and publishers, it has not as yet come out. However, in the present treatment some nomenclature changes and status of some other taxa have been dealt with and (ii) The authors do not agree with the nomenclature of some taxa as proposed by Fraser-Jenkins (1997, 2008), the reasons of disagreement are being given in detail elsewhere. Although the specimens of many taxa listed hereunder are with the authors, reports on others are taken from the literature cited above.

2. ENUMERATION

1. **Family Lycopodiaceae** Mirbel in Lamarck & Mirbel

Huperzia Bernh.

1. *H. hamiltonii* (Spring) Trev.: Lori & Munch (Pithoragarh).
2. *H. pulcherrima* (Wall. ex Hook. & Grev.) Sen & Sen: Champawat, Chokori, Didihat (Pithoragarh).
3. *H. subulifolia* (Wall. ex Hook. & Grev.) Trev.: Sandev & Didihat (Pithoragarh).

Lycopodiella Holub

1. *L. cernua* (L.) Pichi-Sermolli: Pithoragarh, Almora.

Lycopodium L.

1. *L. japonicum* Thunb.: Pithoragarh. (Fig. 1).
2. *L. annotinum subsp. alpestre* (Hartm.) A. Love & D. Love: Kumaon.

2. **Family Sellaginellaceae** Willk.

Selaginella P. Beauv.

1. *S. bryopteris* (L.) Baker: Pithoragarh.
2. *S. chrysocaulos* (Hook. & Grev.) Spring: Throughout Uttarakhand.
3. *S. indica* (Milde) Tryon: Tawaghat (Pithoragarh).
4. *S. involvens* (Sw.) Spring: Throughout Uttarakhand.
5. *S. pallida* (Hook. & Grev.) Spring: East Kumaon.
6. *S. pallidisima* Spring.: Eastern parts of the state.
7. *S. pulvinata* (Hook. & Grev.) Maxim.: Tawaghat (Pithoragarh), Rare.
8. *S. repanda* (Desv. ex Poir.) Spring: In the Didihat region (Pithoragarh).
9. *S. subdiaphana* (Wall. ex Hook. & Grev.) Spring: Common in humus rich forest floors during raining season. (Fig. 2).

10. S. vaginata Spring: In the lower ranges of Uttarakhand.

3. **Family Equisetaceae** L.C. Richard ex A.P.de Candolle

Equisetum L.

1. E. arvense L.: Higher ranges of the state. (Fig. 3)
2. E. diffusum D. Don: Throughout Uttarakhand.
3. E. debile Roxb. ex Vaucher: Throughout Uttarakhand.
4. E. ramosissimum Desf.: Throughout Uttarakhand.

4. **Family Psilotaceae** Kanitz

Psilotum Sw.

1. P. nudum (L.) P. Beauv.: Bhujan, Almora (Based on the report of Pande and Pande, 2003).

5. **Family Botrychiaceae** Nakai

Botrychium Sw.

1. B. lunaria Sw. : Throughout the higher ranges of the region, often confined to pastures.
2. B. daucifolium Wall. ex Hook. et Grev.: Pithoragarh, Champawat .
3. B. multifidum (Gmel.) Ruprecht: Mussoorie, Chamoli Garhwal, Almora, Pithoragarh.
4. B. ternatum (Thunb.) Sw. : Dehradun (Mussoorie) to Pithoragarh.
5. B. lanuginosum Wall. ex Hook. et Greville: Throughout Uttarakhand.
6. B. virginianum (L.) Sw..: Tehri Garhwal and Pithoragarh.

Helminthostachys Kaulfuss

1. H. zeylanica (L.) Hooker: Nainital, Tanakpur (Champawat).

6. **Family Ophioglossaceae** L.

Ophioglossum L.

1. O. gramineum Willd. : Almora (Ranikhet) reported by Y.P.S. Pangtey.
2. O. nudicaule L. : Almora (Ranikhet).
3. O. petiolatum Hooker : Dehradun, Nainital and Pithoragarh.
4. O. polyphyllum Braun ex Seubert : Dehradun.
5. O. reticulatum L.: Dehradun, Chamoli Garhwal, Nainital, Almora, and Pithoragarh.

7. **Family Angiopteridaceae** Fee ex Bonner

Angiopteris Hoffmann

1. Angiopteris indica Desv.: Almora and Pithoragarh.

8. **Family Osmundaceae** Berchtold et J.S. Presl

Osmunda L.

1. O. claytoniana L. *subsp. vestita* (Wall. ex Milde) A. Love & D. Love : Dehradun, Tehri Garhwal, Chamoli Garhwal, Almora and Pithoragarh. (Fig. 4).
2. O. regalis L.: Dehradun, Tehri Garhwal, Chamoli Garhwal, Almora and Pithoragarh.

9. **Family Plagiogyriaceae** Bower

Plagiogyria (Kunze) Mett.

1. P. scandens Mett.: Pithoragarh.

10. **Family Gleicheniaceae** (R. Brown) C. Presl

Dicranopteris Bernhardi

1. D. linearis var. hirta Kaur et Punetha: Common in the eastern part of Kumaon.

11. **Family Grammitidaceae** Ching

Ctenopteris Blume ex Kunze

Figs. 1-4: *1. Lycopodium japonicum, 2. Selaginella subdiaphana, 3. Equisetum arvense, 4.* Vast population of *Osmunda clytoniana.*

1. C. subfalcata Blume ex Kunze: Chamoli, Pithoragarh, Almora.

12. Family Loxogrammaceae Pichi-Sermolli

Loxogramme (Blume) Presl

1. L. involuta (D. Don) Presl: Throughout Uttarakhand.

2. L. parallela Copel.: Almora (Pindari Khati Dwali Road).

3. L. porcata Price: Kumaon, Suring (Pithoragarh). (Fig. 5).

13. Family Polypodiaceae Berchtold et Presl

Arthromeris (Moore) J. Smith

1. A. lehmanii (Mett.) Ching: Chamoli Garhwal and Almora, Pithoragarh.

2. A. mairei (Brause) Ching: Common in Kumaon.

3. A. wallichiana (Spreng.) Ching: Throughout Uttarakhand.

Leptochilus Kaulf.

1. L. ellepticus (Thunb.) Noot. : Jeolikote (Nainital), Didihat (Pithoragarh).

2. L. decurrens subsp. hemionitidea (C. Presl) Fraser-Jenkins: Tawaghat (Pithoragarh).

3. *L. insignis* (Blume) Fraser-Jenkins: Rare in east Kumaon.

Drynaria (Bory) J. Smith

1. *D. mollis* Beddome: Throughout Uttarakhand.
2. *D. propinqua* (Wall. ex Mett.) J. Smith : Throughout Uttarakhand.

Lepisorus (J. Smith) Ching

1. *L. bicolor* Ching: Kumaon, Pithoragarh.
2. *L. x birii* Punetha *et al*., (in preparation).
3. *L. clathratus* (Clarke) Ching: Pithoragarh.
4. *L. contortus* (Christ) Ching: Dehradun, Chamoli Garhwal, Pithoragarh.
5. *L. loriformis* (Wall. ex Mett.) Ching: Chamoli and Pithoragarh.
6. *L. mehrae* Fraser Jenkins: Throughout Uttarakhand.
7. *L. morrisonensis* (Hayata) H. Ito.: Common in the state.
8. *L. nudus* (Hook.) Ching: Throughout Uttarakhand. (Fig. 7).
9. *L. pseudonudus* Ching: Throughout Uttarakhand.
10. *L. scolopendrium* (D. Don) Mehra & Bir: Throughout Uttarakhand.
11. *L. thunbergianus* (Kaulf.) Ching : Throughout Uttarakhand. (Fig. 6).

Microsorum Link

1. *M. membranaceum* (D. Don) Ching: Throughout Uttarakhand.
2. *M. zipelli* (Blume) Ching: Chamoli Garhwal, Almora and Pithoragarh.

Phymatopteris Pichi-Sermolli

1. *P. ebenipes* (Hook.) Pichi-Sermolli: Throughout Uttarakhand.
2. *P. malacodon* (Hook.) Pichi-Sermolli: Throughout Uttarakhand.
3. *P. oxyloba* (Wall. ex Kunze) Pichi-Sermolli: Throughout Uttarakhand.
4. *P. stewartii* (Bedd.) Pichi-Sermolli: Dehradun, Uttarkashi, Chamoli, Nainital, Almora and Pithoragarh.
5. *P. stracheyi* (Ching) Pichi-Sermolli: Dehradun, Chamoli, Nainital, Almora.

Goniophlebium (Bl.) Presl

1. *G. argutum* (Wall. ex Hook.) J.Sm.: Throughout Uttarakhand.

Polypodiodes Ching

1. *P. amoena* (Wall. ex Mett.) Ching : Throughout Uttarakhand.
2. *P. lachnopus* (Wall. ex Hook.) Ching: Throughout Uttarakhand. (Fig. 8).
3. *P. microrhizoma* (Clarke ex Baker) Ching: Throughout Uttarakhand.
4. *P. subamoena* (Clarke) Ching : Dehradun, Tehri Garhwal, Almora and Pithoragarh.

Pyrrosia Mirbel

1. *P. costata* (Wall. ex Presl) Tagawa et Iwatsuki: Dehradun, Almora, Tehri and Pithoragarh.
2. *P. flocculosa* (D. Don) Ching: Throughout Uttarakhand.
3. *P. lanceolata* (L.) Farwell : Chamoli Garhwal and Pithoragarh.
4. *P. mannii* (Gies.) Ching: Nainital, Almora and Pithoragarh.
5. *P. porosa* (Wall. ex C. Presl) Hovenkamp: Throughout Uttarakhand.

14. Family Lygodiaceae Presl

Lygodium Sw.

1. *L. flexuosum* (L.) Sw.: Throughout Uttarakhand (a climbing fern of lower elevations).
2. *L. japonicum* (Thunb.) Sw.: Throughout Uttarakhand.

15. Family Sinopteridaceae Koidzumi

Cheilanthes Sw.

1. *C. albomarginata* Clarke : Throughout Uttarakhand.
2. *C. anceps* Blanford : Dehradun, Tehri Garhwal, Chamoli Garhwal, Nainital, Pithoragarh.
3. *Cheilanthes C. new species* (in preparation) Collected only once from near Lilam (Pithoragarh).
4. *C. bicolor* (Roxb.) Fraser-Jenkins: Throughout Uttarakhand.
5. *C. brevifrons* (Khullar) Khullar: Throughout Uttarakhand.
6. *C. dalhousiae* Hooker: Throughout Uttarakhand.
7. *C. doniana* (S.K.Wu) Fraser Jenkins et Khullar: Pithoragarh.
8. *C. dubia* Hope: Dehradun, Chamoli Garhwal, Tehri Garhwal, Nainital, Pithoragarh.
9. *C. duthiei* Baker : Chamoli Garhwal (Based on the report of Beddome and Hope).
10. *C. grisea* Blanford: Dehradun, Uttarkashi, Chamoli Garhwal and Pithoragarh.
11. *C. rufa* D. Don: Throughout Uttarakhand. (Fig. 9)
12. *C. subvillosa* Hooker: Dehradun, Chamoli Garhwal, Pithoragarh, Tehri Garhwal and Chamoli Garhwal.

Notholaena R. Brown

1. *N. delavayi var. intermedia* (Bak.) C. Chr.: Pithoragarh (Garbyang).
2. *N. marantae* (L.) Desv. : Tehri Garhwal, Uttarkashi, Chamoli Garhwal, Pithoragarh and Almora.

Pellaea Link

1. *P. hastata* (Thunb.) Prantl.: Dehradun, Tehri, Nainital, Almora.
2. *P. nitidula* (Wall. ex Hook) Hook. et Baker: Dehradun, Tehri Garhwal and Pithoragarh.

16. Family Cryptogrammaceae Pichi-Sermolli

Cryptogramma R. Brown

1. *C. brunoniana* Wall. ex Hooker et Greville: Throughout Uttarakhand.
2. *C. stelleri* (Gmel.) Prantl: Kumaon, Pithoragarh, Champawat.

Onychium Kaulfuss

1. *O. cryptogrammoides* Christ : Throughout Uttarakhand.
2. *O. japonicum* (Thnub.) Kunze: Dehradun, Uttarkashi, Almora, Nainital, Pithoragarh.
3. *O. siliculosum* (Desv.) C. Chr.: Dehradun, Chamoli Garhwal, Nainital, Pithoragarh, Almora. (Fig. 10).
4. *O. tenuifrons* Ching : Throughout Uttarakhand.

17. Family Actiniopteridacae Pichi-Sermolli

Actiniopteris Link.

1. *A. radiata* (Sw.) Link : Bhujan (Almora) Based on the repot of Pande and Pande (2003).

18. Family Hypolepidaceae Pichi-Sermolli

Hypolepis Bernhardii

1. *H. polypodioides* (Bl.) Hook.: Throughout Uttarakhand.

19. Family Pteridiaceae Ching

Pteridium Gleditsh ex Scopoli

1. *P. revolutum* (Bl.) Nakai: Throughout Uttarakhand.

20. Family Pteridaceae Ching

Pteris L.

1. *P. aspercaulis* Wall. ex Ag. (*P. pseudoquadriaurita* Khullar *sensu* Khullar): Throughout Uttarakhand.
2. *P. biaurita* L. : Throughout Uttarakhand.
3. *P. cretica* L. : Throughout Uttarakhand.
4. *P. dactylina* Hooker: Dehradun, Uttarkashi, Chamoli Garhwal, Nainital, Almora and Pithoragarh.
5. *P. excelsa* Gaud.: Throughout Uttarakhand.
6. *P. multifida* Poiret : Dehradun (FRI campus by S. Singh).
7. *P. stenophylla* Wall. ex Hook. et Greville: Throughout Uttarakhand.
8. *P. subquinata* Wall. ex Agardh: Almora, Pithoragarh.
9. *P. vittata* L.: Throughout Uttarakhand.
10. *P. wallichiana* Ag. : Throughout Uttarakhand.

21. Family Adiantaceae Ching

Adiantum L.

1. *A. capillus veneris* L. : Throughout Uttarakhand.
2. *A. edgeworthii* Hook. : Throughout Uttarakhand.
3. *A. incisum* Forsskal: Throughout Uttarakhand.
4. *A. philippense* L. : Throughout Uttarakhand.
5. *A. pedatum* L. : Tehri Garhwal, Almora and Pithoragarh.
6. *A. venustum* D. Don: Throughout Uttarakhand.

22. Family Hemionitidaceae Pichi Sermolli

Anogramma Link

1. *A. leptophylla* (L.) Link: Dehradun, Tehri Garhwal, Nainital.

Coniogramme Fee

1. *C. affinis* Wall. ex Hieronymus: Dehradun, Tehri Garhwal and Nainital.
2. *C. fraxinea* (D. Don) Fee ex Diels: Throughout Uttarakhand.
3. *C. intermedia* Hieron.: Dehradun, Pithoragarh, Almora.
4. *C. pubescens* Hieron.: Pithoragarh.
5. *C. serrulata* (Bl.) Fee : Throughout Uttarakhand.

Gymnopteris Bernhardii

1. *G. vestita* (Wall. ex Moore) Underwood: Throughout Uttarakhand. (Fig. 11).

23. Family Vittariaceae Ching

Vittaria J.E. Smith

1. *V. flexuosa* Fee: Throughout Uttarakhand.
2. *V. garhwalensis* Dixit: Chamoli (Chathi, Ukhimath).
3. *V. himalayensis* Ching: Pithoragarh (Kalamuni Munsyari).

24. Family Parkeriaceae Hooker

Ceratopteris Brongn.

1. *C. thalictroides* (L.) Brongniart: Dehradun, Nainital, Almora and Pithoragarh.

25. Family Hymenophyllaceae Link

Trichomanes L.

1. *T. bipunctatum* Poir. ex Lamk: Almora (Pande & Pande 2003), Kumaon (Hope, 1901).

2. *T. campanulatum* Roxb.: Rare at Didihat (Pithoragarh).
3. *T. late-alatum* (v.d. Bosch) Copel.: Throughout Uttarakhand.
4. *T. proliferum* Blume: Almora-Pindari (Hope, 1900).
5. *T. schmidiana* Zencher ex Tascnar: Pindari gorge (Hope 1900, Duthie 1906).
6. *T. striatum* D. Don: Deochula (Pithoragarh).

Hymenophyllum Sm.

1. *H. exsertum* Wall ex Hook.: Throughout Uttarakhand.
2. *H. javanicum* Spreng : Almora (Pindari, Khati), Pithoragarh.
3. *H. tenellum* D. Don: Restricted in the eastern part of Kumaon.

26. Family Cyatheaceae Kaulfuss

Alsophilla R. Brown

1. *A. spinulosa* (Wall. ex Hook.) Tryon: Chamoli and Pithoragarh.

27. Family Dennstaedtiaceae Pichi Sermolli

Dennstaedtia Bernhardii

1. *D. scabra* Wall. ex Moore: Dehradun, Tehri Garhwal, Chamoli Garhwal, Nainital.

Emodiopteris Ching et S.K. Wu

1. *E. appendiculata* (Wall. ex Hook.) Ching et S.K. Wu: Commonly found at Chamoli Garhwal, Almora and Pithoragarh.

Microlepia Presl

1. *M. platyphylla* (D. Don) J. Smith: Tehri Garhwal, Pithoragarh.
2. *M. setosa* (Sm.) Alston : Pithoragarh.
3. *M. strigosa* (Thunb.) Presl: Chamoli Garhwal and Pithoragarh.

Monachosorum Kunze

1. *M. henryi* Christ : Dehradun, Almora and Pithoragarh.

28. Family Lindsaeaceae Pichi Sermolli

Sphenomeris Maxon

1. *S. chinensis* (L.) Maxon: Throughout Uttarakhand.

29. Family Aspleniaceae Mett. ex Frank

Asplenium L.

1. *A. adiantum nigrum* L. : Dehradun, Tehri Garhwal, Uttarkashi, Almora.
2. *A. aitchisonii* Fraser Jenkins et Reichstein: Tehri Garhwal and Almora.
3. A. *bullatum* Wall. ex Mett.: Nainital, Champawat (Lohaghat).
4. *A. capillipes* Makino : Uttarkashi, Almora, Pithoragarh.
5. *A. ceterach subsp. ceterach* L. : Tehri Garhwal, Almora.
6. *A. cheilosorum* Kunze ex Mett.: Chamoli Garhwal and Pithoragarh.
7. *A. dalhousiae* Hooker: Throughout Uttarakhand.
8. *A. ensiforme* Wall. ex Hook. et Greville: Throughout Uttarakhand.
9. *A. fontanum subsp. pseudofontanum* (L.) Bernhardii: Tehri Garhwal, Chamoli and Pithoragarh.
10. *A. gueinzianum* Mett. ex Kuhn: Chamoli Garhwal, Almora and Pithoragarh.
11. *A. khullarii* Viane, Rasbach and Reichstein: Uttarkashi, Almora and Pithoragarh.
12. *A. kukkonenii* Viane, Rasbach and Reichstein: Uttarkashi, Almora and Pithoragarh.
13. *A. laciniatum* D. Don : Throughout Uttarakhand.
14. *A. nesii* Christ: Dehradun, Pithoragarh.

15. *A. phyllitidis* D. Don : Pithoragarh (Ramganga valley, Sandeo, Askot).
16. *A. paucivenosum* (Ching) Bir: Pithoragarh (Ruinguwair).
17. *A. rutamuraria* L.: Pithoragarh (Kali, Kuti and Dhauli valley).
18. *A. septentrionale* (L.) Hoffm. : Tehri Garhwal, Uttarkashi, Chamoli Garhwal, Almora and Pithoragarh.
19. *A. tenuicaule* Hayata: Uttarkashi, Almora and Pithoragarh.
20. *A. tenuifolium* D. Don: Dehradun, Nainital and Pithoragarh.
21. *A. trichomanes* L. : Throughout Uttarakhand.
22. *A. viride* Huds. : Almora and Pithoragarh.
23. *A. unilaterale* Lam. : Dehradun, Nainital, Almora and Pithoragarh.
24. *A. yoshinage* Makino subsp. *indicum* (Sledge) Fraser-Jenkins: Throughout Uttarakhand.
25. *A. yoshinage* Makino *var. yoshinage* : Throughout Uttarakhand.
26. *A. yunnanese* Franch : Dehradun, Uttarkashi, Chamoli Garhwal, Nainital and Pithoragarh.
27. *A. x garhwalense* (Notho subsp.) *garhwalense* Reichst. ex Fraser-Jenkins & Khullar: Garhwal.
28. *A. x garhwalense* (Notho subsp.) *kumaonicum*: Fraser-Jenkins & Khullar: Kumaon.
29. *A. x hunzanum* Reichstein & Fraser-Jenkins: Pithoragarh.

30. Family Onocleaceae Pichi Sermolli

Metteuccia Todaro

1. *M. intermedia* C. Chr.: Uttarakhand (Pangtey and Samant reported it under the *M. orientalis* which according to Fraser-Jenkins (2008) occurs in east Himalaya).

31. Family Woodsiaceae (Diels in Engler and Prantl) Herter

Woodsia R. Brown

1. *W. alpina* (Bolton) Gray: Chamoli Garhwal, Pithoragarh.
2. *W. andersonii* (Beddome) Christ: Tehri Garhwal, Chamoli Garhwal, Almora and Pithoragarh.
3. *W. cycloloba* Handel-Mazzetti: Garhwal (Rammi Bhujgara).
4. *W. elongata* Hooker: Throughout Uttarakhand.
5. *W. glabella* R. Br. ex Richardson: Uttarakhand (Based on Fraser Jenkin's report 2008).
6. *W. lanosa* Hooker: Tehri Garhwal, Chamoli Garhwal, Almora and Pithoragarh.

32. Family Athyriaceae Alston

Athyrium Roth

1. *A. anisopterum* Christ: Dehradun, Tehri Garhwal, Chamoli Garhwal, Nainital, Almora, Pithoragarh.
2. *A. atkinsonii* Beddome: Throughout Uttarakhand.
3. *A. attenuatum* (Wall. ex Clarke) Tagawa: Throughout Uttarakhand.
4. *A. cuspidatum* (Beddome) Kato: Pithoragarh (Didihat, Dunakote).
5. *A. davidii* (Franch.) Christ: Tehri Garhwal, Uttarkashi, Chamoli Garhwal, Almora and Pithoragarh.
6. *A. distans* (D. Don) Moore: Throughout Uttarakhand.
7. *A. x dxitii* Fras.-Jenk.: Nainital.
8. *A. drepanopterum* (Kunze) Braun ex Milde: Tehri Garhwal, Nainital, Pithoragarh.

9. *A. falcatum* Beddome: Dehradun, Tehri Garhwal, Chamoli Garhwal, Nainital, Almora and Pithoragarh.
10. *A. fimbriatum* Moore: Throughout Uttarakhand.
11. *A. flabellulatum* (Clarke) Tardie-Blot: Pithoragarh (Namik).
12. *A. himalaicum* Ching ex Mehra & Bir: Chamoli Garhwal and Pithoragarh.
13. *A. kumaonicum* Punetha: Nainital and Lohaghat (Champawat).
14. *A. mackinnoniorum* (Hope) C. Chr.: Throughout Uttarakhand.
15. *A. micropterum* Fraser Jenkins: Chamoli Garhwal, Uttarkashi, Almora and Pithoragarh.
16. *A. nephrodioides* (Baker) Christ: East Kumaon.
17. *A. otophorum* (Miq.) Koidzumi: Almora, Banlekh (Champawat).
18. *A.x pangteyi* Freser-Jenkins : Rare, Nainital.
19. *A. parasnathense* (Clarke) Ching ex Bir: Nainital (Jeolikote).
20. *A. pectinatum* (Wall. ex Mett.) Moore, Dehradun, Tehri Garhwal, Nainital, Almora and Pithoragarh.
21. *A x pichisermollianum* Fraser-Jenkins: Nainital.
22. *A. puncticaule* (Bl.) T. Moore : Throughout Uttarakhand.
23. *A. rubricaule* (Edgeworth ex Clarke) Bir: Chamoli and Pithoragarh.
24. *A. rupicola* (Edgeworth ex Hope) C. Chr.: Throughout Uttarakhand.
25. *A. schimperi* Moug. ex Fee: Throughout Uttarakhand.
26. *A. setiferum* C. Chr.: Chamoli Garhwal, Nainital, Almora and Pithoragarh.
27. *A. strigillosum* (Moore ex Love) Moore ex Salom: Throughout Uttarakhand.
28. *A. vermae* Fraser-Jenkins, Khullar & Pangtey: Common in district Pithoragarh.
29. *A. wallichianum* Ching: Tehri Garhwal, Chamoli Garhwal, Almora and Pithoragarh.
30. *A. x septrionalioccidentalibhartense* Fraser Jenkins: Dehradun (Mussoorie).

Cornopteris Nakai

1. *C.quadripinnatifida* Kato: Chamoli Garhwal, Kumaon and Almora.

Cystopteris Bernhardii

1. *C. dickieana* Sim. : Dehradun, Uttarkashi, Chamoli Garhwal, Almora and Pithoragarh.
2. *C. fragilis* (L.) Berhn.: Throughout Uttarakhand.
3. *C. montana* (Lam.) Desv.: Almora and Pithoragarh.
4. *C. tenuisecta* (Blume) Mett.: Nainital and Pithoragarh.

Deparia Hooker & Greville

1. *D. acuta* (Ching) Fraser Jenkins: Chamoli Garhwal (Badrinath).
2. *D. allantodioides* (Bedd.) Kato: Throughout Uttarakhand.
3. *D. boryana* (Willd.) Kato: Dehradun, Tehri Garhwal, Chamoli Garhwal, Nainital, Almora, Pithoragarh.
4. *D. japonica* (Thunb.) Kato: Dehradun, Tehri Garhwal, Chamoli Garhwal, Nainital, Almora and Pithoragarh.
5. *D. macdonellii* (Bedd.) Kato: Pithoragarh, Almora.
6. *D. petersenii* (Kunze) Kato: Dehradun, Nainital, Almora and Pithoragarh.

Diplazium Sw.

1. *D. esculentum* (Retz.) Sw.: Throughout Uttarakhand.
2. *D. laxifrons* Rosenst : Common at Dhaj, Sobla in Pithoragarh.
3. *D. longifolium* (D. Don) Moore: Dehradun, Nainital and Pithoragarh.

4. *D. maximum* (D. Don) C. Chr.: Throughout Uttarakhand.
5. *D. sibiricum* (Turcz. ex Kunze) Kurata: Chamoli Garhwal (Trijungi Narayan).
6. *D. sikkimense* (Clarke) C. Chr.: Rare, Uttarakhand (Based on report from Fraser Jenkins, 2008).
7. *D. spectabile* (Wall. ex Mett.) Ching: Dehradun, Nainital, Almora, Pithoragarh and Chamoli Garhwal.
8. *D. squamigerum* (Wall. ex Mett.) Matsum: Chamoli Garhwal and Almora.
9. *D. subsinuatum* (Wall. ex Hook. et Grev.) Tagawa : Didihat (Pithoragarh).

Gymnocarpium Newman

1. *G. fedtschenkoanum* Pojark: Uttarkashi, Chamoli Garhwal and Pithoragarh

33. Family Hypodematiaceae Ching

Hypodematium (Forssk.) Kunze

1. *H. crenatum* subsp. *crenatum* (Forssk.) Kuhn: Throughout Uttarakhand.
2. *H. crenatum* (Firssj) Kuhn subsp. *Loyally* Fraser Jenkins & Khullar: Throughout Uttarakhand.
3. *H. crenatum* (Forssk.) Kuhn notho subsp. x *tiwanae* Fraser Jenkins: Dehradun (Mussoorie, Kempty Falls).

34. Family Thelypteridaceae Pichi Sermolli

Ampelopteris Kunze

1. *A. prolifera* (Retz.) Copeland: Throughout Uttarakhand.

Christella Leveille

1. *C. appendiculata* (Presl) Holttum: Dehradun, Nainital, Almora, Pithoragarh.
2. *C. arida* (D. Don) Holttum: Dehradun, Nainital, Almora, Pithoragarh.
3. *C. dentata* (Forsskal) Brownsey & Jermy: Throughout Uttarakhand.
4. *C. x kumaonica* Holttum: Kumaun (McLeod ex Herb. Beddome. Holttum).
5. *C. multiauriculata* Punetha: East Kumaon, Pithoragarh.
6. *C. papilio* (Hope) Holttum : Dehradun, Chamoli, Nainital, Almora, Pithoragarh.
7. *C. parasitica* (L.) Leveille: Dehradun, Chamoli, Nainital and Almora.

Cyclogramma Tagawa

1. *C. auriculata* (J. Smith) Ching: Dehradun, Chamoli Garhwal, Nainital, Almora and Pithoragarh.

Glaphyropteridopsis Ching

1. *G. erubscens* (Wall. ex Hook.) Ching: Throughout Uttarakhand.

Macrothelypteris (H. Ito) Ching

1. *M. ornata* (Wall. ex Bedd.) Ching: Chamoli Garhwal, Nainital and Pithoragarh.

Metathelypteris (H. Ito) Ching

1. *M. gracilescens* (Blume) Ching: Tehri Garhwal and Pithoragarh.

Phegopteris Fee

1. *P. connectalis* (Michx.) Watt.: Tehri Garhwal, Chamoli, Almora and Pithoragarh.

Pronephrium Presl

1. *P. nudatum* (Roxb.) Holttum: Throughout Uttarakhand.
2. *P. penangianum* (Hook.) P. Chandra: Throughout Uttarakhand.

Pseudocyclosorus Ching

1. *P. canus* (Baker) Holttum & Grimes: Throughout Uttarakhand.

2. *P. tylodes* (Kunze) Ching: Throughout Uttarakhand.

Pseudophegopteris Ching

1. *P. levingei* (Clarke) Ching: Throughout Uttarakhand.
2. *P. pyrrhorhachis* (Kunze) Ching subsp. *distans* Fraser Jenkins: Throughout Uttarakhand.
3. *P. pyrrhorhachi subsp. pyrrorhachis*: Dehradun, Tehri, Almora.

Stegnogramma Blume

1. *S. mollissima* (Kunze) Fraser Jenkins: Dehradun, Tehri Garhwal, Chamoli Garhwal and Almora.

Thelypteris Schmidel

1. *T. palustris* (Salisb.) Schott.: Pithoragarh (Sandeo).

Trigonospora Holttum

1. *T. ciliata* (Wall. ex Benth.) Holttum: Dehradun, Garhwal.

35. Family Peranemataceae Ching

Peranema D. Don

1. *P. cyatheoides* D. Don: Dehradun, Chamoli Garhwal, Almora, Pithoragarh.
2. *P. aspidioides* (Blume) Mett.: Pithoragarh (Kalamuni).

Acrophorus Presl

1. *A. paleolulatus* Pichi Sermolli: Chamoli (Above Trijungi Narayan).

36. Family Dryopteridaceae Ching

Arachniodes Blume

1. *A. rhomboidea* (Wall. ex Mett.) Ching : Pithoragarh, Banlekh (Champawat).
2. *A. aristata* (Forst. f.) Tindale: Pithoragarh (Suwalekh near Pithoragarh).

Cyrtomium Presl

1. *C. caryotideum* (Wall. ex Hook. & Greville) Presl: Throughout Uttarakhand.
2. *C. macrophyllum* (Makino) Tagawa: Pithoragarh, Almora.
3. *C. nervosum* Ching & Shing: Pithoragarh (Gini, Tejum).

Dryopsis Holttum & Edwards

1. *D. manipurensis* (Bedd.) Holttum et P.J. Edwards: Kalamuni (Pithoragarh).

Dryopteris Adanson

1. *D. acutodentata* Ching: Uttarkashi, Tehri Garhwal, Pithoragarh.
2. *D. barbigera* (Moore ex Hooker) O. Ktze: Throughout Uttarakhand.
3. *D. basisora* Christ: Chamoli (Monghu).
4. *D. blanfordii* (Hope) C. Chr.: Dehradun, Tehri Garhwal, Uttarkashi, Chamoli, Pithoragarh.
5. *D. caroli-hopei* Fraser Jenkins: Throughout Uttarakhand.
6. *D. chrysocoma* (Christ) C. Chr.: Throughout Uttarakhand.
7. *D. cochleata* (Buch.-Ham. ex D. Don) C. Chr.: Throughout Uttarakhand.
8. *D. conjugata* Ching: Uttarkashi, Tehri Garhwal, Pithoragarh.
9. *D. flemingii* Fraser Jenkins: Uttarkashi, Almora, Buddhi.
10. *D. gamblei* (Hope) C. Chr.: Tehri Garhwal, Pithoragarh and Almora.
11. *D. hirtipes* (Kunze) C. Chr.: Tehri Garhwal.
12. *D. juxtaposita* Christ: Throughout Uttarakhand.
13. *D. khullarii* Fraser Jenkins: Tehri Garhwal, Chamoli Garhwal (Trijungi Narayan).

14. *D. komarovii* Kossinsky: Almora (Pindari, Sraju, East Ramganga valley).
15. *D. lepidopoda* Hayata: Dehradun, Tehri Garhwal, Uttarkashi, Pithoragarh.
16. *D. neorosthornii* Ching: Uttarkashi, Chamoli Garhwal, Pithoragarh.
17. *D. panda* (Clarke) Christ: Tehri, Uttarkashi, Chamoli, Almora, Pithoragarh.
18. *D. pulvinulifera* (Beddome) O. Ktze: Nainital, Pithoragarh.
19. *D. ramosa* (Hope) C. Chr.: Throughout Uttarakhand.
20. *D. redactopinnata* Basu and Panigrahi: Throughout Uttarakhand. (Fig. 13).
21. *D. serratodentata* (Beddome) Hayata: Tehri Garhwal, Uttarkashi, Chamoli Garhwal, Pithoragarh.
22. *D. sparsa* (Buch.-Ham. ex D. Don) Kunze: Throughout Uttarakhand.
23. *D. stenolepis* (Bak.) C. Chr.: Pithoragarh (Dhaj).
24. *D. stewartii* Fraser Jenkins: Dehradun, Tehri Garhwal, Uttarkashi, Pauri Garhwal, Chamoli Garhwal.
25. *D. subimpressa* Loyal: Chamoli, Almora, Pithoragarh.
26. *D. sublacera* Christ in Lectome: Dehradun, Tehri Garhwal, Uttarkashi, Chamoli Garhwal, Almora and Pithoragarh.
27. *D. wallichiana* (Spreng.) Hylander: Throughout Uttarakhand.
28. *D. woodsiisora* Hayata: Tehri Garhwal, Pithoragarh.
29. *D. xanthomelas* (Christ) C. Chr. : Throughout Uttarakhand.
30. *D. yoroii* Serizawa: Pithoragarh (Kali valley, Jipti).
31. *D. zayuensis* Ching & S.K. Wu: Uttarkashi (Janki Chatti).

Polystichum Roth

1. *P. backerianum* (Atkins. & Clarke) Diels: Tehri Garhwal, Chamoli Garhwal, Almora and Pithoragarh.
2. *P. discretum* (D. Don) J. Smith: Throughout Uttarakhand.
3. *P. duthiei* (Hope) C. Chr.: Uttarkashi, Almora, Pithoragarh.
4. *P. lachenense* (Hook.) Beddome: Tehri Garhwal, Chamoli Garhwal, Almora, Pithoragarh.
5. *P. lentum* (D. Don) Moore: Throughout Uttarakhand.
6. *P. longipaleatum* Christ: Dehradun, Tehri Garhwal, Almora, Pithoragarh.
7. *P. luctuosum* (Kunze) Moore: Dehradun, Pithoragarh.
8. *P. manmiense* (Christ) Nakaike: Throughout Uttarakhand.
9. *P. mehrae* Fraser Jenkins & Khullar: Throughout Uttarakhand.
10. *P. mucronifolium* (Blume) Presl: Nainital.
11. *P. neolobatum* Nakai: Dehradun, Tehri Garhwal, Chamoli Garhwal, Almora and Pithoragarh.
12. *P. nepalense* (Spreng.) C. Chr. : Throughout Uttarakhand.
13. *P. obliquum* (D. Don) Moore: Throughout Uttarakhand.
14. *P. piceopaleaceum* Tagawa: Throughout Uttarakhand.
15. *P. prescottianum var. prescottianum* (Wall. ex Mett.) Moore: Throughout Uttarakhand. (Fig. 12).
16. *P. prescottianum* (Wall. ex Mett.) Moore var. *casteneum* Clarke: Garhwal.
17. *P. shensiense* Christ: Uttarkashi, Chamoli, Pithoragarh.
18. *P. squarrosum* (D. Don) Fee: Throughout Uttarakhand.

19. *P. stimulans* (Kunze ex Mett.) Beddome: Throughout Uttarakhand.
20. *P. thomsonii* (Hooker Fil.) Beddome: Throughout Uttarakhand.
21. *P. wilsonii* Christ: Tehri Garhwal, Uttarkashi, Chamoli, Almora, Pithoragarh.
22. *P. yunnanense* Christ: Dehradun, Tehri Garhwal, Uttarkashi, Pauri Garhwal, Chamoli Garhwal, Nainital, Almora, Pithoragarh.
23. *P. x flemingii* Fraser Jenkins: Dehradun, Pithoragarh.
24. *P. x jamunae* Fraser Jenkins: Uttarkashi.

37. Family Tectariaceae Panigrahii

Tectaria Cavanilles

1. *T. coadunata* (J. Smith) C. Chr.: Throughout Uttarakhand.
2. *T. polymorpha* (Wall. ex Hook.) Copel.: Garhwal, Almora, Pithoragarh.

38. Family Lomariopsidaceae Alston

Elaphoglossum Schott ex J. Smith

1. *E. marginatum* (Wall. ex Fee) Moore: Almora (Pindari, Kharpatta, Dharanghar).
2. *E. stelligerum* (Wall. ex Baker) Moore: Pithoragarh.

39. Family Nephrolepidaceae Pichi-Sermolli

Nephrolepis Schott

1. *N. auriculata* (L.) Trimen: Throughout Uttarakhand.

40. Family Oleandraceae Pichi Sermolli

Oleandra Cavanilles

1. *O. wallichii* (Hook) Presl: Throughout Uttarakhand.

41. Family Davalliaceae Mett. ex Frank.

Araiostegia Copeland

1. *A. beddomei* (Hope) Ching: Dehradun, Tehri Garhwal, Chamoli Garhwal, Almora, Pithoragarh.
2. *A. hookeri* (Moore ex Beddome) Ching: Dehradun, Garhwal, Almora, Pithoragarh.
3. *A. pulchra* (D. Don) Copeland: Throughout Uttarakhand.
4. *A. squamata* (Decne.) Freser-Jenkins: West Himalaya (Based on the report of Fraser-Jenkins, 2008).

Davallia Smith

1. *D. trichomanoides* Blume: Almora, Pithoragarh.

Leucostegia Presl

1. *L. truncata* (D. Don) Fraser-Jenkins : Throughout Uttarakhand.

Paradavallodes Ching

1. *P. membranulosum* (Wall. ex Hooker) Ching: Garhwal, Nainital, Pithoragarh.
2. *P. multidentatum* (Hooker & Baker) Ching: Chamoli Garhwal, Almora.

42. Family Blechnaceae Copeland

Blechnum L.

1. *B. orientale* L.: Dehradun, Tehri Garhwal, Pithoragarh.

Woodwardia J. Smith

1. *W. unigemmata* (Makino) Nakai: Throughout Uttarakhand.

43. Family Azollaceae Wettstein

Azolla Lamarck

1. *A. pinnata* R. Br.: Rare, collected from Hachila and Hat (Pithoragarh).

5

6

7

8

9

10

11

12

13

Figs. 5-13: 5. *Loxogramme porcata*, 6. *Lepisorus thumbergianus*, 7. *L. nudus*, 8. *Polypodiodes lachnopus*, 9. *Cheilanthes rufa*, 10. *Onychium siliculosum*, 11. *Gymnopteris vestita*, 12. *Polystichun prescotianum*, 13. *Dryopteris redactopinnata*.

REFERENCES

Dixit, R.D. and Ramesh, K. (2002). Pteridophytes of Uttarakhand (A check list). Bishan Singh Mahendra Pal Singh, Dehra Dun.

Fraser-Jenkins, C.R. (1997). *New Species Syndrome in Indian Pteridology and the Ferns of Nepal.* International Book Distributors, Dehra Dun.

Fraser-Jenkins, C.R. (2008). *Taxonomic Revision of 300 Indian subcontinental Pteridophytes.* International Book Distributors, Dehradun.

Kholia, B.S. (1992). *Taxonomic Studies on the Pteridophytic Flora of Pithoragarh District of Kumaon (N.W. Himalaya),* Ph.D. Thesis, Kumaon University, Nainital.

Kholia, B.S. and Punetha, N. (1995). Notes on taxonomy and distribution of some Pteridophytes in Kumaon (W. Himalaya). *J. Indian Bot. Soc.* 74**:** 183-187.

Khullar, S.P., Pangtey, Y.P.S., Samant, S.S., Rawal, R.S. and Singh, P. (1991). *Ferns of Nainital.* Bishan Singh Mahendra Pal Singh, Dehra Dun.

Khullar, S.P. (1994). *An Illustrated Fern Flora of the West Himalaya* .I. International Book Distributors, Dehra Dun.

Khullar, S.P. (2000). *An Illustrated Fern Flora of the West Himalaya* .II. International Book Distributors, Dehra Dun.

Pangtey, Y.P.S. and Punetha, N. (1987). Pteridophytic flora of Kumaon Himalaya: An updated list. In. Y.P.S. Pangtey and S.C. Joshi (Eds.). *Western Himalaya: Environment Problems and Development,* 1: 389-412. Gyanodaya Prakashan Nainital.

Pande, P.C. (1990a). An annotated check list of the fern-allies of Kumaon Himalaya. *Indian Fern J.* **7:** 58-65.

Pande, P.C. (1990b). A census of Kumaon ferns (north western Himalaya). *Indian Fern J.* **7:** 140-195.

Pande, H.C. and Pande, P.C. (2002a). *Pteridology in Western Himalaya.* Bishan Singh Mahendra Pal Singh, Dehra Dun.

Pande, H.C. and Pande, P.C. (2002b). *An Illustrated Fern Flora oF Kumaon Himalaya* .II. Bishan Singh Mahendra Pal Singh, Dehra Dun.

Pande, H.C. and Pande, P.C. (2003). *An Illustrated Fern Flora oF Kumaon Himalaya* .I. Bishan Singh Mahendra Pal Singh, Dehra Dun.

Punetha, N. and Kaur, S. (1987). Pteridophytic flora of Pithoragarh district of Kumaon (WH). *J. Econ. Tax. Bot.,* 9: 269-286.

Diversity of Lower Plants (2014): pp. 264-278
Editors: Rajan Kumar Gupta & Mukesh Kumar
IK International Publishing House Pvt. Ltd.

21

Distribution Pattern of Terricolous Lichens in Garhwal Himalayas (Chopta-Tungnath Tract) with Reference to Morphological and Environmental Variables

HIMANSHU RAI[1***], RAJAN K. GUPTA*, D.K. UPRETI** AND PRAMOD NAG***

**Department of Botany, Pt. L.M.S Govt. P.G. College, Rishikesh, Dehradun, Uttarakhand–249 201*
***Lichenology Laboratory, National Botanical Research Institute, CSIR, Lucknow, Uttar Pradesh–226 001, India*
****Department of Botany, Mahendra Multiple Campus, Tribhuvan University, Nepalgunj, Banke–00977, Nepal*
*E-mail: *himanshurai08@yahoo.com*

ABSTRACT

India is a rich centre of lichen biodiversity. Soil inhabiting lichens (Terricolous lichens), are potential group of lichen species which can be calibrated for ecological indicator studies, as their direct growth on the soil make them very sensitive to any change in the topographic and climatic environment.

In the present study soil-inhabiting lichens of Chopta-Tungnath tract in Garhwal Himalayas, are defined on the basis of ecological and morphological characters using hierarchical bootstrap-clustering, and the various habitat and climatic variables are analyzed for their effect on the determination of structure and distribution pattern of terricolous lichen community, using Correlation-Regression analysis.

Of the three macrohabitats identified along the elevation gradient maximum soil lichen diversity was observed at middle elevation (3000-3400 mt.) A total of nine lichen species belonging to four families, seven genera and five morphological groups/thallus type (Squamulose, Squamulose-Fruticose, Leprose, Fruticose and Foliose) were identified. Squamulose was dominating morphological group, followed by fruticose, dimorphic, foliose and leprose forms. Soil lichens exhibit maximum associations with mosses followed by angiosperms, only few soil lichens were found growing on ground. Among the environmental variables soil pH and minimum and maximum temperature were positively related to lichen diversity whereas relative humidity and elevation negatively influenced soil lichen community. Cluster analyses revealed six lichen groups on the basis of their maximum distribution at middle elevations, dominance of moisture absorbing lichen groups (fruticose and foliose) and maximized associative growth with mosses.

Keyword: Terricolous lichens, Chopta, Tungnath, Cladonia, Cluster analysis.

[1]Communicating Author

Fig. 3. The three macrohabitat types of Chopta-Tungnath habitat complex based on dominant vegetation along the elevation gradient—(A) Oak-Rhododendron mixed forest (2700-3000 mt.); (B) Rhododendron rich middle altitude subalpine forest (3000-3400 mt.); (C) and (D) Alpine grassland—Bugyal, with exposed rocks, (3400-4001 mt.)

Table 1: Attributes of macrohabitats with reference to species abundance of terricolous lichens in Chopta-Tungnath landscape

Microhabitats	Elevation range (in metre)	Average elevation (in metre) AM ± SE*	Lichen spp. abundance (% Frequency)	Genera of soil lichens recorded/macrohabitat
Oak-Rhododendron mixed forest (ORMF)	2700-3000	2862.5 ± 62.5	34.21	*Cetrelia olivetorum, Cladonia pyxidata, Heterodemia pseudospeciosa, Lepraria* spp.,*Physconia detersa, Stereocaulon foliolosum, Stereocaulon pomiferum*
Rhododendron rich middle altitude subalpine forest (RMSF)	3000-3400	3261 ± 11.4	60.53	*Cladonia furcata,Cladonia pyxidata, Everniastrum cirrhatum, Lepraria spp., Stereocaulon foliolosum*
Higher altitude alpine grassland (HAGL)	3560-3890	3706 ± 9.62	5.26	*Cladonia pyxidata, Stereocaulon foliolosum*

*Arithmetic Mean ± Standard error

Lichens were examined in the laboratory using a stereomicroscope, a light microscope morphologically and anatomically, and chemically with the help of 'spot tests', UV light and standardized thin-layer chromatography (TLC) following procedure of John A. Elix *et al.,* 1993 and Orange A. *et al.,* 2001. Owing to their difficult taxonomy, species of *Lepraria* were treated at the generic level. Reference material in the form of voucher specimens are deposited in the lichen herbarium (LWG) at the National Botanical Research Institute (NBRI) of Lucknow, Uttar Pradesh, India. Identification of lichen species was done using relevant keys and monographs (Ahti, 2000; Awasthi, 2007). The lichen chemistry of soil extracted from the samples was studied with reference to soil pH. Data regarding frequency of number of lichen species in all the three macrohabitats, their growth form diversity, associations with other vegetation, macro-scale and micro-scale (soil pH) environmental variables were carefully recorded.

4. DATA ANALYSIS

To determine the natural grouping in the soil lichens, lichen species were hierarchically classified by bootstrap-cluster analysis using paired group algorithm and Euclidean distances as similarity measure, on three criteria: Lichen morphological groups, distribution in the three macrohabitats and associations with other vegetations, using abundance data (% frequency) (Hammer, 2001). The effect of various environmental variables on community structure of soil lichens was studied by analyzing four macro-scale (elevation, maximum temperature, minimum temperature and relative humidity-Rh) and one micro-scale (soil pH) environmental variable in the three stratified macrohabitats of the study area using Linear correlation and Linear regression (least square method) analysis.

5. RESULTS

5.1 Composition of Soil Lichen Community

Of the three macrohabitats identified along the elevation gradient, Rhododendron rich middle altitude subalpine forest–RMSF-3000-3400 mt., harboured maximum diversity of soil lichens (63.08%) followed by Oak-Rhododendron mixed forest—ORMF-2700-3000 mt. (34.06%) and higher altitude alpine grassland- HAGL -3400-4001 mt.(5.2%) .

A total of nine lichen species belonging to four families, seven genera and five morphological groups/thallus type (Squamulose, Squamulose-Fruticose, Leprose, Fruticose and Foliose) were identified (Table 2, Fig 4). Family **Cladoniaceae** represented by *Cladonia pyxidata* (52.57%) and *Cladonia furcata* (2.6%) was the most dominant family followed by **Stereocaulaceae** represented by *Stereocaulon foliolosum* (23.65%), *Stereocaulon pomiferum* (2.6%) and *Lepraria* spp. (10.52%); **Parmeliaceae** represented by *Cetrelia olivetorium* (2.6%), and *Everniastrum cirrhatum* (2.6%) and **Physciaceae** represented by *Heterodermia pseudospeciosa* (2.6%) and *Physconia detersa* (2.6%) (Table 2, Fig. 5).

Squamulose thallus (*Cladonia pyxidata)* was dominant (34.2%) morphological group, distributed maximum at the middle altitudes (3202 mt.) followed by fruticose thallus (26.3%) with highest concentration in the middle altitudes (3202 mt.), dimorphic thallus (18.39%) distributed from 2862 mt. to 3706 mt. with highest concentration in middle altitudes (3202 mt.), foliose thallus (10.52%) with highest concentration in lower elevations (2700-300 mt.), and leprose (10.52%) thallus with even distribution in lower to middle altitudes (2700-3400 mt.) (Tables 3, 4).

Table 2: Terricolous lichens in the three macrohabitats of Chopta-Tungnath landscape

Terricolous lichens	Family	Morphological group	Macrohabitat types (% Frequency) (*n* = *38*)		
			ORMF	RMSF	HAGL
Cetrelia olivetorum	Parmaliaceae	Fo	2.6	0	0
Cladonia furcata	Cladoniaceae	Fr	0	2.6	0
Cladonia pyxidata	Cladoniaceae	Sq, Dm	10.5	39.47	2.6
Everniastrum cirrhatum	Parmeliaceae	Fo	0	2.6	0
Heterodermia pseudospaciosa	Physciaceae	Fo	2.6	0	0
Lepraria spp.	Stereocaulaceae	Lp	5.26	5.26	0
Physconia detersa	Physciaceae	Fo	2.6	0	0
Stereocaulon foliolosum	Stereocaulaceae	Fr	7.9	13.15	2.6
Stereocaulon pomiferum	Stereocaulaceae	Fr	2.6	0	0

Lp = Leprose, Fo = Foliose, Sq = Squamulose, Fr = Fruticose, Dm = Squamules as primary thallus bearing erect fruticose body as secondary thallus; ORMF = Oak-Rhododendron mixed forest; RMSF = Rhododendron rich middle altitude subalpine forest; HAGL = Higher altitude alpine grassland.

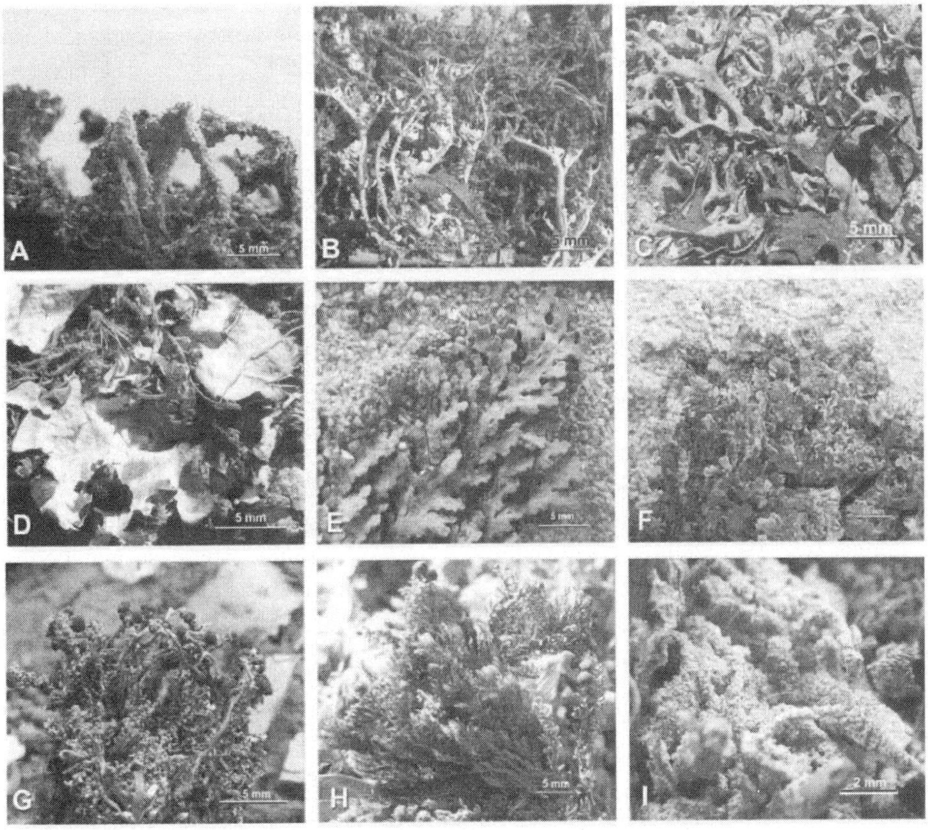

Fig. 4. A- *Cladonia pyxidata* (L.) Hoffm.; B-*Cladonia furcata* (Huds.) Schrad.; C-*Everniastrum cirrhatum* (Fr.) Hale ex.; D-*Cetrelia olivetorum* (Nyl.) W. L. Culb. & C. F. Culb.; E-*Heterodermia peudospeciosa* (Kurok.) W. L. Culb.; F-*Physconia detersa* (Nyl.) Poelt.; G-*Stereocaulon foliolosum* Nyl.; H-*Stereocaulon pomiferum* P. A. Duvign.; I-*Lepraria* spp.

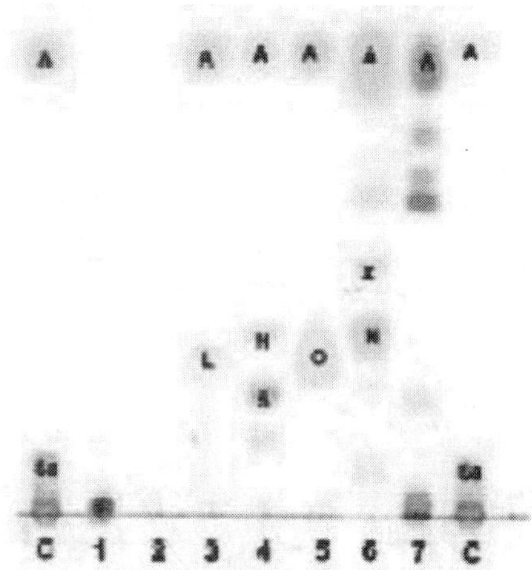

Fig. 5. TLC profile of terricolous lichens of Chopta-Tungnath in solvent system A: 1- *Cladonia pyxidata* (L.) Hoffm.; 2-*Physconia detersa* (Nyl.) Poelt.; 3-*Stereocaulon foliolosum* (Nyl.); 4-*Stereocaulon pomiferum* (Nyl.); 5-*Cetrelia olivetorum* (Nyl.) W. Culb. & Culb.; 6-*Heterodermia pseudospeciosa*; 7-*Lepraria caesioalba* var groenlandica.; C-*Parmelinella wallichiana* (Taylor) Elix & Hale. (A : Atranorin; F : Fumarprotocetraric Acid; L : Lobaric Acid; S : Stictic Acid; Sa: Salazinic Acid; O : Olivetoric Acid; Z : Zeorin)

Table 3: Values of lichen community and environmental variables in the three macrohabitat types of Chopta-Tungnath landscape.

Variables	Macrohabitat types (*n* = *38*)		
	ORMF (*n* = 13)	RMSF (*n* = 23)	HAGL (*n* = 2)
Lichen community variables : (% Frequency)			
Lichen growth forms:			
Leprose	5.26	5.26	0
Fruticose	10.5	13.15	2.63
Foliose	7.89	2.63	0
Squamulose	5.26	28.94	0
Dimorphic	5.26	10.5	2.63
Associations (with other vegetation):			
Mosses	26.31	26.31	5.26
Angiosperms (grasses, etc.)	5.26	31.57	0
No association (on ground)	2.63	2.63	0
Environmental Variables:			
Macroscale Variables (AM ± SE):			
Elevation (Meter)	2862.5 ± 62.5	3202 ± 11.4	3706 ± 9.62
Temperature-maximum (°C)	20.5 ± 0.14	16.9 ± 0.7	9.93 ± 0.06
Temperature-minimum (°C)	3.6 ± 0.3	2.4 ± 0.1	0.6 ± 0.2
Relative Humidity (%)			
Micro-scale variable (s):	89.7 ± 0.15	71.9 ± 0.06	80.5 ± 0.29
pH	6.3 ± 0.1	6.8 ± 0.1	5.2 ± 0.2

ORMF = Oak-Rhododendron mixed forest; RMSF = Rhododendron rich middle altitude subalpine forest; HAGL = Higher altitude Alpine grassland.

*Arithmetic Mean ± Standard error

Table 4: Terricolous lichen species recorded in Chopta-Tungnath landscape and their attributes with reference to morphological groups, distribution in the three macrohabitats and association with other vegetation

Terricolous Lichens	Family	Dominance category	Morphological groups (% Frequency) (*n* = *38*)					Macrohabitat types (% Frequency) (*n* = *38*)			Associations (% Frequency) (*n* = *38*)		
			Lp	Fo	Sq	Fr	Dm	ORMF	RMSF	HAGL	Ms	Ag	NAs
Cetrelia olivetorum	Parmaliaceae	X	0	2.6	0	0	0	2.6	0	0	2.6	0	0
Cladonia furcata	Cladoniaceae	X	0	0	0	2.6	0	0	2.6	0	2.6	0	0
Cladonia pyxidata	Cladoniaceae	A	0	0	34.2	0	18.42	10.5	39.47	2.6	21.1	31.5	0
Everniastrum cirrhatum	Parmeliaceae	X	0	2.6	0	0	0	0	2.6	0	0	0	2.6
Heterodermia pseudospaciosa	Physciaceae	X	0	2.6	0	0	0	2.6	0	0	2.6	0	0
Lepraria spp.	Stereocaulaceae	C	10.5	0	0	0	0	5.26	5.26	0	5.26	5.26	0
Physconia detersa	Physciaceae	X	0	2.6	0	0	0	2.6	0	0	2.6	0	0
Stereocaulon foliolosum	Stereocaulaceae	B	0	0	0	21.1	0	7.9	13.15	2.6	15.8	2.6	2.6
Stereocaulon pomiferum	Stereocaulaceae	X	0	0	0	2.6	0	2.6	0	0	2.6	0	0

A = Most Dominant, B = 2nd most dominant, C = 3rd most dominant, X = present in one macrohabitat and in one sample; Lp = Leprose, Fo = Foliose, Sq = Squamulose, Fr = Fruticose, Dm = Squamules as primary thallus bearing erect fruticose body as secondary thallus; ORMF = Oak-Rhododendron mixed forest; RMSF = Rhododendron rich middle altitude subalpine forest; HAGL = Higher altitude alpine grassland; Ms = Associative growth with mosses, Ag = Associative growth with angiospermic vegetation, NAs = non-associative growth on ground.

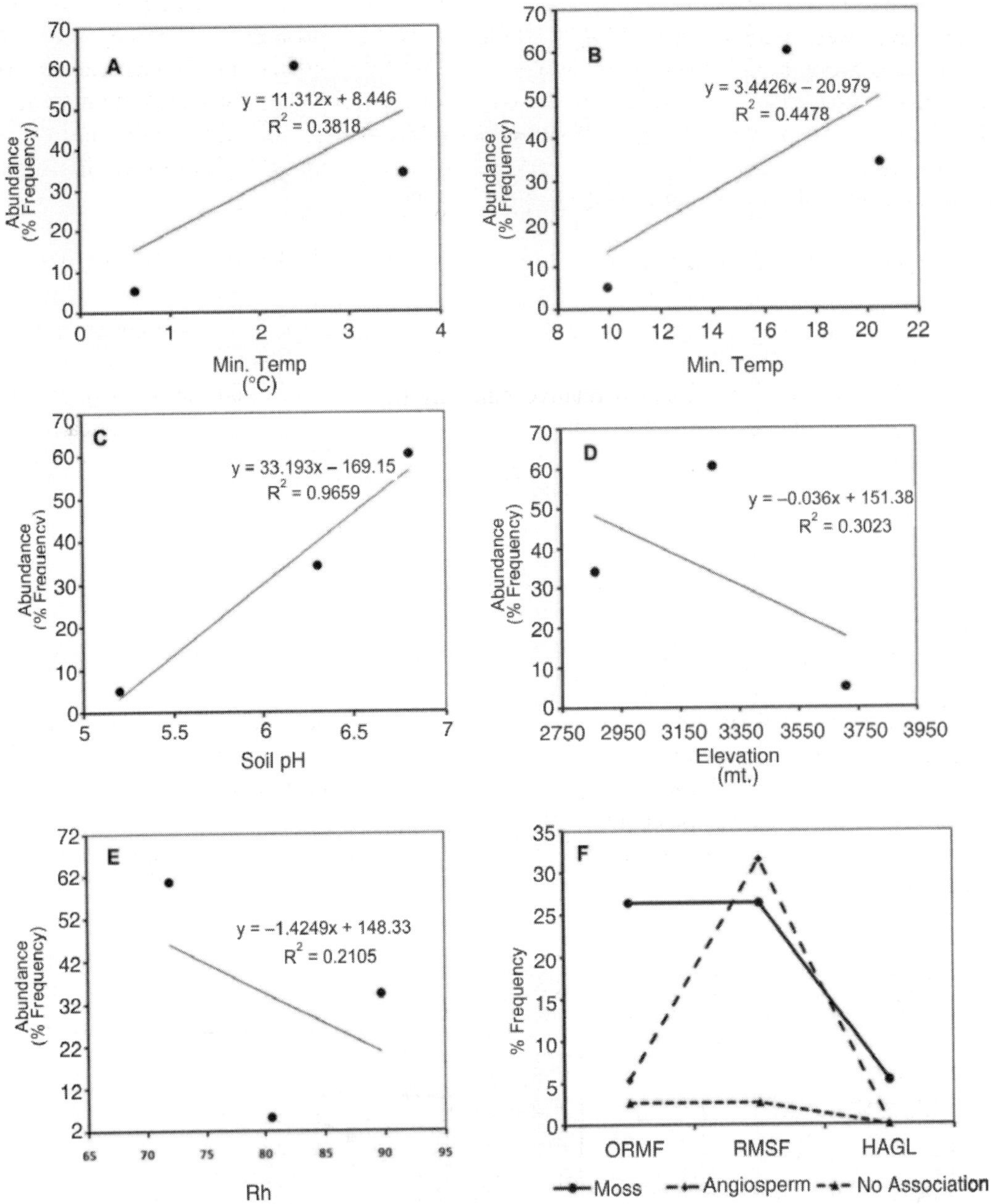

Fig. 6. A, B, C, D, E, F: Effect of environmental variables (Minimum Temperature, Maximum Temperature, Soil pH, Elevation, and Relative humidity (Rh) on abundance of soil lichen species in Chopta-Tungnath landscape; Association pattern of soil lichen species with other vegetation on the three macrohabitats of Chopta-Tungnath landscape (ORMF = Oak-Rhododendron mixed forest; RMSF = Rhododendron rich middle altitude subalpine forest; HAGL = Higher altitude alpine grassland.)

The analysis of data of associative growth of soil lichens revealed that maximum associations were with mosses (57.88%) followed with angiosperms (36.83%), only 5.26% soil lichens were found growing on ground without any association. Though mosses were consistently associated with lichens in all the three macrohabitats their maximum frequency (26.31%) was at middle elevations (3200 mt.) (Table 4, Fig. 6-F). Associative growth with angiospermic vegetation though dominated (31.57%) at middle elevations, this type of association was absent at higher altitude.

5.2 Effect of Environmental Variables on Soil Lichen Community

Macroscale and microscale environmental variables critically affect the soil lichen community composition. At 5% significance level soil lichen diversity was positively correlated to soil pH (p = +0.98) and minimum (p = +0.61) and maximum (p = + 0.66) temperature whereas it was found negatively correlated to relative humidity (p = – 0.45) and elevation (p = –0.54). Soil pH (R^2 = 0.96), and maximum temperature (R^2 = 0.45) were the most significant drivers of soil lichen diversity followed by minimum temperature (R^2 = 0.38), elevation (R^2 = 0.30) and relative humidity (R^2 = 0.21) (Fig. 6-A, B, C and D).

5.3 Lichen Groups

Six lichen groups were defined (Table 4, Fig. 7). Group A consisted of a single species (*Cladonia pyxidata*), occurred in all the three macrohabitats (i.e., ORMF, RMSF & HAGL), exhibiting squamulose and dimorphic thallus, and associated with mosses and angiospermic vegetation. Group B consisted of only one species (*Stereocaulon foliolosum*), occurred in all the three macrohabitats (i.e., ORMF, RMSF and HAGL), exhibiting fruticose thallus, and associated mostly with mosses and angiospermic vegetation and also in one instance found growing on ground without any association. Group C consisted of a single Leprose species (*Lepraria*), occurring at lower to middle altitude macrohabitats (ORMF and RMSF), and associated with mosses and angiospermic vegetation. Group D consisted of a two-fruticose species (*Stereocaulon pomiferum* and *Cladonia furcata*), occurring in one of the HAGL macrohabitats—ORMF and RMSF, respectively associated with mosses only. Group E and F consisted of four foliose species

Distribution in the three macrohabitats	Morphological groups	Associations	Lichen Group	Species
Present in all three macrohabitats	Squamulose & Dimorphic	Moss & Angiosperms	A	*Cladonia pyxidata*
	Fruticulose	Moss, angio & no asso.	B	*Stereocaulon foliolosum*
Present in 2 of the 3 macrohabitats	Leprose	Moss	C	*Lepraria* spp.
Present in one of the 3 macrohabitats	Fruticulose		D	*Stereocaulon pomiferum* *Cladonia furcata*
	Foliose		E	*Physconia detersa* *Cetrelia olivetorum* *Heterodermia pseudospaciosa*
		No Association	F	*Everniastrum cirrhatum*

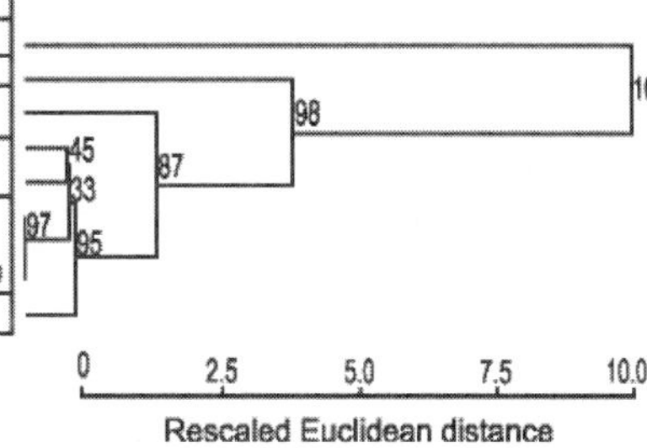

Fig. 7. Groups of soil lichens resulting from bootstrap-cluster analysis (Bootstrap No = 100) based on ecological (distribution in three macrohabitats along the elevation gradient and association with other vegetation) and morphological characters (thallus morphological groups).

(*Physconia detersa*, *Cetrelia olivetorum*, *Heterodermia pseudospaciosa* and *Everniastrum cirrhatum*), ocuurring in one of the three macrohabitats, mostly in lower to middle altitude habitats, and associated with mosses usually, only *Everniastrum cirrhatum* was found occurring on ground without any association.

6. DISCUSSION

Though soil lichens form a very small group in the overall lichen community of the area (1.2%) (Negi 2000), the present study reveals that this small group of lichens can be a very efficient indicator of ecosystem health in relation to various anthropogenic pressures in Himalayan habitats.

Soil lichen species richness pattern in the three macrohabitats revealed that maximum species concentration was in middle altitudes (RMSF-3560-3890 mt.), which eventually decreased at higher altitudes (HAGL, 3560-3890 mt.). This disparity is due to high use of the lower altitude *Quescus* forest (ORMF, 2700-300), for lopping, cutting and uncontrolled grazing throughout the year, whereas though middle and higher altitude macrohabitats are also open for grazing, etc. but only in summers (Negi, 2000). The drop in lichen species at higher altitude can be attributed to the change in topography of the area, increasing exposed rocks and boulders which eventually limit soil rich habitats.

Dominance of squamulose lichen species followed by fruticose, dimorphic and foliose species, over leprose species in the terricolous niche of the study area, is due to the more efficient atmospheric moisture absorbing capacity of foliose and fruticose morphological forms (Lalley & Viles, 2005).

Association with mosses and angiospermic vegetation (small herbs and grasses) in nearly all the macrohabitats is due to the dominance of these plant groups on the ground strata of the vegetation the only vegetation surviving at the higher altitudes of the landscape.

Environmental variables influenced the terricolous lichen community of the area decisively. Among the macro-scale variables temperature regime (minimum-maximum temperature), was found positively influencing the soil lichen community; elevation and relative humidity (Rh) negatively influenced the terricolous lichens, which can be attributed to the absence of tree cover on the higher altitudes. At micro-scale level soil pH was found significantly related to soil lichen diversity (Eldridge & Tozer, 1997; Eldridge & Koen, 1998; Ponzetti & McCune, 2001), where lower acidic pH (6.3-6.8) sustains higher concentration of soil lichens than higher acidic pH (5.2).

Based on the ecological and morphological characters, soil-associated lichens of the Chopta-Tungnath tract were clustered into three major groups, species distributed in all the three macrohabitats, two of the three macrohabitats and one of the three macrohabitats, which were subsequently subdivided into two, one and three sub-groups. In the study area, moisture absorbing groups (squamulose, dimorphic, fruticose and foliose) and moss associated growth decided the natural clustering of terricolous lichens.

ACKNOWLEDGEMENTS

The authors are grateful to Director, NBRI and Uttarakhand State Council for Science and Technology for necessary lab facilities and technical assistance.

REFERENCES

Ahti, T. (2000). Cladoniaceae-Flora neotropica monograph 78. *The New York Botanical Garden Press.*

Awasthi, D.D. (1991). A key to the microlichens of India and Nepal and Sri Lanka, *Bibl Lichenol.* **40:** 1-337.

Awasthi, D.D. (1998). A key to the macrolichens of India and Nepal. *J. Hattori. Bot. Lab.* **65:** 207-302.

Awasthi, D.D. and Ahti, T. (2007). A compendium of the macrolichens from India, Nepal and Sri Lanka. Bishen Singh Mahendra Pal Singh, Dehra Dun, India. pp. 1-580.

Belnap, J. and Gillette, D.A. (1998). Vulnerability of desert biological soil crusts to wind erosion: the influence of crust development, soil texture and disturbance *J. Arid. Environ.* **39:** 133-142.

Beymer, R.J. and Klopatek, J.M. (1992). The effects of grazing on cryptogamic crusts in Piyon-Juniper woodlands in Grand Canyon National Park. *Am. Midl. Nat.* **127:** 139-148.

Bjelland, T. (2003). The influence of environmental factors on the spatial Eldridge, D.J. and Koen, T.B. (1998). Cover and floristics of microphytic soil crusts in relation to indices of landscape health. *Plant Ecol.* **137:** 101-114.

Eldridge, D.J. (1998). Trampling of microphytic crusts on calcareous soils and its impact on erosion under rain-impacted flow. *Catena* **33:** 221-239.

Eldridge, D.J. and Leys, J.F. (2003). Exploring some relationships between biological soil crusts, soil aggregation and wind erosion. *J. Arid Environ.* **53:** 457-466.

Eldridge, D.J. and Tozer, M.E. (1997). Environmental factors relating to the distribution of terricolous bryophytes and lichens in semi-arid eastern Australia. *Bryologist.* **100:** 28-39.

Evans, R.D. and Belnap, J. (1999). Long-term consequences of disturbance on nitrogen dynamics in an arid ecosystem. *Ecology* **80:** 150-160.

Gadgil, M. and Meher-Homji, V.M. (1990). Ecological diversity; in *Conservation in developing countries: Problems and prospects* (Eds.) Daniel, J.C. and Serrao, J.S. (Bombay: Bombay Natural History Society) pp. 175-198.

Greig-Smith, P. (1983). *Quantitative plant ecology,* 3rd edition (London: Blackwell).

Hammer, Øyvind, Harper, David, A.T., and Paul D. Ryan, (2001). Past: Paleontological Statistics Software Package for Education and Data Analysis. *Palaeontologia Electronica*, Vol. 4, issue 1, art. **4:** 9.

Hawksworth, D.L. (1988). Coevolution of fungi with algae and cyanobacteria in lichen symbioses. in Pirozynski, K.A. and Hawksworth, D.L. (Eds). *Coevolution of fungi with plants and animals. Academic Press, New York, NY.*

John, E. Elix and Karen, D. Ernst-Russel, (1993). A catalogue of standardized thin layer chromatographic data and biosynthetic relationships for Lichen substances (2nd edition.) *Australian National University*, Canberra.

Krebs, C.J. (1989). *Ecological methodology* (New York: Harper and Row).

Lalley, J.S. and Viles H.A. (2005). Terricolos lichens in the northern Namib Deserts of Namibia: distribution and community composition, *The Lichenologist.* **37(1):** 77-91.

Lalley, J.S., Viles, H.A., Henschel, J.R. and Lalley, V. (2006 a). Lichen-dominated soil crusts as arthropod habitat in warm deserts. *J. Arid. Environ.* **67:** 579-593.

Lalley, Jenifer S., Viles, Heather A., Copeman, Neil, and Cowley, Catherine. (2006 b). The influence of multi-scale variables on the distribution of terricolous lichens in a fog desert. *J. Vegetation Science* **17:** 831-838.

Negi, H.R. (1999). Co-variation in diversity and conservation value across taxa: a case study from Garhwal Himalaya, *Ph.D. thesis, Indian Institute of Science, Bangalore.*

Negi, H.R. (2000 a). Species richness and turnover of moss communities in western parts of Nanda Devi Biosphere Reserve; *Int. J. Ecol. Environ.* **26:** 1-18.

Negi, H.R. (2000 c). On the patterns of abundance and diversity of Macrolichens of Chopta-Tungnath in Garhwal Himalaya, *J. Biosci.* **4:** 367-378.

Negi, H.R. and Gadgil, M. (1996). Patterns of distribution of Macrolichens in western parts of Nanda Devi Biosphere Reserve; *Curr. Sci.* **71:** 568-575.

Negi, H.R. and Upreti, D.K. (2000 b). Species diversity and relative abundance of lichens in Rumbak catchment of Hemis National Park in Ladakh; *Curr. Sci.* **78:** 1105-1112.

Orange, A., James, P.W. and White, F.J. (2001). Microchemical methods for the identification of Lichens, *British Lichen Society.*

Pinokiyo, A., Singh, K.P. and Singh, J.S. (2008). Diversity and distribution of lichens in relation to altitude within a protected biodiversity hot spot, north-east India. The Lichenologist **40(1):** 46-62.

Pirintsos, S.A., Diamantopoulos, J. and Satmou, G.P. (1995). Analysis of the distribution of epiphytic lichens within homogeneous *Fagus sylvatica* stands along an altitudinal gradient (Mount Olympos, Greece), *Vegetatio* **116:** 33-40.

Ponzetti, J.M. and McCune, B. (2001). Biotic soil crusts of Oregon's shrub steppe: Community composition in relation to soil chemistry, climate and livestock activity. *Bryologist.* **104:** 212-225.

Sheard, J.W. (1968). Vegetation Pattern in a Moss: Lichen Heath Associated with Primary Topographic Features on Jan Mayen, *Bryologist*, Vol. 71, No. 1, pp. 21-28.

Sundriyal, R.C. (1992). Structure, productivity and energy flow in an alpine grassland in the Garhwal Himalaya; *J. Veg. Sci.* **3:** 15-20.

Upreti, K. and Büdel, B. (1990). The Lichen genera *Heppia* and *Peltula* in India *J. Hattori. Bot. Lab.* **65:** 279-248.

Will-wolf, S., Scheidegger, C. and McCune, B. (2002). Methods for monitoring biodiversity and ecosystem function-Monitoring scenarios, sampling strategies and data quality. In: Nimis, P.L., Schedegger, C. and Wolseley, P.A. (Eds.). *Monitoring with Lichens- Monitoring Lichens., Kluwer Acdemic Pblishers, Netherlands.* 163-181.

Wolseley, P.A., Moncrieff, C. and Aguirre-Hudsen, B. (1994). Lichens as indicators of environmental stability and change in tropical forests of Thailand, *Global Ecol. Biogeogr. Lett.* 116-123.

Zaady, E. and Bouskila, A. (2002). Lizard burrows association with successional stages of biological soil crusts in an arid sandy region. *J. Arid. Environ.* **50:** 235-246.

Diversity of Lower Plants (2014): pp. 279-296
Editors: Rajan Kumar Gupta & Mukesh Kumar
IK International Publishing House Pvt. Ltd.

22

Bryophytes and Ecosystem

YATEESH MOHAN BAHUGUNA[1*], SUMEET GAIROLA[1], D.P. SEMWAL[2], P.L. UNIYAL[2] AND A.B. BHATT[1]

[1]*Department of Botany, HNB Garhwal University, Srinagar Garhwal-246 174, Uttarakhand*
[2]*Department of Botany, University of Delhi, Delhi-110 007*
Email: ymbahuguna@gmail.com

ABSTRACT

Bryophytes are a diverse group of land plants of small stature but have large ecological impact. There are some 23,000 described species worldwide, making it the largest group of land plants except for the flowering plants. The group includes three phylogenetically distinct lineages: mosses, hornworts, and liverworts. The bryophytes are generally considered a "key" group in our understanding of how the modern land plants (the embryophytes) are related to each other phylogenetically and how they came to conquer the hostile land environment. Bryophytes are able to occupy variety of habitats in diverse ranges of growth forms. They are integral part of natural environment of forest ecosystems. Although the bryophytes display much species diversity, a major limitation in the use of bryophytes as study organisms has been the lack of basic floristic, ecological, and alpha-taxonomic knowledge of the plants in many regions.

Keywords: Liverworts, Hornworts, Mosses, Habitat, Growth forms, Ecology, Threats.

1. INTRODUCTION

Bryophytes comprise a very distinct group among the land plant vegetation. They comprise nearly 23,000 species and form the most diverse group of terrestrial plants. They evolved as an unparallel diversity in size and structure. In India, the bryophytes are represented by about 2850 taxa (Singh and Hajra 1996) and Western Himalaya, Eastern Himalaya and Western Ghats represent hotspots of biodiversity and acquire numerous endemic bryophytes. They are able to grow in extreme conditions and form major component in Arctic, Antarctic and alpine environment.

[1*]Communicating Author

Most of the bryophytes are of small size, but some of them attain a height up to half metre or a little more. They are able to store large amount of water, nutrients and carbon in their biomass. In most ecosystems, especially in peatlands, bryophytes function as carbon sinks, which is a matter of great concern when considered with the rise of global carbon dioxide level. A remarkable adaptation of bryophytes is their ability to remain alive for a long period without water even under high temperature, and then resume photosynthesis within seconds after being moistened by rain or dew. Most bryophytes are ectohydric, i.e., ability to absorb water, inorganic nutrients and mineral elements directly from the atmosphere rather than the soil and substratum.

Bryophytes are widely recognized as the oldest living land plants (Shaw and Renzaglia 2004). The features of bryophytes relate them to the green algae and both appear to share a common ancestor. Bryophytes possess short-lived sporophytic and dominant gametophytic phase. Bryophytes have a basal phylogenetic position among the extant land plants, remnant lineages surviving from spectacular radiation of the land plants in the Devonian period (400 million years ago). Recent phylogenetic reconstruction of relationships suggests hornworts as the basal group to higher land plants; moss, and liverworts forming a monophyletic sister clade. In adapting to the erratic sub-aerial supply of water, bryophytes in general adopted the alternative strategy of evolving desiccation tolerance, photosynthesizing and growing during moist period and suspending metabolism during times of drought.

The growth and sexual reproduction in bryophytes depend on external water and are therefore favoured by a humid microclimate. They have remarkable capacity of regeneration from any fragment of plant. Various modes of reproduction play an important role in the life cycle of bryophytes especially in stands within high disturbance (During 1997). Rainforest bryophytes that hang in long festoons from the trees rely on torrential rain to transport their sperms from tree to tree. The small short mosses move their sperms with the help of dewdrops in early morning. Spores of mosses travel a long distance with the help of winds, even moving between continents through the jet streams. The spore walls are highly protective and some spores are reported to remain viable for up to 40 years.

Bryophytes are able to grow in stressful conditions like cold, drought, shades and in nutrient poverty conditions and a dominating role in such ecosystem by subsequently influencing factors *viz*., soil temperature, nutrient input, etc. Generally, the soil under bryophytes mats become cool and moist, which affect nutrient cycling of ecosystem by catching atmospheric deposition and leaching from dripping above ground. Bryophytes play an efficient role to filter the nutrients reaching to soil by absorbing them directly from the atmosphere in liquid phase. Bryophytes protect the soil against erosion due to their netted and webbed protonemata and gametophores to cover the exposed substrata and help in increasing water-holding capacity of the soil. The role of bryophytes in an ecosystem is governed by four properties, viz., their ability to establish soils, to trap and hold moisture, to exchange cations and, to tolerate desiccation. These characters are enhanced by their ability to spread by frequent branching. Mosses like *Atrichum*, *Nardia*, *Pogonatum*, *Pohlia* and *Trematodon* are inhibitors of soil erosion because of their trample-resistant structure and their regenerative capacity.

Bryophytes have ecological association with microorganisms, protozoans, rotifers, nematodes, earthworms, molluscs, insects, spiders and many other invertebrates (Gerson, 1982) as well as other plants and fungi. Bryophytes provide food, shelter and nesting material for small animals and invertebrates and indirectly serve as a matrix for a number of varieties of interactions among all the organisms. Bryophytes furnish a favourable substratum and seed bed for the establishment of seedlings of higher plants. They provide congenial habitat for nitrogen-fixing Cyanophyceae, e.g., *Nostoc*, *Scytonema*, etc. (Basilier, 1979).

Mosses alone account about 75% of the annual accumulation of phosphorus. Some time bryophytes reduce mineral availability to other organisms and become successful competitor. During precipitation, mosses are almost perfect sink for some elements. This process is governed by high cation exchange capacity of many mosses. This characteristic phenomenon takes place by large net negative charge carried extra cellularly on the cell walls of these mosses, play a role in absorption of some nutrient cations, e.g., Mg^{++} and K^{+}. These ions firstly sequester by the negative charges and later pass from the cell wall into the living cells. Some bryophytes that can grow in salt marshes appear to be best representative where annual precipitation is high; *Schistidium maritimum* can tolerate salinity through their low permeability to cation as well as an active extrication of Na from the cells.

Pollutants reduce the level of gametangia thus sexual reproduction reduces. Some taxa such as *Ceratodon purpureus* and *Funaria hygrometrica* produce sporophytes under highly polluted conditions but differ in reproductive behaviour as it shows differences in tolerance (Gilbert 1968). Sergio (1987) observed that *Tortula laevipila* produces few sporophytes in urban population in comparison to rural area. The highest mercury contents 12,100 Hg g^{-1} has been reported from the basal segments of *Jungermannia vulcanicola* growing in an acidic stream (Satake *et al.* 1983). Atmospheric pollutants along with some minerals directly enter the cell of the bryophytes. The ultimate survival of bryophytes is critically dependent on the preservation of their natural habitats.

Table 1: Summary of the numbers of the species of the plant taxonomic classes which are listed as extinct, extinct in the wild (the species has been extirpated from its natural habitat), or globally threatened with extinction:

	Extinct	*Extinct in the wild*	*Threatened*
Bryopsida	2	0	36
Anthocerotopsida	0	0	2
Marchantiopsida	1	0	42
Coniferopsida	0	1	140
Ginkgoopsida	0	0	1
Magnoliopsida	69	14	5099
Liliopsida	1	2	291
Total	73	17	5611

Adapted from Hilton-Taylor, 2000

2. DIVERSITY IN MORPHOLOGICAL FORMS

Bryophytes are divided into three groups viz., Liverworts, Hornworts and Musci (Plates 1 and 2). The ultrastructure of the cell and the molecular biology confirmed that bryophytes itself have three separate evolutionary patterns which are crowned as Liverworts (Hepaticopsida),

Hornworts (Anthocerotopsida) and Mosses (Bryopsida) (Shaw and Goffinet 2000). Further studies pertaining to detailed analysis of land plant relationship, Kenrick and Crane (1997) proposed that these three groups of bryophyte represent a remarkable structural slab, in plant evolution recognized by their "monosporangiate" life cycle.

Plate 1. Bryophytes, integral part of Forest Environment.
A: Frog adapted to surroundings of carpets of mosses. B: Liverwort as a nesting medium for reptiles eggs.
C: Mosses growing on the fallen tree trunk. D: *Meteoriopsis formosana* growing on *Quercus* sp.
E: *Meteorium buchananii* hanging from tree branches. F: Bryophytes growing in water stream.

2.1. Liverworts

The estimated number of liverwort species range from 6000 to 8000. Liverworts are represented by about 850 species belonging to 141 genera and 52 families (Singh 2001). Leafy forms (Jungermaniopsida) are represented by nearly 85% of liverwort species that show enormous amount of morphological, anatomical and ecological diversity. Plants with leafy shoot systems are the most common growth form in this class, e.g., *Cololejeunea*, *Frullania*, *Jubulopsis* and *Radula*. Thalloid forms, e.g., Metzgeriales and Marchantiales are widely distributed in moist, humid, shady, terrestrial, semi aquatic places, especially in high altitude, dense sub alpine and wet temperate forests. The sporophyte of liverworts grows within gametophyte plant body until their capsules are ready to dehisce. The dispersal of spores is by the twisting motion of elaters and by splitting of sporophyte into four segments.

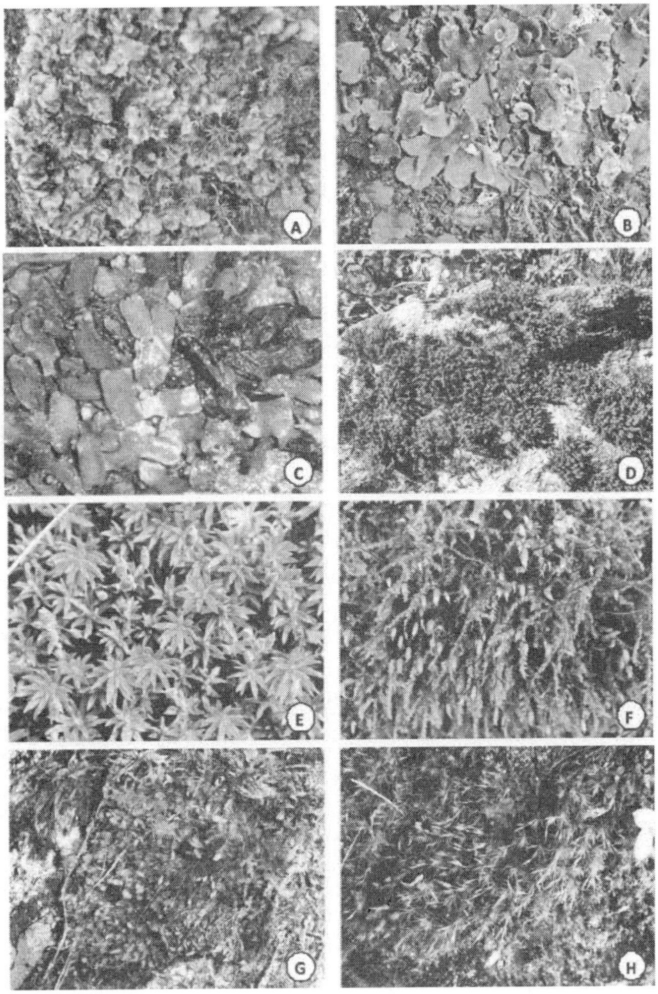

Plate 2. Some liverworts and mosses growing in their natural habitats.

A: *Blasia pusilla* B: *Dumortiera hirsuta*
C: *Targionia indica* D: *Grimmia donniana*
E: *Pogonatum microstomum* F: *Stereophyllum anceps*
G: *Bryum coronatum* H: *Dicranum crispifolium*

2.2. Hornworts

Hornworts consists of approximately 100-150 species in the world (Renzaglia and Vaughn 2000). They resemble with some liverworts only in case of unspecialized thalloid gametophyte. Hornworts possess *Nostoc* colonies ventrally in their thallus. This alga shows a symbiosis nature which provides organic nitrogen for metabolism of the hornworts thallus and the thallus provide food (carbohydrate) and shelter. They possess a cylindrical horn-like sporangia (Sporophyte). The liberation of spores from the sporophyte takes place gradually over a long period. Their spores are dispersed by water movement and not by wind. Hornworts differ from all other land plants in having only one large, algal-like chloroplast in each thallus cell.

2.3. Mosses

Mosses display greatest species diversity with estimates of the number of species ranging from 10,000 to 15,000. In India mosses are represented by 2300 species in 330 genera. Mosses are one of the dominant plant communities of Himalaya at higher elevation growing in humid and cool conditions and some time give more than 50% of active biomass (Groombridge 1992). The leaves in most cases are spirally arranged; nerve is usually present. The capsules possess unique structure called peristome for effective spore dispersal. The *Sphagnum* mosses are one of the most ecologically and economically important groups of bryophytes.

3. HABIT AND HABITAT DIVERSITY

Generally, bryophytes are smaller than most other vascular plants. *Dawsonia superba* is the tallest moss, which attains a height up to 70 cm; *Polytrichum commune* grows in wet conditions and attains a height up to 50 cm. The other pendent mosses as *Meteoriopsis* spp. and the aquatic moss *Fontinalis* spp. may attain a height up to 1 metre. Some of the bryophytes are quite small; the smallest tiny plants of *Cephaloziella* are visible only under microscope and some mosses as species of *Ephemerum* and *Micromitrium*. Moist evergreen forests have a great variety of microhabitat. Bryophytes are significant components of temperate and tropical forests where they are found as carpet on the humid soil, boulders, living and dead logs, hanging from tree branches, and on leaves. The distribution of bryophytes is influenced first by microclimatic factors, i.e., rainfall and temperature, latitude and altitude (Sveinbjörnsson and Oechel 1992) and by micro environmental conditions like shade, humidity, humus and temperature (Alpert 1991). The bryophyte vegetation may be effected by side factors, e.g., age of the soil, rocks, composition of forest soil, moisture content (Sillett and Neitlich 1996) and by substratum like its pH and its humus status (Batty *et al.* 2003).

Bryophytes have an especial nature to grow on certain habitats such as rocks of preferred type, barks of especial type, decomposed logs and stumps, on such soil, which have a suitable moisture and humus, exposure and pH. Moss carpet on the forest floor provides suitable substrata for seed germination and seedling growth of higher plants. Many species of *Sphagnum* are aquatic some time they form floating mates. *Riccia fluitans* is also an aquatic bryophyte. Some bryophytes are saprophytic in nature as *Buxbaumia* and *Cryptothallus*, which generally grow on the dead organic matter, *Cryptothallus* totally lack chlorophyll and nutrients are absorbed by the associated mycorrhiza. The special character of the bryophytes is that they absorb water within no time and get fresh when they are dried in unfavourable conditions, thus they are also named–"Resurrection plants".

The epiphytic bryophytes are found in tropical rainforests, with plenty of water and nutrients. Thus, the broad leaves forest create a good environment for these lower plants. Epiphytic bryophytes occur on irregular surface of bark and twigs and depression on branches as well as require fissures for their establishment, which provide a place for them as well as humus and moisture. In alpine regions, a rich vegetation of turf forming bryophytes occurs and the species diversity is very high within a limited space. It has also been observed that where organic material rich soil is present, there is the richer bryoflora, along with the vascular plants. In arid and semi-arid zones, bryophytes survive in unfavorable conditions in the form of protonema, rhizoidal tubers and gemmae or the apical part of the gametophore as latent (this apical part survive with the help of scales, which provide mineral nutrition and moisture). A few highly characteristic species also grow in other salty habitat. Bryophytes such as *Sphagnum* species are capable of modifying their environment and thus affect the life of other organisms.

4. BRYOPHYTES OF SPECIEAL HABITATS

4.1. Epiliths (Saxicolous)

Some bryophytes grow only on rocks possess some special adaptation and requirements, perhaps for a more permanent habitat, need of less water and their adaptive capability to grow on such substrata. Some of these epilithic bryophytes are *Grimmia* sp., *Gymnostomum* sp., *Porella platyphylla*, *Seligeria* and *Tortella* species.

4.2. Xerophytic

Mosses are able to colonize the mobile sand. During unfavourable conditions, they are buried in the soil; and may reappear shortly after a shower of rain or humidity.

4.3. Disturbed Sites

Some species grows on freshly disturbed soil and on road cutting, etc., are pioneers e.g., *Barbula*, *Bryum*, *Funaria* and *Tortula*.

4.4. Calcicoles and Calcifuges

Calcicoles are calcium loving and calcifuges include calcium repelling taxa. The calcicoles are restricted to rocks and on soils containing high amount of calcium carbonate. On the other hand, calcifuges taxa live on substrata with an acid reaction. Bates (1978) reported a 3-4 fold greater calcium exchange capacity in the calcicoles taxa. *Fissidens cristatus*, *Grimmia anodon*, *Gymnostomum* sp., *Hydrogonium* sp., *Philonotis* sp., etc., grow on calcicoles habitat. *Andreaea rothii* grow on calcifuges habitat which has a large concentration of iron and aluminum. Calcifuges taxa are *Marsupella emarginata* and *Scapania undulata* that grow only in low solute content in water. The Calcicoles taxa *Chiloscyphus polyanthos*, *Cratoneuron filicinum*, *Rhynchostegium riparioides* and *Thamnobryum alopecurum* commonly grow in base-rich water but found in low Ca^{++} concentration.

4.5. Fire Mosses

Fires deposit a large amount of ashes containing high percentage of potassium, phosphates and magnesium along with nitrogen and other organic matter which greatly change the soil texture. The pH rises from 5-6 to 8-10. After an interval of time, some species of bryophytes grow on these burn sites. The pioneer moss community is found there, e.g., *Funaria hygrometrica* rapidly dominate along with *Bryum argenteum* and *Ceratodon purpureus*.

4.6. Coprophils

The coprophilous taxa are obligate and are most frequent in nutrient poor substrata, grow only on decomposed animal excreta, e.g., many Splachnaceae are adapted to live on such substrata. *Splachnum* spp., are exclusively coprophilous.

4.7. Epixylic Bryophytes

In humid forest, on dead logs and wood, a carpet forming taxa of pleurocarpous mosses is known as epixylic bryophytes. Thallose and foliose hepatics, e.g., *Aneura*, *Cephalozia*, *Lophocolea*, *Reboulia*, *Riccardia* and *Symphyogyna* appear on the dead wood. The bryophytes grow on dead wood especially in temperate forest, on dead trees, fallen logs and on stumps, forming major structural features of ecological importance. Such bryophytes provide a suitable habitat for various organisms, which helps as a source of mineral nutrients and energy and influence soil deposition (Harmon *et al.* 1986). Some epixylic bryophytes drought sensitive and these communities show the successional changes in the physiochemical properties of the wood during its decay, e.g., change in texture, density, pH and water holding capacity (Vellak and Paal, 1999, Crites and Dale, 1998). Dead tree decay stage, trunk dimension, species of the tree and microclimate are the main factors for the inhabitation of these communities (Heilmann-Clausen and Christensen, 2003).

4.8. Epizoic Bryophytes

Some bryophytes grow on the bodies of living insects and molluscs in the tropical forest. In very humid mossy montane forest at an altitude of 2000-3000 feet, a community of *Cololejeunea*, *Daltonia angustifolia*, *Metzgeria*, *Microlejeunea* and *Odontolejeunea* is reported to live on the backs of certain weevils, e.g., *Gymnopholus* (genus of beetles) and other genera (Gressitt *et al.* 1968).

5. GROWTH FORMS

The adaptation by some bryophytes according to habitat and their environment is a significant character. It is determined and classified by their shoot growth pattern and morphology and referred as "growth-forms" (DuRietz 1931). Gimingham and Robertson (1950) classified the growth forms based only on their morphology. The study of these growth forms is a valuable tool in many fields of bryophytes ecology. The morphological structure of the gametophore shows a positive correlation with its environmental factors. Mostly it is very uncertain to place a moss or a moss assemblage under one or other life forms. It is observed that one and same type of moss can show two different life forms, e.g., in *Plagiomnium undulatum* the sterile shoots are fan-like while the fertile shoots look like a dendroid. *Bazzania trilobata* forms tall turfs according to environmental conditions. They are of following types:

5.1. Turfs

Turfs is composed of many shoots, which forms a compact upright shoot. Branches from the main shoots are similar and parallel to the main stem. These mosses grow on exposed site. The turfs are of following types:

I. **Tall Turfs:** Tall turf grows on temperate forest surface. The upright shoots attain a height generally more than 5 cm. The water is conducted through capillary action in between the shoots. Some tall turf taxa are *Dawsonia, Dicranum, Polytrichum, Sphagnum, Timmia*, etc.

(a) Tall Turfs with erect branches: In this type, the shoots are generally simple or branched with vertical growth converted by lateral innovations, e.g., *Dicranum majus, Polytrichum commune, Tortula ruralis*, etc.

(b) Tall Turfs with divergent branches: The divergent branches are of limited growth. The erect shoots are normally over 2 cm high with short lateral, scattered or whorled branches, e.g., *Campylium stellatum, Sphagnum* sp., etc.

II. **Short Turfs:** They attain a height up to 2 cm and grow closely together vertically after ripening of the sporogonia by means of regenerative shoots. They grow on mineral soils and on rocks. These are very spreading turfs and remain for more than a few years e.g., *Barbula* sp., *Bryum argenteum, Ceratodon* sp., *Gymnomitrion* sp., *Trichostomum* sp., etc.

III. **Open Turfs:** Their shoots are short, clearly separate from one another, mostly having persistent protonema or rhizoid strand, e.g., *Diphyscium foliosum, Polytrichum* sp. and *Trematodon* sp.

5.2. Canopy Forms

Tree-like structure forms an erect leafy canopy. The erect portion of the shoot is unbranched below with scale leaves and normal leaves, which are found abundant at the tip forming canopy. These are of two types:

I. Unbranched Dendroid: Mostly they are found in terrestrial habitat. The leafy branches are present only at the uppermost part of thc main shoot, e.g., *Climacium* sp., *Dawsonia* sp. and *Pogonatum* sp.

II. Branched Dendroid: They are found in epiphytic or terrestrial habitat. They are frequently branched, erect, leafy shoot, e.g., *Hypnodendron* sp., *Plagiomnium undulatum* and *Rhodobryum dendroides*.

5.3. Cushions

In this pattern shoots radiating from a central point form a rounded tuft. The cushions are predominant in open sites. They are of two types:

I. **Large Cushions:** This type of cushion attains a diameter of about 5 cm, e.g., *Amphidium* sp., *Andreaea* sp., *Leucobryum glaucum, Orthotrichum* sp. and *Schistochila* sp.

II. **Small Cushions:** This type of cushion attains a diameter less than 5 cm, e.g., *Calymperes* sp., *Grimmia* sp., *Plagiopus* sp.

5.4. Mats

Mats are formed of creeping shoots and their branches, which are firmly affixed, to the substratum with the help of rhizoids. The mats frequently occur on forest floor and are of following types:

1. **Rough Mats:** These mats are formed when main shoots attach to substratum with numerous short erect laterals and the branches are in many directions. It can fix quickly with the substratum, e.g., *Brachythecium rivulare, Brachythecium albicans, Sematophyllum* sp. and *Trachypodopsis* sp.
2. **Smooth Mats:** These types of mats are formed by closely interwoven branches lying mainly in the same plane as the main shoot, e.g., *Frullania tamarisci, Hypnum cupressiforme, Isopterygium* sp., *Radula* sp., and many Hookeriaceae and Lejeuneaceae.
3. **Thread Forms:** These are delicate, creeping irregularly and sparingly branched and sometimes form open mats. It is when these species occur among other vegetation, e.g., *Hypnum cupressiforme* var. *ericetorum, Lophocolea bidentata.*
4. **Thalloid Mats:** These thalloid bryophytes form the overlapping dorsiventral thalli, e.g., *Conocephalum conicum, Pellia epiphylla, Dumortiera hirsuta*, members of Marchantiaceae and Metzgeriaceae.

5.5. Wefts

Interwoven, frequently branched creeping shoots form this pattern and rhizoids loose on the substratum. This type is found in the forest floor in temperate and humid climate, e.g., *Rhacopilum* sp., *Thuidium* sp. and *Trichocolea* sp., etc.

5.6. Pendulous

In this form the base of the stem is attached to the tree branches and main shoot and branches hang downwards, found in temperate and humid forest, e.g., *Cryptoleptodon* sp., *Meteoriopsis* sp., *Meteorium* sp. and members of Phyllogoniaceae.

5.7. Fans

The branches and main shoot of gametophore occur in a single plan and appear like a fan, e.g., *Bryopteris* sp., *Plagiochila* sp. and members of Neckeraceae.

5.8. Tails

These bryophytes are shade loving, creeping in nature; grow on tree trunk and vertical rocks. Thus, they form tail-like structures. They are slightly branched and commonly have radial leaves, e.g., *Cyathophorella, Leucodon, Plagiochila*, etc.

5.9. Feather Forms

Secondary stems of some bryophytes are pinnately branched and are placed dorsiventrally, epiphytic in nature, e.g., *Bryopteris* sp., *Callicosta bipinnata, Hypopterygium* sp. and *Neckeropsis* sp.

5.10. Annual Forms

Some bryophytes stop their growth, which has produced gametangia and dies after the production of sporogonium without any production of generative shoots. Such bryophytes are known as pioneer bryophytes and occur on open mineral soils, e.g., *Buxbaumia*, *Calobryum*, *Phascum*, *Riccia*, *Sphaerocarpos*, etc.

5.11. Protonemal Bryophytes

These bryophytes have the persistent protonema and leafy shoots, which are minute and scattered, e.g., *Micromitrium* and *Protocephalozia*, etc.

5.12. Bracket Mosses

These mosses exhibit horizontal shoots and mainly grow as epiphytes, e.g., *Leiomela* and *Spiridens*.

5.13. Moss Balls

The moss cushions on small stones in very cold climate are over turned and form a ball-like structure, it is due to wind, freeze-thaw and movements of the soil or ice, e.g., *Drepanocladus* sp.

6. ECOLOGY OF BROPHYTES

Bryophytes play important role in an ecosystem in many ways: (i) Bryophytes have a great capacity to stabilize soil, particularly mosses are very effective and successful soil binder and nutrient trapper. (ii) They have high water holding capacity and ability to tolerate desiccation. Bryophytes are spread by apical growth and even after their subsequent decay, the apex grows in a mat in favourable condition which helps to hold the water and soon the young ones; cover the soil area. (iii) They form a moist wet ground to form a cushion; which ultimately helps to grow the other vascular seedlings later. This maintains the high humidity regime within such forest. (iv) Filtering takes place through fall and runoff water through the bryophyte mat and peat using cation exchange system. This is an additional significance of the forest ecosystems elsewhere in the world (Bates 1990).

Bryophytes play an important role in the ecosystem of the temperate and tropical forest with special reference to the extreme environmental conditions. Many species have a capability to sustain against drought, frost and scarcity of nutrition. Some bryophytes are able to fix nitrogen in association with cyanobacteria. They are the best host to provide shelter, food and suitable place for the invertebrates. Bryophytes have an ability to change the rough surfaces by their growth, to a substratum adding humidity and humus that becomes suitable for colonization to seed plants. These initial bryophytic colonies are replaced by the tolerant and competitive species. This tolerance is the significant character of bryophytes by which they survive successfully. Epiphytic bryophytes play a very important role in the water balance and humus formation in gymnospermous forest. The bryophytes which grow on the canopy of the tree differ physiologically from those growing at the tree base. Bryophytes at tree canopy have increased respiratory system due to higher illumination than those that grow on the base. Bryophytes are found on the nutrient rich tree bases, ridges of the bark, irregular surface on twigs and exposed surfaces on branches and trunks.

Some bryophytes such as *Riccia himalayensis*, *Funaria hygrometrica* and *Physcomitrium cyathicarpum* grow throughout the year in the cultivated fields. The associations of lichens and bryophytes with symbiosis of cyanobacteria play an important role in nitrogen cycling. Bryophytes absorb mineral elements directly from the atmosphere rather than the substratum. Nitrogen and phosphorus are the most important minerals for the growth of these nonvascular plants and these minerals have limiting role in natural and semi-natural environments (Vitousek and Howarth 1991). *Sphagnum* accumulates nutrients direct from atmosphere in their living cells (Sala *et al.* 2000). A rapid growth of moss colonization takes place when barren rocks are covered with soil. Species of *Bryum* have an especial quality to grow fast on this soil, which make a thick cushion. The litter of the forest when decomposes helps to grow some mosses which later on helps in the germination of seeds of higher plants. The mixed oak-conifer forest provides a suitable platform of the mixed litter on which form a thick mossy carpet which controls the soil and nutrient erosion by holding the water.

Vellak and Paal (1999) stated that bryophytes are the most sensitive indicators for evaluating human impact in forest. Many bryophyte species are may survive only in forest without considerable human impact (Cooper-Ellis 1998). These species are known as hemerophobic *sensu* (Linkola 1916) and they usually disappear first after forest cutting or drainage. Forest management has been intensified in recent years; some forest types have been put under stronger economic pressure and by which their area has decreased drastically (Paal 1998). It is not yet clear if forest management affects different forest types similarly. Soil conditions are responsible for bryophyte diversity and rarity. Predicting species diversity from environmental conditions have benefited in the last few years, from the increasing availability of geographic information system (GIS) to combine information on species richness and habitat condition (Cowley *et al.* 2000). Soil condition and landscape features, control bryophytic diversity and rarity at a regional scale. Conservation of bryophytes is important as their ecological importance is critical in human habitats and their scientific use is great because of their presence at particular area can give suitable information about taxa and minerals.

7. BRYOPHYTES ASSOCIATION WITH ANIMALS

The degree to which different taxonomic groups show resemblance in diversity patterns has attracted increased attention, yet such studies on stream biota are lacking. Environment correlates the species richness patterns of bryophytes, micro-invertebrates and fishes in streams. Bryophytes and micro-invertebrates displayed the strongest degree of resemblance, but even this relationship had a relatively weak predictive power. The growth forms of bryophytes play an important role to give shelter to other organism. Thus, they form an attractive habitat for many invertebrates, which get shelter, food from them even in adverse conditions, e.g., *Frullania* and *Herbertus* give shelter to a number of rotifers, nematodes, invertebrates and algae in their lobules. Some aquatic mosses are the best habitats for laying eggs of snails. Insects suck juice from the cells of gametophores and sporangia. Spider, mites, millipedes, centipedes and ants make their shelter within these mosses. This activity of invertebrates within bryophytes helps in the mineral cycling which promote the growth of this vegetation. These invertebrates also help in cross-pollination (transfer of antherozoids from antheridia to archegonia) and help in spore dispersal, e.g., especially in coprophilous taxa, Splachnobryaceae. Some insects carry the spores from mature sporangium to the nitrogen substratum where this spore can germinate and grow. Many birds use mosses for the construction of their nests. The shoots of pleurocarpous mosses collected by birds for nest construction, as these mosses are lighter in weight in comparison to other vegetation and can be easily woven into desired pattern and provide insulation. In reference to alpine regions, bryophytes form the suitable habitat for lichens.

8. DESICCATION TOLERANCE

The bryophytes have more desiccation tolerance than other plants. They have an ability to revive from the drought and resume photosynthesis and respiration within minutes or hour when they get hydrated. Vertical pattern and their flourishing in a forest are governed by the balance between water and nutrient availability and depend on the tolerance to periods of desiccation. Bryophytes and lichens exhibit a type of desiccation tolerance, which seems a primitive character (Oliver *et al.* 2000). Some mosses have a cuticle which prevents water loss but most of the leafy mosses can prevent water loss by changing the shape of leaves and curling of branches, e.g., *Tortula ruralis* have a capability to tolerate drought condition up to ten months and functions normally within a few hours of rehydration. The youngest part of the shoot is capable to tolerate freezing temperature. In semiarid climate, the bryophytes grow near the pebble base where there is availability of water. Bryophytes can survive from desiccation by limiting the damage incurred up to a repairable level, a capacity to maintain its physiological reliability in the dried conditions and assemble restore system upon rehydration. They can survive even the loss of 80 to 90% of their cell water. *Andreaea rothii* grows on hard acidic mountain rocks; it can revive its photosynthesis even after 12 months of drought. Desiccation tolerant species are of two types (i) Homoiochlorophyllous (ii) Poikilochlorophyllous. The bryophytes, lichens and some other plants have a significant physiological adaptation in which they are able to control their metabolism during the periods of drought and the metabolism revive itself when water is available to them.

9. CHEMICAL CONSTITUENTS

The biochemical information on bryophytes is consistent and provides detailed data on relationship among bryophytes as well as on affinities with other groups of plants. The terpenoids and flavonoids are useful chemosystematic markers at different taxonomic levels. The sesquiterpenoids mostly occur in liverworts and are located in the oil bodies. Asakawa (1997) reported 65 types of sesquiterpenoids from liverworts. Bryophytes have many inorganic compounds, primary and secondary metabolites and other compounds like terpenoids, flavonoids and other aromatic compounds. *Conocephalum* contains monoterpenoids like pinene, camphene, sabinene, myrcene. *Frullania* having monoterpenoids like B-pinene and camphene, *Jungermannia* contains monoterpenoids pinene, camphene, myrcene, terpinene, limonene, Y-terpinene, terpinolene and p-cymene. Sesquiterpenoids like drimenol, cuparene are found in *Bazzania*, *Porella* and *Scapania*. Squalene is found in all liverworts and *Racomitrium*. Diplotene, a monoterpinoid, is found in all mosses. Hopenes, Betulic and Ursolic acid are found in all bryophytes. The sesquiterpenoids reported in liverworts belong to drimanes, bisabloanes, cadinanes, humulanes, caryphyllanes, dumorterol, germacranes, elemanes, maalianes, aromadendranes, pinguisanes and lactones.

Diterpenoids are reported only in liverworts. At least 18 diterpenes have been reported in liverworts of which labdanes and kauranes are most common. Triterpenes are reported in all groups of bryophytes. About 40 types of triterpinoids have been reported in bryophytes; at least 20 of them are moss specific. These include amyrin, friedelanol, friedeline, hopene, fernene, lupeol, squalene, taraxerol, taraxerone and betulic and ursolic acid (Mues 2000). Fourteen steroids have been isolated from bryophytes. These include sitosterol, pytosteroids, campesterol and stigmasterol are most common. Campesterol is more common in liverworts while rest of steroids occur mainly in mosses.

Flavonoids are broadly distributed in liverworts and mosses but are not common in hornworts. The important flavones reported in bryophytes are apigenin, glucuronide, acacetin, luteoin, chrysoeriol, scoparin and tricetin. Bryophytes also contain a wide range of benzoic and cinnamicacids and derivatives of phenylglycosides, phenylathers, bibenzyles, acetophenones, lignas, napthallenes and alkanes and some of them are important constituents of the epicuticular waxes, which act as protective coatings. Liverworts contain frequently n-paraffins ranging from C 10 to about C 30.

10. ECOLOGICAL IMPORTANCE

Bryophytes have variety of uses. Some of important uses of Bryophytes are listed below:

10.1. Pioneer Colonizers in Succession

Bryophytes colonize on such barren lands which have poor nutrients and where no other plants may grow. After a long interval, these bryophyte colonies built up an organic layer on that barren land which helps for the growth of microorganisms. These microbes change the mineral status of the substratum and the site become suitable for the establishment of other vegetation. The epiphytic bryophytes colonies growing on tree trunk and branches absorb water during high moisture. This result in increase in weight and the branches break down. Thus, a new succession is formed with the change in moisture, illumination regime and with the decomposition of the wood.

10.2. Pollution and Heavy metal Indicator

Bryophytes are bioindicators of air, water pollution and accumulators of heavy metals. Communities of mosses, lichens and liverworts reduce in size over a period in such disturbed environmental conditions (air pollution). Mosses disappear from such polluted areas accept a few tolerant species, viz., *Bryum*, *Ceratodon*, *Dicranoweisia*, *Funaria*, *Hyophila* and *Tortula*. Relationship between the distribution of communities, Index of Atmospheric Purity (IAP) values and sulphur dioxide levels can provide a clear picture of degree of pollution. Even a low concentration of SO_2 inhabits the flourishing of the mosses, as protonemata are especially sensitive to the pollutant. Bryophytes are very sensitive to hydrogen fluoride, even a low concentration of HF (0.001 to 0.1 ppm) and show symptoms of injury. *Pylaisiella* and *Orthotrichum* are shown to change the colour of leaves due to chlorophyll degeneration when they are exposed to HF polluted area (Le Blanc *et al.* 1971). *Sphagnum* have a peculiar character to bind radioactive compounds through cation exchange and also used for purifying contamination of waste water. Some species of *Anomodon*, *Dicranum*, *Eurhynchium*, *Leucodon*, *Mnium*, *Rhynchostegium* and *Thuidium* are able to accumulate Cesium. Leafy liverworts are found to be more sensitive to ionizing radiations. Species of *Brachythecium*, *Buxbaumia* and *Grimmia* are able to concentrate much strontium than that found in their substrata. Some terrestrial bryophytes accumulate uranium. High levels of pollutants like fluoride, sulphur dioxide, acidified rain and heavy metals are responsible to inhibit sporophyte growth in many moss species. Aquatic bryophytes are the best monitoring agents for heavy metal pollution as they are able to concentrate these heavy metals and can release these metals only after decomposition. The high concentration of cadmium in some bryophytes shows a distinct change in pigmentation and growth rate of these lower plants. In *Marchantia* and *Funaria*, the zinc concentration (> 50 ppm) reduce the spore germination. Species of *Bryum*, *Dicranella* and *Polytrichum* are able to tolerate high levels of zinc (55000 ppm), cadmium (610 ppm) and copper (2700 ppm) in their tissue.

10.3. Bryophytes as Site Indicators

Bryophytes have great capability to retain some minerals in their substrata than those of vascular plants. So they are used as a tool for geobotanical studies. Species of *Barbula*, *Gymnocolea*, *Merceya*, *Mielichhoferia*, *Scopelophila*, *Solenostoma* prefer to grow in high copper (320-770 ppm) bearing substrate, and also occur on zinc, iron and lead which exist as sulfides. Many taxa of *Campylopus*, *Barbula*, *Brachythecium*, *Bryum* and *Gymnostomum* are known to grow on gypsum (pH 4.9-7.8) and are often encrusted with this mineral (Tewari and Pant 1994). Some bryophytes build up tufa deposits. Calcareous tufas are formed through precipitation of calcium present in the surrounding soil. Some species have a relationship with the seed plants; because the specific seed plant community provides a favourable microenvironment to the bryophytic vegetation. Thus, it is useful in characterizing a site in some areas, the particular bryophyte vegetation can inform about the specific community even after the vascular plant vegetation has been destroyed. They are also able to indicate the past existence of type of vegetation and can be used as indicator to make plantation of the particular taxa on that area. Some bryophytes are able to grow in such soil, which has high copper concentration, and they accumulate copper from the soil, e.g., *Solenostoma* grows on the seepage water from copper mines (Brown and House 1978). Mosses easily accumulate K, Ca and Mg from rainfall. The patches of *Hyophila*, *Oxystegus* and *Zygodon* indicate iron hematite bearing substrates at the particular site. Bryophytes grow in sharply defined ecological range and they are also a good indicator of environmental condition, indicators of soil quality. *Polytrichum* is an indicator of acidic soil. Taoda (1977) listed a few indicators of soil pH as *Atrichum undulatum* (4.5-6.0), *Marchantia polymorpha* (6-7.5), *Tortula rhyzophylla* (6.1-7.4), etc.

11. THREATS TO BRYOPHYTES

Developmental activities create loss of habitat to both macro and micro level. The main cause of threatening to these lower plants are removal of forest to make it to agricultural land and other means as for shelter, rehabilitation, road, dam, etc. All these activities are fully responsible for the loss of bryophytic habitat. Global warming makes climate dry loss of humidity and drier microclimate are responsible for the loss of bryophytes. The unplanned forest management further enhances the loss of these valuable lower plants, viz., the forest fire either natural or by man, causes a subsequent drought, which effects the growth of bryophytes and other plants. According to Hilton-Taylor (2000), 36 species of Bryopsida, 2 of Anthocerotopsida and 42 of Marchantiopsida are threatened and 2 species of Bryopsida and 1 species of Marchantiopsida have become extinct. Following are some of the causes which are threatening diversity of the bryophytes:

1. Habitat destruction.
2. Forest clearance.
3. Land use change.
4. River valley projects.
5. Anthropogenic disturbances.
6. Exotic species invasion.
7. Climate change.
8. Road construction.

In recent years, we have witnessed a sharp decline in biodiversity, mainly due to urbanization and other anthropogenic activities, which results in loss or fragmentation of natural habitats. In the Himalayan region, disturbances in the form of removal of small fraction of forest biomass through grazing, lopping, and surface burning are common. These factors posed a serious threat to many species of bryophytes. In Himalaya large amount of mosses are illegally extracted from the forest for variety of purposes. Since conservation strategies are meant only for attention-getting 'keystone' plant or animal species, their *in situ* protection can only be assured by helping to establish sufficient large nature reserves needed by these keystone species.

12. PROSPECTS OF BRYOPHYTES

In reference to present day condition, Bryophytes have an important role with respect to environmental conditions. It is proved that these are the first colonies of the terrestrial habitats and represent a bridge between the Pteridophytes on one hand and the Algae on the other. The species richness of Bryophytes is very high and placed next to Angiosperms and has a great capability to grow even in adverse conditions, while other vascular plants are not able to do so, bryophytes are the best pollution indicators. However, this lower group of plants received lesser attention in this country and remained neglected in exploration, due to their less direct economic potentials. At present, the impact upon these lower plants is more adverse due to global problems, which is governed by several factors like; global warming, shifting of monsoon, drought, landslides, earthquakes, manmade pollution and habitat distraction by broad constructions and post effect of tourism. More bryo-exploration is essential for the unexplored areas and necessary steps should be taken up for their conservation. India one of the 12-megabiodiversity countries of the world, possesses a large area and a variety of phytoclimatic conditions which contribute to great diversity of the flora. Pande (1958) divided these zones in to 7 bryogeographical regions namely the Western and Eastern Himalaya, Punjab and west Rajasthan, Gangetic plains, Central India, Deccan Plateau and the Western and Eastern Ghats. Bryophytes are ecologically significant and play a key role in ecosystem dynamics. They cover the barren soil and conserve the soil and nutrients, provide habitats for invertebrates and maintain water balance in the forest. Bryophytes contain some specific compounds, which possess antibiotic and antimicrobial properties.

REFERENCES

Alpert P. 1991. Microtopography as habitat structure for mosses on rocks. *In*: Bell SS, Mccoy D.E. and Mushinsky H.R. (Eds.), *Habit structure. The physical arrangement of objects in space*. Chapman and Hall, London. U.K. pp. 120-139.

Asakawa Y. 1997. Heterocyclic compounds found in bryophytes. *Heterocycles* **46:** 795-848.

Basilier K. 1979. Moss-associated nitrogen fixation in some mire and coniferous forest environments around Uppsala, Sweden. *Lindbergia* **5:** 84-88.

Bates J.W. 1978. The influence of metal availability on the bryophytes and macrolichen vegetation of four rock types on Skye and Rhum. *J. Ecol.* **66:** 457-482.

Bates J.W. 1990. Interception of nutrients in wet deposition by *Pseudoscleropodium purum*: an experimental study of the uptake and retention of potassium and phosphorus. *Lindbergia* **15:** 93-98.

Batty K., Bates J.W. and Bell J.N.B. 2003. A transplant experiment on the factors preventing lichen colonization of oak bark in southeast England under declining SO_2 pollution. *Canad J. Bot.* **81:** 439-451.

Brown DH and House K.L. 1978. Evidence of a copper tolerant ecotype of the Hepatic, *Solenostoma crenulata. Ann. Bot.* **42:** 1383-1392.

Cooper-Ellis S. 1998. Bryophytes in old-growth forests of western Massachusetts. *J. Torrey Bot. Soc.* **125:** 117-132.

Cowley M.J.R., Wilson R.J., Leon-Cortes J.L., Guttierez D., Bulman C.R. and Thomas C.D. 2000. Habitat-based statistical models for predicting the spatial distribution of butterflies and day-flying moths in a fragmented landscape. *J. Appl Ecol.* **37:** 60-72.

Crites S. and Dale M.R.T. 1998. Diversity and abundance of bryophytes, lichens and fungi in relation to woody substrate and successional stage in aspen mixed wood boreal forests. *Canad J. Bot.* **76:** 641-651.

DuRietz G.E. 1931. Life forms of terrestrial flowering plants. *Acta Phytogeogr Suec* **3:** 1.

During H. 1997. Bryophyte diaspore banks. *Adv. Bryol* **6:** 103-134.

Gerson U. 1982. Bryophytes and invertebrates. *In*: Smith A.J.E. (Ed.) *Bryophyte Ecology*. Chapman and Hall, London. pp. 291-330.

Gilbert O.L. 1968. Bryophytes as indicators of air pollution in the Tyne valley. *New Phytol* **67:** 15-30.

Gimingham C.H. and Robertson E.T. 1950. Preliminary investigation on the structure of bryophytic communities. *Trans. British Bryl Soc.* **1:** 330-344.

Gressitt J.L., Samuelson G.A. and Vitt D.H. 1968. Moss growing on living Papuans forest weevils. *Nature* **217:** 765-767.

Groombridge B. (Ed.) 1992. *Global Biodiversity: Status of the Earth's Living Resources*. Chapman and Hall, New York.

Harmon M.E., Franklin J.F., Swanson F.J., Sollins P., Gregory S.V., Lattin J.D., Anderson N.H., Cline S.P., Aumen N.G., Sedell J.R., Lienkaemper G.W., Cromack K. and Cummins K.W. 1986. Ecology of coarse woody debris in temperate ecosystems. *Adv. Ecol. Res.* **15:** 133-302.

Heilmann-Clausen J. and Christensen M. 2003. Fungal diversity on decaying beech logs implications for sustainable forestry. *Biodivers Conserv* **12:** 953-973.

Hilton-Taylor C. (comp.) 2000. 2000 IUCN Red List of Threatened Species. IUCN, Gland. (Also available at http:// www.redlist.org).

Kenrick P. and Crane P.R. 1997. *The Origin and Early Diversification of Land Plants: A Cladistic Study.* Washington. D.C.: Smithsonian Institution Press.

Le Blanc F., Comeau G. and Roa D.N. 1971. Fluoride injury symptoms in epiphytic lichens and mosses. *Canad J. Bot.* **49:** 1691-1698.

Linkola K. 1916. Studien über den Einfluss der Kultur auf die Flora in den Gegenden nördlich vom Ladogasee. I. *Acta Soc. Fauna Flora Fennica* **45:** 1-432.

Mues R. 2000. Chemical constituents and biochemistry. *In*: Shaw A.J. and Goffinet B. (Eds.). *The Biology of Bryophytes.* Cambridge Univ. Press, U.K.

Oliver M.J., Tuba Z. and Mishler B.D. 2000. The evolution of vegetative desiccation tolerance in land plants. *Plant Ecol.* **151:** 85-100.

Paal J. 1998. Rare and threatened plant communities of Estonia. *Biodivers Conserv* **7:** 1027-1049.

Pande S.K. 1958. Some aspects of Indian Hepaticology (Presidential address). *J. Ind. Bot. Soc.* **37:** 1-26.

Renzaglia K.S. and Vaughn K.C. 2000. Anatomy, development and classification of hornworts. *In*: Shaw A.J. and Goffinet B. (Eds.), *Bryophyte Biology.* Cambridge University Press, U.K. pp. 1-19.

Sala O.E., Chapin F.S.I., Armesto J.J., Berlow E., Bloomfield J., Dirzo R., Huber-Sanwald E., Huenneke LF, Jackson R.V., Kinzig A., Leemans R., Lodge D.M., Mooney H.A., Oesterheld M., Leroy Poff N., Sykes M.T., Walker B.H., Walker M. and Wall D.S. 2000. Global Biodiversity Scenario for the Year 2100. *Science* **287:** 1770-1774.

Satake K., Soma M., Seyama H. and Uehiro T. 1983. Accumulation of Mercury in the liverwort *Jungermannia volcanicola* Steph., in an acid stream Kashiranashigawa in Japan. *Arch Hydrobiol* **99:** 80-92.

Sergio C. 1987. Epiphytic bryophytes and air quantity in the Tejo Estuary. *Symposia Biologist Hungaria* **35:** 795-814.

Shaw A.J. and Goffinet B. 2000. *Bryophyte Biology.* Cambridge University Press, London.

Shaw J. and Renzaglia K.S. 2004. Phylogeny and diversification of bryophytes. *A.m. J. Bot.* **91:** 1557-1581.

Sillett S.C. and Neitlich P.N. 1996. Emerging themes in epiphyte research in Westside forests with special reference to cyanolichens. *Northwest Sci.* **70:** 54-60.

Singh D.K. 2001. Diversity in Indian Liverworts: their status, vulnerability and conservation. *In*: Nath V and Asthana A.K. (Eds.), *Perspectives in Indian Bryology.* Dehradun pp. 235-300.

Singh D.K. and Hajra P.K. 1996. Floristic diversity *In*: Gujral G.S. and Sharma V. (Eds.), *Changing Perspective of Biodiversity Status in the Himalayas.* New Delhi. pp. 23-38.

Sveinbjörnsson B. and Oechel W.C. 1992. Controls on growth and productivity of bryophytes: environmental limitations under current and anticipated conditions. *In*: Bates J.W.* and Farmer A.M. (Eds.), *Bryophytes and lichens in a changing environment.* Claredon Press, Oxford, U.K. pp. 77-102.

Taoda H. 1977. Bryophytes of urban ecosystem. *In*: Numata M. (Ed.), *Interdisciplinary studies of urban ecosystems in the metropolis of Tokyo.* Japan.

Tewari S.D. and Pant G.B. 1994. *Bryophytes of Kumaun Himalaya.* Bishen Singh Mahendra Pal Singh, Dehradun, India.

Vellak K. and Paal J. 1999. Diversity of bryophyte vegetation in some forest types in Estonia: a comparison of old unmanaged and managed forests. *Biodivers Conserv* **8:** 1595-1620.

Vitousek P.M. and Howarth R.W. 1991. Nitrogen limitation on land and sea: how can it occur? *Biogeochemistry* **13:** 87-115.

Diversity of Lower Plants (2014): pp. 297-316
Editors: Rajan Kumar Gupta & Mukesh Kumar
IK International Publishing House Pvt. Ltd.

23

Biodiversity and Conservation Needs in Uttarakhand Himalaya: A Case Study of District Rudraprayag

G. P. Raturi* and A. B. Bhatt

Department of Botany, HNB Garhwal University, Srinagar Garhwal, Uttarakhand
*Email: *Girja_prasad2002@yahoo.com*

ABSTRACT

Uttarakhand Himalaya constitutes one of the unique biological regions of the world. Fast land-use and land-cover change has serious implications for the biodiversity of the region and need concern for protection. Prioritizing any area for biodiversity conservation requires the complete knowledge and information about the factors affecting biodiversity. All of them have to be taken into account, while prioritizing the area for conservation. The chief factor responsible for biodiversity loss in Uttarakhand Himalaya, is the dependency of local people on forests for daily needs while illegal grazing and uprooting of medicinal plants from the alpine pastures for means. Second important factor adversely influencing the diversity is developmental activities. It is expected that the results of present study would facilitate the planners and policy makers to arrive at sustainable planning strategies not only in a newly carved district like Rudraprayag but also at the state level.

1. INTRODUCTION

Himalayan range is unique on this planet with its rich biodiversity, scenic grandeur and abode of many famous rivers, viz., Brahmaputra, Ganga, Indus, Sharda, Sutlej, Tista, Yamuna, etc. The Himalayan region is of utmost importance to India, not only ecologically and geographically but politically too. The developmental needs of the Himalaya have recently been realized which

*Communicating Author

has attracted the attention of country planners towards this region. Increasing attention has been paid towards the understanding of the developmental problems of the region. With the result, small hill states have been formed in Himalayan region, Uttarakhand is the recent one of them.

The Uttarakhand Himalaya represents one of the most glorious and characteristic vegetation all over the world. Marked variations are noticeable both in quality and quantity of vegetation and habitat conditions. Garhwal Himalaya situated in the western part of Central Himalaya lies between latitude 29°26′-31°28′N and longitude 77°49′-80°06′ E with altitude ranging from 250-7817 m above sea level. It has an area about 30090 sq. km, spreading 220 km in east-west, 235 km in north-south and is well known for its diverse land forms, panoramic sights, perennial streams, lush green valleys, rich faunal and floristic diversity. Besides, wide ranges of geographical features such as snow covered peaks, deep gorges, glacial moraines and perennial water streams further add to its grandeur.

Since the beginning of human civilization, mankind has gained subsistence in a close relationship with nature. Large pressure of growing population coupled with industrial activities have increased the demand for food, fodder and fuel wood which in turn, have inevitably led to rapid change in forest cover particularly, in developing countries. Likewise, over exploitation and unscientific management of forests in India have caused much reduction in the forest land which is only 64.1 million ha, accounting 19.47 per cent of the total geographical area of the country (The State Forest Report, 1991). The National Forest Policy, 1988, stipulates that "the national goal should be to have a minimum of one-third of the total land area of the country under forest or tree cover". While against this policy, the forest area contributes 75.18 million ha which is only 22.8 per cent of the total geographical area of the country. Planning for development of natural resources without endangering the environment is a crucial issue, the world is facing today (Kachhwaha, 1985; Ives, 1985; Sharma, *et al.* 1989; Khoram and John, 1991). Information on the rate and kind of change in the use of land resource is essential for proper management, planning and regularizing the use of such resource (Shankaranarayanan and Sen, 1977). Land use data are needed in the analysis of environmental processes and problems, which must be understood if living conditions and standards are to be improved or maintained at least at current level (Anderson, *et al.* 1976).

The growing population and limited availability of land necessitates proper utilization of available land through scientific land use planning. Data on land use pattern, their spatial distribution and changes are the prerequisites for making developmental plans (Dhinwa, *et al.* 1992). Thus the planning and management strategies of natural resources hold key for development of any region. Information with respect to forest cover, their type, classes, density quality and distribution is of prime need for eco-balance and planning-cum-management aspect (Pant and Roy 1992). Remote sensing technology has emerged as an efficient and powerful tool in providing reliable information on various natural resources essential for planning (Roy, *et al.* 1991).

It is anticipated that present investigations would constitute a stepping stone on which detailed studies would be carried out not only in the district but also at the level of Uttarakhand State to facilitate the state planners for better planning and implementation of their developmental policies.

2. BRIEF ACCOUNT OF RUDRAPRAYAG DISTRICT

The Rudraprayag district of Uttarakhand lies between the latitude 30° 12′ 58 ″-30° 56′ 47″ N and longitude 78° 50′ 07″ -79° 22′ 34″ E, covering Survey of India top-sheet No. 53J and 53N. The altitudinal range of study area lies between 500-6940 m above sea level and covers two sections of the Himalaya, viz., Lesser Himalaya and Greater Himalaya. The district is bounded northwest by Uttarkashi, southwest by Tehri, southeast by Pauri and east by Chamoli districts respectively (Figure 1).

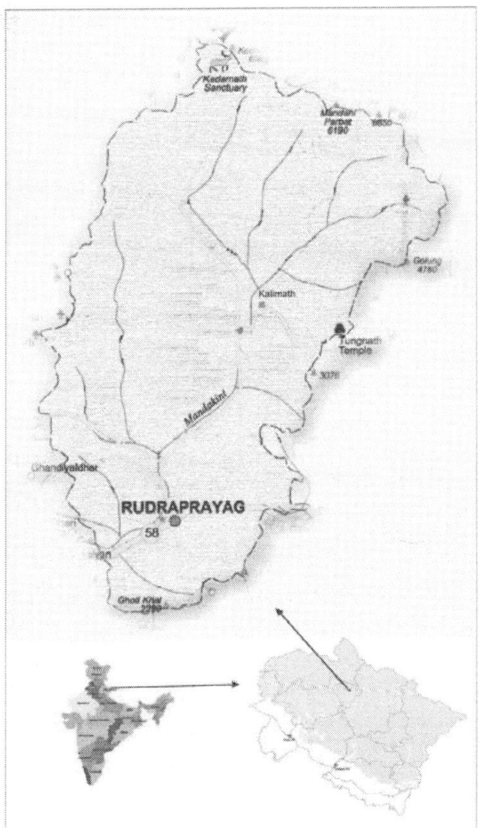

Fig. 1. Location map of Rudraprayag District

The Rudraprayag is the ninth largest district of newly carved state of Uttarakhand, well known for its holy shrines of Kedarnath, Madhyamaheswar and Tungnath. The district in many ways is a land of extremes and lies chiefly along the Mandakini valley presenting rich landscape diversity. Rudraprayag district was formed by carving out some parts of Chamoli, Tehri and Pauri district in September 1997. The district is divided into two tehsils (Okhimath and Rudraprayag) and three blocks (Agastyamuni, Jakholi and Okhimath). The total area of the district is 2,328 km^2 with a population of 2,27,461 and the population density of the district is 120 capita km^2 (Anonymous, 2001).

The Rudraprayag district is a land of profound serenity for the pilgrims of Hindu devotees. According to the Hindu mythology out of "Panch Kedar" three (Kedarnath, Tungnath and Madhyamaheswar) are in the district. Other famous shrines of the district are Trijuginarayan and Kalimath. Rather than the religious spirit, the district is abounding in forests, meadows and drainage system and characteristic vegetation pattern. Deoriya Tal and Gandhi Sarovar (Chaurabari tal) are two high altitude lakes of the district.

Geomorphology

The Himalaya gifted with amazing geomorphic diversity offers an exceptionally suitable field for the study. Most of the geomorphic features in the region are the result of polycyclic, endogenic and exogenic processes of varying intensities through time (Naithani, 1997).

Physiography

Like other parts of Himalaya, topography of the region is irregular with rich vegetation cover. The high ridges and steep slopes are the characteristics of the region. The mountains of Garhwal Himalaya are north, northeast to south, southwest and north, northwest to south and southeast in directions. Between these huge mountain ridges, rivers rising from trans-Himalaya enter this region with one or few tributaries.

Peaks, lakes and thermal springs

The peaks, high altitude lakes and geothermal springs of the district have its own characteristic features. The famous peaks are Kedardome (6940 m), Bharatekhunta (6578 m) and Brahmagupha (5040 m). The important lakes are Devariya Tal (near Okhimath, 2410 m), Chaurabari Tal or Gandhi Sarovar (3970 m) (Plate 1) and Vasuki Tal (4350 m). The geothermal springs are at Gaurikund (3298 m, 53°C) and Madhyamaheswar (4396 m, 53°C).

Glaciers

The Kedarnath glacier is the only prominent glacier located in the district. It is about 14 km long with average width of 500 m. The glacier is the snout of the Mandakini River, a tributary of Alaknanda (Plate 2).

Drainage system

Mandakini is the principal river of the district. It holds a direction of flow generally southwards. Among its tributaries, Son river joins Mandakini at the right bank near Sonprayag, while the Kali Ganga and Madhu Ganga form confluences near Kalimath and Nala, respectively at the left bank of the river. Further downstream the Saur rivulet joins Mandakini at the left bank while Gabini, Laster, and Damar gad (rivulets) on the right bank at Agastyamuni. Continuing its flow in southern direction, Mandakini joins Alaknanda to form a confluence at Rudraprayag following a course of about 68 km.

Chaurabari Tal or Gandhi Sarovar (3970 m)

Devariya Tal (near Okhimath, 2410 m)

Plate 1. High altitude lakes

Geology

The geological setting of the central Himalaya, viz., Himachal, Garhwal, Kumaon and Nepal is similar. Garhwal Himalaya comprises the litho-tectonic zone of the Outer, Lesser, Greater and Tethys (Tibatian) Himalaya. The Rudraprayag district of Garhwal Himalaya falls in the Higher and northern part of Lesser Himalaya. It comprises two main tectonic units, viz., Garhwal Group/Chail Nappe and Central Crystalline Group. These tectonic units are separated from one

Kedardome (6940 m), a high altitude peak

Snout of the Mandakini River, a tributary of Alaknanda

Plate 2. Kedarnath glacier

another by main Central Thrust (MCT) which is a well-known major tectonic contact in the Kund area (Joshi *et al.* 2001).

Soil

In general, the nature and type of soil vary with topography and climate of the district. Mostly the soil is clay loam and sandy loam mixed with pebbles. Soil colour is in different brown shades due to accumulation and decomposition of various amount of litter. Soil pH ranged 5.9 – 6.1 which indicates the slight acidic nature of soil mainly because of high precipitation resulting in leaching. Water holding capacity is in high range for all forest types of the district. The textural type is classified as clayey loam, loam and sandy loam in different forest types.

Agriculture

Two type of agricultural land occurring in the district are irrigated and non-irrigated. Irrigated land is generally located in the valley portion along with the drainage/river course while the non-irrigated land is located on the gentle slopes.

Minerals

In the Uttarakhand Himalaya, there has not been any specific survey conducted to explore the mineral type, quantity and their location of availability till present. Magnetite and Phospharite were found economically useful. Some minerals of the district are known traditionally and used locally since ancient time. The common minerals and their location is given below (Negi, 1995).

Asbestos	Kandara and Okhimath
Arsinopeirite	Makku
Antimony	Mohankhal
Copper	Pinglapani (Qunjagad) and Dandapani (Mohankhal)
Iron	Sankari.

Climate

Meteorological condition of the district is most important factor in its natural setting. Slope, aspect, elevation, mountain peaks and ridges, forest and proximity to glacier are the factors of minor and major variation regulating the climatic conditions. The climatic condition in the district varies from subtropical in lower altitude to cold alpine and glacial in high altitude. High altitude regions of the district experience severe cold and moist while lower elevations and valleys are moderately warmer round the year. Winter is strong with snow while the summer is charming and comfortable. In total, climate of the district is fascinating and enjoyable particularly in the summers.

Temperature

It is a well-known fact that the increase in altitude is directly proportional to the temperature. According to Ozenda there is a decrease of 0.55°C for every 100 m rise in elevation if the slope and aspect remain regular and uniform. The mean temperature of the district is utmost in July coupled with high humidity. The temperature is lowest in January. Temperature during December to January fluctuates between 8.85 – 13.21°C and June to August varies between 27.59 – 31.71°C. Remaining months are enjoyable with moderate temperature.

Rainfall

The amount of rainfall is associated with temperature and altitude. Fall in temperature and increase in altitude influence rainfall. During the study period the rainfall was frequent because of its location in high mountain zone of Himalaya. The precipitation data as recorded by the Central Water Commission and Forest Department show the maximum precipitation in the month of August to be 600 mm and minimum in November (1.5 mm). The precipitation varies between 35.0 – 486.0 mm during rest of the year (Figure 2).

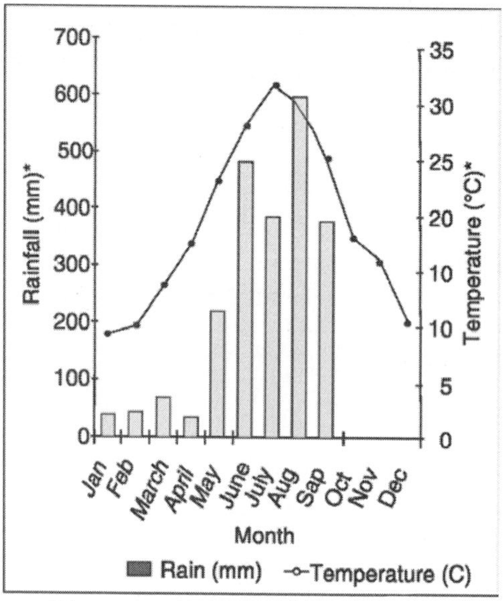

*Mean annual rainfall and temperature

Fig. 2. Climatic data of Rudraprayag district.

Snowfall

Snowfall is directly related to the altitude and temperature, as the altitude rise and temperature declines the frequency of snowfall increases. Duration of snow cover depends upon the aspect of the slope. Snow cover is thick in the north facing shady slopes than sunny slopes. Frequency of snowfall is 3-4 time in a year at an altitude between 2000-2500 m and very little at low lying area. Between the altitudes 2500-3300 m the snowfall is 7-8 time a year and between 3300-4500 m it happens more frequently. Above this altitude there is perpetual snow.

Socio-economic Status

Development of any district reflects its socio-economic status. Population of any area affects the social status, resource sharing and the developmental activities. The Rudraprayag district has well-developed road network of national and state high ways and other connecting road and paths. According to the Census 2001 (Anonymous, 2001), population of the district is 2, 27,461 of which 1, 07,425 are male and 1, 20,036 female, and contributes 2.68% of state population. The population density of the district is 120 capita per km^2 and sex ratio 1117/1000 male. The chief source of economy lies in the tourism and government services or personal business during the summer season when pilgrims and naturalists from the India and abroad visit the region. Females are generally engaged in domestic as well as agricultural activities round the year. The district is profoundly rich in its folk and ancient traditions like its other counter parts in Himalayan region.

Bio-resources

Bio-resources play prominent role in conserving, regulating and stabilizing the environment of the area. Bio-resources of the district include the forests, medicinal plants, minor forest products, etc. Alpine meadows and wild life occupy special ecological niches.

Flora

The flora of district is fairly well known (Naithani, 1967, 84-85). The vegetation has an altitudinal distribution and can be broadly classified into temperate forest, subtropical forest and alpine meadows/moraine zone.

Temperate forest

Temperate forests of Himalaya are evergreen forest occurring in high altitude at 1500-3300 m and are characterized by cold climate. The forest types are further classified on the basis of, physiognomy and reconnaissance survey into three classes, viz., Oak-Rhododendron forest, temperate mixed forest and conifer-I forests (Plate 3). These forests are localized at the higher elevation of the district and mostly occur in Mandakini catchment area of Okhimath block. On the satellite imagery, the forest type exhibits dark red to blackish red tone with varying size and medium smooth to slight mottled texture. Forest patches are irregular in distribution. The forest type covers an area of 43% of the district.

(i) ***Oak-Rhododendron forest:*** This type is restricted to the upper reaches of mountain. The forests are moderately distributed and degraded in the vicinity of village due to lopping for fodder and fuel. Large patches are exceptionally found in the landscape because of fragmentation (Quinn *et al.* 1988). Top stratum of the forest is composed of *Quercus leucotrichophora, Q. glauca, Q. semicarpifolia, Rhododendron arborium, Myrica esculenta, Lyonia ovalifolia Pinus roxburghii, P. wallichiana, Dalbergia sericea, Neolitsea cuipala Aesculus indicus, Cupressus torulosa, Prunus cerasoides, Betula utilis, Toona hexandra Taxus baccata* and *T. wallichiana*, etc. The shrub layer comprises *Hypericum cernuum, Cotoneaster microphyllus, Berberis asiatica* and *Rubus ellipticus.*

Oak-Rhododendron forest

Temperate mixed forest

Plate 3. Temperate mixed forest

(ii) ***Temperate mixed forest:*** It occurs at the same elevation of 1500-3300 m and along cold and moist climate. The forest is located in rarely accessible area of the valley and catchments of the mountain. These forests are relatively undisturbed and support a diversified flora. *Quercus leucotrichophora, Q. semicarpifolia, Q. glauca, Dalbergia sericea, Pyrus pashia, Myrica esculenta, Lyonia ovalifolia, Pinus wallichiana, Bauhinia variegata, Cinnamomum tamala, Neolitsea cuipala, Pinus roxburghii, Pyrus pashia, Rhododendron arboreum, Aesculus indica, Cupressus torulosa, Prunus cerasoides, Juglans regia, Betula utilis, Toona hexandra Taxus baccata* and *T. wallichiana*, etc., mostly occupy top canopy and *Carrisa congesta, Rubus ellipticus, Princepia utilis,* etc. cover the second storey of temperate mixed forest.

(iii) ***Cedrus-blue pine forest:*** The forest type occurs in the areas having elevation range 1800-2200m in temperate zone of Himalaya. The tree stratum is composed of *Cedrus deodara, Cupressus torulosa, Pyrus pashia, Pinus wallichiana, P. roxburghii, Aesculus indica, Rhododendron arboreum,* etc. while *Cotoneaster microphyllus, Prinsepia utilis, Pyracantha crenulata, etc.* cover shrub stratum.

Subtropical forest

Subtropical forests of the district are localized on the lower elevation (500-1500 m) and mostly dominated by *Pinus roxburghii*. On the basis of the physiological study and other related factors like, topography, climate, and reconnaissance survey, the forest is again classified into Chir pine forest and subtropical mixed forest (Plate 4). These forest types are generally restricted at lower elevations in relatively warmer climate of Agastyamuni and Jakholi block. The total area covered by the subtropical forest is 26% of the district.

Chir pine forest

Subtropical mixed forest

Plate 4. Subtropical forest

(i) ***Chir pine forests:*** These represent the pure formation of chir pine forest. *Pinus roxburghii* constituted the canopy and *Berberis asiatica, Carrisa congesta, Pyracantha crenulata, Rhus parviflora,* etc. occupy the second storey of such forests.

(ii) ***Subtropical mixed forests:*** These are located in the lower elevation of 500-700 m experiencing a warmer climate of the area. Subtropical mixed forest is in degraded form due to high pressure of fuel and fodder. As the name indicates, the species of these forests are mixed so that no particular species dominated the vegetation. Tree stratum of the forest is composed of *Pinus roxburghii, Mallotus phillippensis, Toona hexandra, Lagerstroemia parviflora, Ougeinia oojeinensis, Acacia catechu, Aegle marmelos, Bauhinia purpurea, Cassia glauca, Cassia fistula, Dalbergia sissoo, Eugenia jambolina, Haldinia cordifolia, Kydia calycina, Shorea robusta, Terminalia alata, T. belerica, etc.* and shrub stratum by *Berberis asiatica, Carrisa congesta, Cotinus coggygria, Rhus parviflora* and *Woodfordia fruticosa, etc.*

Alpine pasture

Alpine pastures of Rudraprayag are located at the high elevation (above 3300 m) between timberline and snow line. Grasses slope with sedge and shrubs characterize this zone. The two zones consisted of dwarf shrubs like *Juniperus indica, J. squamata*, *Rhododendron anthopogon, R. lepidotum* and important herb species included *Aconitum heterophyllum, Anaphalis royleana, Anemone obtusiloba, A. rivularis, Caltha palustris, Geranium wallichianum, Iris kemaonensis, Morina longifolia, Polygonum affine, P. macrophyllum, Potentilla atrosanguinea, P. cuneata, Primula denticulata, Ranunculus hirtellus, Salix lindleyana, Saussurea obvallata*, *Saxifraga paranasifolia, Selinum candollii,* and *Trachidium roylei,* etc. Total life span of the flora is six months in the summers; rest of the year the region is covered by snow.

Fauna

The characteristic of natural vegetation and specific environmental factor is to sustain a variety of wild lives since distant past. However, modern developmental activities have denuded the vegetation and eliminated the majority of fauna of the district. Some of the common wild fauna of the district is presented in Table 1 with zoological, English and vernacular name.

Table 1: Common fauna of district Rudraprayag

Zoological name	*English name*	*Local name*
Mammals		
Canis aureus Linn.	Asiatic Jackal	Siyar
*Macaca mulatta (*Zimmer.) Villosa	Monkey	Bander
Moschus chrysogastres	Musk Deer	Kastura Mriga
Muntiacus muntjak Zimmer.	Barking Deer	Kakar
Naemorhedus goral Hardwicke	Goral	Ghural
Panthera pardus (Linn.) *fusca* Meyer	Leopard	Tendua
Panthera tigris-tigris Linn.	Tiger	Bagh
Presbytis entellus Dufresne.	Black Monkey	Langur
Selenarctos thibetanus Cuvier.	Himalayan Bear	Reechh
Vulpes bengalensis Shaw.	Indian Fox	Lomadi

Contd.

Table 1: Contd...

Birds		
Aquila heliacea Temm.	Imperial Eagle	Garur
Columba intermedia Strickl.	Blue rock Pegion	Jungle Kabutar
Corvus marcrorhynchas Adams.	Jungle Crow	Kawa
Corvus splendens Viell.	House Crow	Kawa
Francholinus francholinus Bons.	Black Patridge	Titer
Icinaetus malayrnsis Tern.	*Black eagle*	*Cheel*
Lophophorus impejanus Lath.	Monal Pheasant	Monal
Streptopelia decaocto Lath.	Ringed Dove	Ghugati
Reptiles		
Callotes versicolor Linn.	Garden Lizard	Chipara
Hemidactylus brooki Gray.	House Gecko	Chipkali
Varanus monitor Linn.	Indian Monitor	Golu
Fishes		
Barilius bendelisis Hamilton.	Dhaur	Hill Trout
B. varga Hamilton.	Phatkwa	Trout
B. varna Hamilton.	Phatkwa	Trout
Crossocheilus latitus Hamilton.	Sunhera	Sunhera
Cyprinus specularis *Lacepede.*	*Mager carp*	
Garra gotyla gotyal Gray.	Gondal	
Glyptothorax cavia Hamilton.	Siplya	
Noemacheilus zonatnus Mc.Clelland.	Gadiyal	
Pseudecheneis sulcatus Mc.Clelland.	Mungria	
Schizothorax plagiostromus Heckel.	Mahseer	Snow trout
S. richardsonii Gray and Hard.	Mahseer	Snow trout
S. sinuatus Hackel.	Mahseer	Snow trout
Tor tor Hamilton.	Mahseer	Mahseer
T. putitora Hamilton.	Mhaseer	Mahseer

Table 2: Some overexploited fauna and their present status

Zoological name	*Local name*	*Threatened category*
Mammals		
Moschus chrysogastres	Kastura Mrig	Endangered
Naemorhedus goral Hardwicke	Ghural	Vulnerable
Muntiacus muntjak Zimmer.	Kakar	Vulnerable
Panthera tigris Linn.	Bagh	Endangered
Birds		
Lophophorus impejanus Lath.	Monal	Endangered
Ithaginis cruentus Hardwicke	Jangli Murga	Endangered
Francholinus francholinus Bons.	Titar	Vulnerable
Fish		
Tor tor Hamilton.	Mahseer	Endangered

Increasing rate of population and their dependency on forests also increased lopping, logging and felling of the trees for fuel wood, fodder, green manure, grazing and building materials. These factors along with expanding agricultural land constituted the second prime problem for biodiversity conservation.

Likewise, the aquatic biodiversity has also declined to a minimum mark, because of the illegal fishing coupled with the use of dynamites and other explosives. These illegal activities not only kill all fishes of various age groups but also damage other fauna as well. Such

Table 3: Overexploited medicinal plants and their status in the region

Name of the plant species	*Local name*	*Threatened category*
Aconitum atrox (Bruhl.) Mukerjee.	Bis, Mitha	Endangered
Aconitum violaceum Jacq. ex. Stapf.	Dudhiya Atis	Endangered
Aconitum heterophyllum Wall. ex. Royle	Atis, Ativis.	Endangered
Angelica archangelica Linn.	Choru	Endangered
Angelica glauca Edgew.	Choru	Rare
Betula utilis D.Don.	Bhojpatra	Endangered
Dactylorhiza hatagirea D. Don.	Hatthazari	Rare
Ephedra gerardiana Wall. ex Stapf.	Somlata	Vulnerable
Iris kemaonensis D. Don. Ex. Royle.	Iris	Rare
Juniperus indica Bertol.	Padmak	Vulnerable
Meconopsis aculeata Royle.	Kalyari	Endangered
Nardostachys jatamansi DC.	Jatamansi,	Endangered
Orchis latifolia Plur.non. Linn.	Salap	Endangered
Podophyllum hexandrum Royle.	Ban kakari	Endangered
Picrorrhiza kurrooa Royle.	Kutki	Vulnerable
Potentilla fulgens Wall. ex. Hook. F.	Bajardanti	Endangered
Rheum emodi Wall. ex. Meissn.	Archa,	Rare
Rhododendron anthopogon Linn.	Burans	Rare
Salix karelinii Turcz.	Garhbhains	Rare
Saussurea obvallata DC. Edgew.	Braham-kamal	Endangered
Saussurea lappa (Dcne.) Sch-Bip.	Kuth	Vulnerable
Viola biflora Linn.	Banafsa, Birali	Dundi

Source: Gupta *et al.* 2000.

practices can be seen in every valley with a stream or a rivulet in the district. Till today no body has ever thought about the loss of aquatic biodiversity in Himalaya, which is of equal importance. Table 4 presents a list of the fauna of the district which have been considered vulnerable and their very survival is threatened.

Table 4: Landslide prone area of Rudraprayag District (Joshi *et al.,* 2001).

Location	*Type of movement*	*Slope condition*	*Favourable Lithology*	*Remarks*
Nagjagai slide	Debris flow	Moderate	Sericite-chlorite-biotite schist	Due to heavy rain, road cutting, toe erosion pulverized weathered rock type
Lawara slide	Shallow rotational	Moderate	Biotite schist	Triggered first time in 1992 due to heavy rains, presence of under ground stream, toe erosion jointed and weathered rock
Devidhar slide	Rock cum debris flow	Moderate to steep	Garneteiferous mica schist	Unfavourable discontinuities, water seepage, highly weathered rocks and jointed rocks and shear zone
Barasu slide	Debris subsidence	Moderate	Amphibolite and schist	Presence of shear zone, water seepage, highly weathered rock mass and toe cutting

Contd.

Table 4: Contd.

Sansari slide	Rock fall	Steep	Mica schist and granitic gneiss	Presence of highly fractured and jointed rock, shear planes within the bedding plane and near to MCT and road cutting
Sirsoli slide	Rock and debris fall	Steep	Granitic gneiss	Highly fractured and jointed rock, shear planes and road cutting
Kalimath slide	Debris slide	Steep (>65^0)	Schist, mylonite	Road cutting, highly fractured and disintegrated rock mass
Temriya slide	Debris flow	Steep to moderate	Mica schist and granitic gneiss	Cloud burst followed by heavy rain, jointed-fractured rock mass
Bhiri slide	Debris slide	Moderate	Gneissic	Road cutting, toe erosion, high rain fall and presence of fractured-jointed rock mass
Bansbara slide	Rock fall	Steep (>45^0)	Gneissic rock	High rain fall, movement along joint planes and presence of fractured rock at the scar at top
Chandra-puri slide	Debris slide	Moderate to steep	Gneissic and quartzitic	Cloud burst, presence of tension cracks at the top, highly fractured and jointed rocks

Landslides

Landslides are frequent in the district and are triggered most frequently during the monsoon period when the current and volume of flowing water is high that collectively acts as an important catalyst for initiating landslides (Plate 5). At the higher elevations of the area, both frost action and snow contribute to activate landslides. The most important factor causing destructive landslides are the excessive use of explosives in road construction. Many of these roads have been unimaginatively planned and their construction has been cruelly executed, leading to destabilization of hillsides and production of huge amount of debris (Joshi *et. al,* 2001). The main landslide sensitive zones of the district and their location, type of movement, slope conditions and mechanism by which these activate are presented in Table 4.

Table 5: Area estimation for vegetation of Rudraprayag District

S. No.	*Forest Types*	*Area*	*Percent area*
1	Oak-rhododendron forest	530.29	23.27
2	Temperate mixed forest	430.89	18.91
3	Cedrus-bluepine forest	7.08	0.31
4	Chir-pine forest	402.83	17.68
5	Sub-tropical mixed forest	209.74	9.20
6	Alpine pastures	106.06	4.65
7	Landslide area	10.94	0.48
8	Agriculture land	133.64	5.86
9	Snow covered area	444.21	19.49
10	River/water bodies	3.33	0.15

Landslides due to high current and volume of river in rainy season

Landslides due to excessive use of explosives in road construction

Plate 5. Landslides

3. DISCUSSION

Physiography like mountainous ranges, valleys, gorges and ravines of Rudraprayag district represent the complex vegetation in the form of forest and meadows with respect to varied topography, soil and climate. Forests have served mankind by providing food, shelter and fodder while alpine meadows "Bugyal" provide medicinal plants for health care. These have given birth to an affectionate relationship between nature and human beings since the beginning of life. The natural philosophy can be observed even today in remote areas where modern culture has yet to permeate with its communication and transportation science.

The present investigation was aimed to study the forest vegetation of Rudraprayag district with the help of remote sensing data, field knowledge, ground survey and observations. Temperate forests of the study area are evergreens occupying high altitudes between 1500-3300 m and are characterized by cold climate. The forest types are further classified on the basis of quantitative analysis, physiognomy, reconnaissance survey and ground truth verification into three classes, viz., oak-rhododendron, temperate mixed forest and cedrus-blue pine forest. Oak-rhododendron forests generally occur in upper ridges of Guptakashi, Sakri, Deosal, Phata, Jamu, Badasu, Sonprayag, Narayankoti, Gaurikund, Rambara, Okhimath, Tala, Chopta, Malla-Kaliphat Reserve Forest, Deobariatal, Hariyalidevi, etc. Temperate mixed forests are localized along the valleys associated with the Madakini river, Son, Kali and Madhyamaheshwar ganga of Rudraprayag district whereas the cedrus-blue pine forests occur in Ghimtoli, Naini Devi, and in association with *Pinus wallichiana* at lower altitudes in Kandai, Kharpatiya, Kuadi, Kotkhal, etc.

Subtropical forests of the district are restricted on the lower elevation (500-1500 m) of Agastyamuni and Jakholi block and mostly dominated by *Pinus roxburghii*. Chir-pine forests are localized in Agastyamuni, Tilwara, Jakholi, Koti, Mayali, Ratura, Gadora, Kandai, Pokhri, Lamari, Punar, Kapin, Rudraprayag, Simli, and Jawari villages/townships. Subtropical mixed forests are mostly occurring in Bachansuin patti of Agasyatmuni block of Narkota, Khankra, Phatapur and other localities at lower altitude of the district.

Alpine pastures are located on the high elevation (above 3300 m) between timberline and snowline. Kedarnath, Tungnath, Madhyamaheshwar alpine meadows characterized by grassy slope with grass, sedge and shrubs. Total favourable period for germination of the flora is six months of summers; rest of the year the region is covered by snow.

High peaks of the district are usually roofed by snow. The famous peaks are, Bharatkhunta, Brahmagupha, Chaukhambha and Kedardome. The peaks are steep and rocky, either covered with perpetual snow or the snow of the peaks has melt in early summers due to steepness. Ground truth verification of some the area was done from a distance with the help of binocular because of the inaccessibility in approach.

In Garhwal Himalaya, floating population crowd due to pilgrims, mountaineering expeditions, trekking parties and other category of tourists, have resulted in the enhancement of constructional activities during the last two decades. Thus there has been a mushroom growth of several new lodges, hotels and huts all along the highways. The damage to mountain environment in some place due to excessive anthropogenic activities has come to an alarming level. The tourists do indiscriminate plucking of flowers, twigs and even whole plants simply because they are lured by the beauty of the flowers. Tourists also carry the packed food materials and beverages in cans, polythenes, polybags and plastic bottles and throw it elsewhere after use at high altitude region (Plate 6). All these activities are polluting the environment adversely affecting the biodiversity of the region at a larger scale. Kedarnath and Tungnath area are the burning examples where one can experience a drastic decline in the biodiversity during last two decades.

4. CONCLUSION AND RECOMMENDATION

Construction of new roads in the remote areas using high power explosives and widening of roads followed by felling of trees is another significant reason for the loss of biodiversity in the area. The damage to the ecological balance is mostly man-made and is caused by the human negligence and over exploitation.

Pollutants thrown by tourists in the form of cans, polythene and polybags

Mushroom growth of lodge, hotels and huts

Plate 6. Anthropogenic factors affecting Himalayan environment

Developmental trends of the area are also related with social, political and economic change. Such developmental activities are getting multiplied at a tremendously high rate with human expectations. The result is that several rural areas are developing to small townships and colonization continues. These processes have lead to the exploitation of surrounding vegetation at high rate. The key points valuable on account of their applicability throughout the region for conservation of plant wealth are mentioned below:

5. SUGGESTIONS AND AGENDA FOR ACTION TO CONSERVE BIODIVERSITY

- To ensure sustainable use of natural resource for encountering the basic need of the present and future generation without damaging the environment.

- To ascertain the implementation of conservation policy, involvement and support of local people in the planning and management of the forest resources.
- To develop and design the information system to promote the knowledge of local people regarding the sustainable use of forest resources. This is important in conservation strategies.
- To take step to prevent grazing, illegal and mismanaged exploitation of medicinal plant and, develop natural nurseries for the production of these plants. The areas which are dominated by medicinal plants, i.e., Kedarnath, Tungnath, Madhya Maheshwar, Vasuki tal and other alpine zones need more protection.
- To promote the *ex-situ* conservation of important medicinal plant/herb species in controlled environment or in the field as arboreta, that are depleting due to different anthropogenic activities. These medicinal plant/herbs can be important from commercial point of view, if grown and multiplied scientifically.
- For controlling the landslide and to check the toe erosion, construction of masonary toe wall, masonary breast walls with proper weep holes and filters are essential.
- As a general principal construction of dams, tunnels, multistoried building and roads particularly in hill area should be planned on the basis of landslide hazard zonation maps.

Dependency on forest is an essential prerequisite to check the conservation of forest based biodiversity. Applying the views and knowledge of the scientist and specialist outside the natural science and academics, we can save our vegetation wealth. Government must apply the policies recommended by the scientists for sustainable utilization and development of natural resources with the involvement of local people and NGOs can protect Himalayan environment more effectively and efficiently. We are the first generation with the tools and knowledge to detect how the terrestrial sphere is changed; at the same time we are the last with the opportunity to affect the course of many of these changes. Information systems and database are needed to exploit the data already available and to assist in the planning and execution of further survey and inventories and, analysis/interpretation of information. Extended adaptability and integration of database is important for strategic planning and operational levels for effective conservation programmes.

ACKNOWLEDGEMENTS

We gratefully acknowledge the Department of Science and Technology (DST), New Delhi for financial support in form of a research project sanctioned to Prof. A. B. Bhatt.

REFERENCES

Anderson J.R., Hardy E.E., Roach J.T., and Wirmer R.E. (1976) a land use and land cover classification system for use with remote sensing data. Professional paper no. 964 usgs reston verginia, 28 pp.

Anonymous (2001) Census Report 2001.

Dhinwa P.S., Pathak S.K., Sastry S.V.C., Rao M., Majmumdar K.L., Chotani M.L., Singh J.P. and Singh R.L.P. (1992) Land-use change analysis of Bharatpur district using gis. J. Indian Soc. Rem. Sens. **20** (4): 237-250.

Gupta Sangeeta, Raturi G.P. and Bhatt A.B. (2000) Observations on the rarity patterns in flowering plants of U.P. Himalaya and their relevance for Conservation Planning. In: *K.G. Mukherji, B.P. Chamola and A.K. Sharma* (Eds.), Glimpses in Botany, A.P.H. Publication Corp, New Delhi, 145-168 pp.

Ives J.D. (1985) himalayan environmental regeneration: an overview – what are problems and how can they be tackled: environmental regeneration in the Himalayas: concepts and strategies, Central Himalayan Environment Association, Nainital, India. 1-11.

Joshi V., Naithani A and Negi G.C.S. (2001) Study of landslides in Mandakini River valley, Garhwal Himalaya, India, gala *n°* ***16***: 145-156.

Kachhwaha T.S. (1985) Temporal monitoring of forestland for change detection and forest cover mapping through satellite remote sensing. Proc. sixth Asian Conf. on Rem. Sens., Hyderabad. 77-85 pp.

Khorram S. and John A.B. (1991) A regional assessment of land use/land cover types in Sicily with Landsat T.M. data. *Int. j. Rem. Sens.*, **12** (1): 69-78.

Naithani A.K. (1997) Morphotectonics and Landslide Problems in area between Kedarnath and Okhimath, Garhwal Himalaya. Ph.D. thesis, H.N.B. Garhwal University, Srinagar Garhwal.

Nathani, B.D., 1967. Plant collection with the Kedarnath parvat expedition: *Bull. Bot. Sur. India,* 11: 224-233.

Nathani, B.D., 1984-85. Flora of Chamoli District. BSI Howrah, Bisen Singh Mahendra Pal Singh (reprint), Dehradun.

Negi, D.S., 1995. Mandakini ghati ka samakalit vikash: A.k. adhayan. Ph.D. thesis in Geography, H.N.B. Garhwal University Srinagar, Garhwal India.

Pant, D.N. and Roy, P.S., 1992. Mapping of tropical dry deciduous forest and land use in part of Vindhyan range using satellite remote sensing. *J. Ind. Soc. Rem. Sens.,* 20(1): 9-20.

Roy, P.S., Maharana, S.C., Prasad, S.N. and Singh, I. J., 1992. Vegetation analysis and study of its dynamics in Chandaka Wildlife Sanctuary (odisha) using Aerospace Remote Sensing. *J. Ind. Soc. Rem. Sens.,* 20: 223-235.

Shankaranarayanan, K.A. and Sen, A.K., 1977. The methodology of mapping land resources for rational utilization (Chapter 9): In: Desertification and its control ICAR, New Delhi. pp. 77-80.

Sharma, K.D., Singh, Surendra, Singh Nepal, and Bohra D.N., 1989. Satellite remote sensing for detecting temporal change in the grazing lands. *J. Ind. Soc. Rem. Sens.,* 17(4): 55-59.

The State Forest Report, 1991. The state forest report, Forest Survey of India (Ministry of Environment and Forest). Govt. of India Press, Dehradun.

Diversity of Lower Plants (2014): pp. 317-324
Editors: Rajan Kumar Gupta & Mukesh Kumar
IK International Publishing House Pvt. Ltd.

24

Diatoms: Occurrence, Structure and Application

RACHNA NAUTIYAL

[1]*Department of Zoology, Govt. P.G. College, Dakpathar, Vikasnagar, Dehradun, 248125*
Email: rachnanautiyal@ymail.com

ABSTRACT

"Every other breath of oxygen that we breathe comes from diatom". Diatoms a microalgae, occupy 2/3 sphere of earth, is known for its unicellular and unique cell wall, made up of silica. They are typically a few microns in diameter, ten times smaller than the width of a human hair. Diatoms are the old known forms live either as single cells or unite in colonies of intricate shapes. Most of the diatoms are photosynthetic, both planktonic and benthic diatoms are restricted to living within the photic zone. Centric diatoms are mostly planktonic, floating near the sea surface, while pennate diatoms are mostly benthic, living on the water floor, or attached to floating objects. They cannot do without silicon, make perfect use of it and the consumption becomes largest in the propagation season. They widely use readily soluble organic substances and so, can be sensitive indicator of a degree of the organic environment pollution including that due to man. Besides some important functions, e.g., oxygen regulation, pollution indications, primary production, assessment of palaeo-climatic conditions, carbon fixation, they act as a major source for silicon used in various nano-technological applications

Keywords: Diatoms, Silica, Ecological indicators, Nanotechnological applications

1. INTRODUCTION

Our knowledge of the diatoms today has accumulated to the point where approximately 2,00,000 species have been described from various parts of the globe. Diatoms are unicellular,

*Communicating Author

microorganism having chloroplast, occur in all major types of waters and on sediments or solid surface that are wet. Being able to live in any aqueous medium, diatoms are ubiquitous and can be found all over the world from the poles to the equator and in extreme climates. They are nearly all photosynthetic and make their own food as primary producers in the world ocean. It is often said that diatoms live inside a glass house. In fact their hallmark is the unique type of cell well composed of silica. Through the holes/pores in the wall, the living cell can communicate with the outside world and get its food, oxygen and carbon dioxide for photosynthesis.

Because their cell wells are readily preserved and because they settle to the bottom of a lack or ocean after death, they can preserve in such number in the sediments that they can tell us the story about the water body in which they grew and in what sequence and where they may have evolved. The complete genome has been sequenced from two diatoms ***Thalassiosira pseudonana,*** centric diatom and ***Phaeodactylum tricornutum*** a pennate diatom genome reveals a 'mosaic' nature with genes derived from plant, animal and bacterial lineages.

2. HISTORY

Molecular clocks have been made to estimate of their origin. This calculation show that the diatom probably evolved about 230 million years ago. These first diatoms probably did not live in glass house that they live in now. Instead, they were likely to have been naked flagellated cells. The very first diatom deposit has been found in Korea from a terrestrial deposit. It has been hypothesized that when the ocean flooded the land, pools of water were left behind with the early flagellated diatom in them. These diatoms in the early cretaceous are cells with vary unusual, thick and ornate cell wall with strong mechanisms to link the cell in to chains. They look very different from modern diatoms. These diatoms are so heavy that it is likely that they lived in near-shore marine, turbulent environments as semi benthic or epiphytic or epilithic. They first appeared more than 180 million years, or before the early Jurassic period. Origin may be related to the end Permian mass extinction (~250 Ma), after which many marine niches were opened. At some early stage in their evolution the diatom established the ability to metabolize silica, abundant in their surrounding environment and to use it to build their cell wall from silica. Countless descriptions have been made by workers. Macro and microprocesses of diatom morphogenesis are receiving special attention today because of their potential in nanotechnology.

3. STRUCTURE

'Diatoms' (Greek: '' (dia)" = "through" + '' (temnein)" = "to cut", i.e., "cut in half") are a major group of eukaryotic algae. They are three dimensional. The unicellular diatoms are made up of two valves like Petri dish overlapping each other. The upper valve is called epitheca and lower as hypotheca, called as frustules (re 1). They also have intricate perforations and ornaments such as spines or ribs. The frustules allow light to pass through so that the conspicuous golden-brown chloroplasts can capture light energy for photosynthesis. Diatoms range from 1-500 μm in size. Biochemical studies demonstrate that diatom biosilica is a composite material containing zwitterionic proteins (silaffins) and long-chain polyamines in addition to silica which acts as nucleation points for silica deposition involved in making their shells. The silica particles are about 10 nm in diameter, and sport mesoscale (around 100 nm to 1 μm) features such as ridges, protuberances and pores and each species has a unique shell design. Chloroplasts are the most noticeable internal component of almost all diatoms, giving the diatom its colour and photosynthetic capabilities. Diatom chloroplasts are usually yellowish-brown in colour, ranging between yellowish-green and dark brown. This coloration is due to the presence of photosynthetic pigments such as chlorophyll, beta carotene, and fucoxanthin. The pigment chlorophyll is also a key component of terrestrial plants. Without this pigment, neither plants nor diatoms grow very well. The cell wall is one of the major reasons why these algae are today a favourite tool of modern ecological and evolutionary researchers, because the fossils are often well preserved in lake and marine systems.

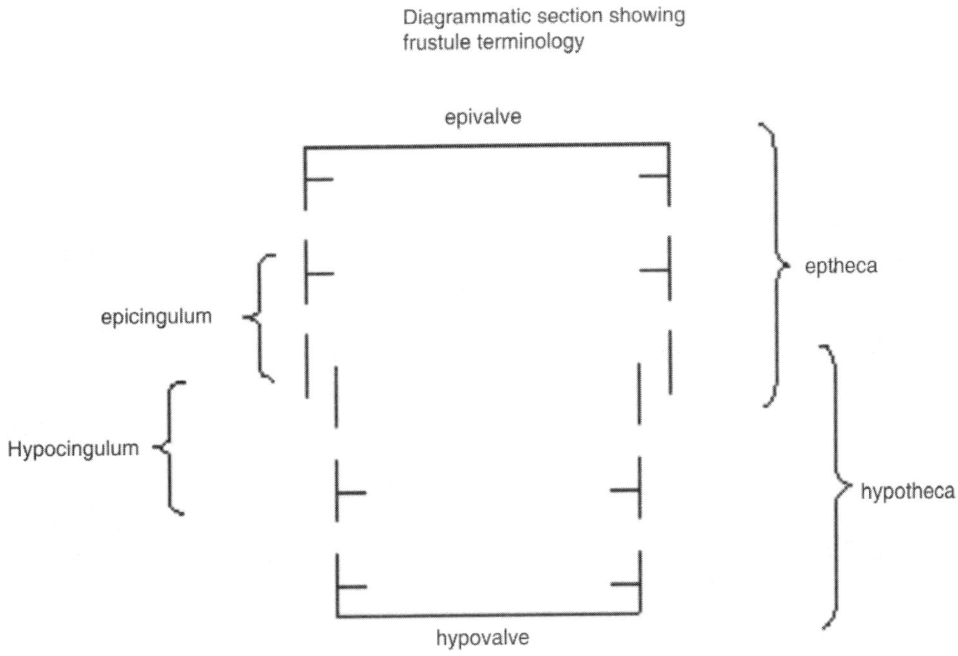

Frustules are made like two overlapping boxes

4. CLASSIFICATION

Diatoms are unicellular organisms of the kingdom Protista. Based on the organization of the pattern of ribs and pores on the valve face, for descriptive purpose diatoms fall into two broad groups, centric diatoms and pennate diatoms. The relationship among the diatoms has been investigated by analyzing their genome. These result shows that there are not two groups of diatom as we think from the pattern on their cell well but instead there are three groups: two centric groups and one Pennate group. Each group has a different type of sexual reproduction. The classification of heterokonts is still unsettled, and they may be treated as a division biology (or phylum), kingdom biology, or something in-between. Accordingly, groups like the diatoms may be ranked anywhere from class (usually called **Bacillariophyceae**) to division (usually called **Bacillariophyta**), with corresponding changes in the ranks of their subgroups. The diatoms are also sometimes referred to as **Class Diatomophyceae**. Diatoms are traditionally divided into two orders Order **centrale** diatoms, which are radially symmetric and **pennate** diatoms, which are bilaterally symmetric. The former are paraphyletic to the latter. A more recent classification divides the diatoms into three classes: centric diatoms, pennate diatoms without a raphe, and pennate diatoms with a raphe. It is probable there will be further revisions as understanding of their relationships increases.

5. HABITAT

Diatoms belong to a large group called the heterokont, including both autotroph (e.g. golden algae) and heterotrophs (e.g. water moulds). Their yellowish-brown chloroplasts are typical of heterokonts, with four membranes and containing biological pigments such as fucoxanthin. Individuals usually lack flagella, but they are present in gametes and have the usual heterokont structure, except they lack the hairs (mastigonemes) characteristic in other groups. Most diatoms are non-motile, although some move via flagellation. As their relatively dense cell walls cause them to readily sink, planktonic forms in open water usually rely on turbulent mixing of the upper layers by the wind to keep them suspended in sunlit surface waters. Some species actively regulate their buoyancy with intracellular lipids to counter sinking. Almost all living diatoms require sunlight to survive and photosynthesize, limiting them to the uppermost 200 meters of the water column. This sunlit region of the water column is referred to as the photic zone. Since the entire surface of the ocean and river is exposed to sunlight at least part of the year, diatoms live practically everywhere at the water surface.

Diatoms are classified into two categories according to their lifestyle: planktonic or benthic. Since all diatoms are photosynthetic, both planktonic and benthic diatoms are restricted to living within the photic zone. Centric diatoms are mostly planktonic, floating near the sea surface, while pennate diatoms are mostly benthic, living on the water floor, or attached to floating objects. They are non-motile, fixed or depend on water current velocity and live either as single cells or unite in colonies of intricate shapes of chain, tubes, stars, fans, ribbon, pellicles and many more and hence, are highly polymorphic. They cling together by means of a gelatinous substance to form zigzag rows of ribbon shaped colonies and perhaps sensitive to microgravity.

6. REPRODUCTION

Under the right conditions, diatoms can reproduce very rapidly and populations of a small species can double each day. Diatoms normally reproduce asexually by binary fission, where

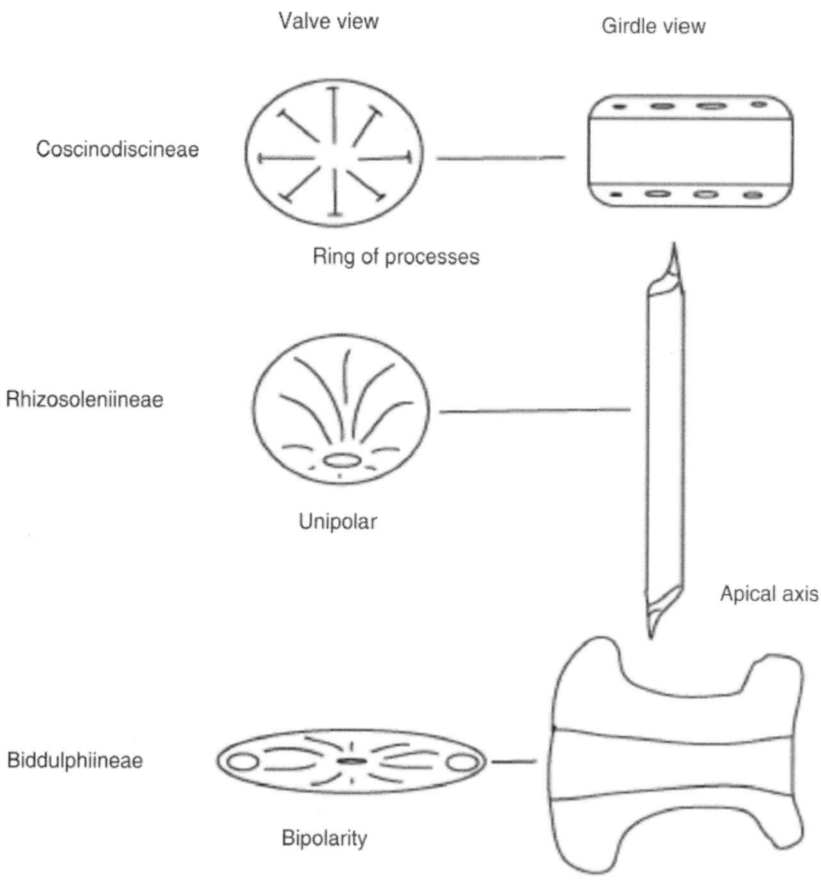

Schematic diagram of centric diatom suborders

one 'mother cell' splits into two daughter cells. The two halves of the mother frustule become the 'lids' for the daughter frustules- the 'box' parts are made inside, as the daughters separate. One daughter cell will therefore be the same size as the parent, and the other will be slightly smaller. Over time, a lineage of diatoms will get smaller and smaller, although a few species avoid this problem by having stretchable frustules. When cells get too small, they discard their old frustules and build new, much larger ones. This process can be connected to sex. In many centrics, some individuals will break up into many small sperm-like cells. These 'fertilise' other diatoms, which then produce new, large frustules. In pennates, 'sex' involves two adult cells lining up alongside each other, dividing and then swapping one daughter cell each. The new pairs of daughter cells fuse, giving two cells that are each half of each 'parent'. They then produce new, large frustules and glide away.

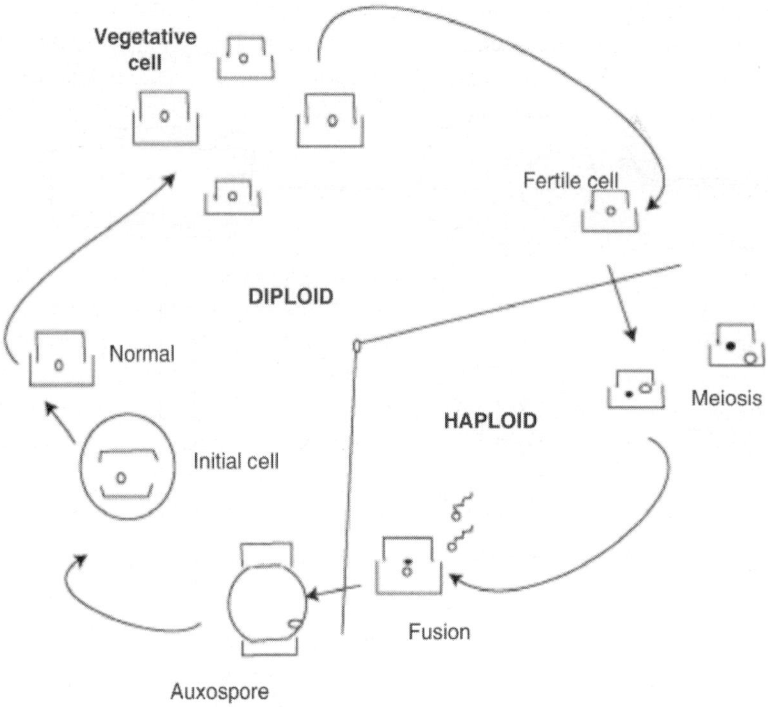

Diagram showing simplified life cycle of a typical diatom. The decrease in the average cell size of a diatom population requires at a certain point the restoration of cell size by the production of an auxospore in which the cell sheds its siliceous frustule. The resulting organic walled cell then produces new maximal sized frustule within itself. The new first cell may differ from the normal vegetative cell in girdle straucture, valve outline and process pattern.

7. IN NATURE

Decomposition and decay of diatoms leads to organic and inorganic (in the form of silicates) sediment, the inorganic component of which can lead to a method of analyzing past marine environments by coring's of ocean floors or mud bay, since the inorganic matter is embedded in deposition of clay summary and silts and forms a permanent geological record of such marine strata.

They propagate quite fast and rate of fission vary for different species and depend on the environmental factors, such as geographic latitude, season, lighting, temperature, and nutrients content. The diatom requires many diverse chemical elements for their normal growth and reproduction. They cannot do without silicon, which constitutes their shells and which eventually limits their quantitative growth. Diatoms with rough cell consume much more silicon then thin cell ones. The complete genome has been sequenced from two diatom, both of them show that the diatom are half animal and half plant. This is a reflection of their evolutionary history because million of years ago a flagellated animal like unicell engulfed a unicellular alga and stole its plastids and kept them for them self so that they would always have a food supply. This is called endosymbiosis and that's why part of the gene from the diatom belongs to the animal host and part belongs to the alga it engulfed. The diatom can make urea just like animal do.

8. APPLICATIONS

Technological and Industrial

For many years diatom has been utilized in a wide variety of industries in the form of their fossil remain. All live cells contain oil reserves and if this becomes incorporated in the sediments before becoming dispersed in the water, large quantity of oil may become trapped beneath the sediments surface and eventually sealed in the rock and compacted to form petroleum. Most of such deposits in the world today come from the rapid burial of massive blooms of diatom and other microalgae. The raw material, Diatomite, Diatomaceous earth or Kieselgur (German) has a number of valuable properties: High porosity, hygroscopicity, low, coefficient of heat conductivity, etc. As building material, diatomite was used in the ancient world. Today diatomite has many uses. It is used for producing light fireproof heat and sound isolating bricks slabs. Generally, it serves as a very fine polish for diamonds or teeth and other abrasives. It also acts as an inert stabilizer for cream, pastes, rubber, and asphalt sulfur cement and paints and as accommodating base, in pills, powders, ointments, etc. Another important use in the form of fine filters that are used through out the wine and brewing industries. Perhaps most famously, Alfred Nobel used diatomaceous earth as a stabilizer in the manufacture of dynamite-nitro-glycerin is absorbed and stabilized by the glass walls of the diatom. Diatomite is used in surgery and balneology. It is an excellent substitute for cotton and disinfectant in bandaging wounds (practiced during World War II), treating extensive burns, dermatitis, eczemas, and even surgical tuberculosis. Diatomaceous mud has a fine effect on skin never ending. Increasingly, diatom plays an important role in the global economy. Cultures of living diatom are now been used for the production of food or valuable biochemical compounds. Diatom also have a used as individuals or in their ecological association –in **forensic** work. Any crime involving natural water may well concern diatoms. Because of their finely-resolved habitat specificity, the diatom found on a body can often point precisely to the scene of a drowning.

Ecological

Diatoms have been used for aquatic ecosystem assessment around the world. Diatoms most often have been used to diagnose levels of stressors, such as organic contamination, lake acidification, climate change, and nutrient concentrations. We define stressors as the habitat alterations and contaminants that are managed to protect and restore valued ecological attributes. Diatom indicators of stressors complement actual measurement of stressors by providing another perspective on stressor condition. For some highly variable stressors (e.g., nutrient concentrations), diatom indicators can be more precise than a 1-time measurement of water chemistry because they integrate stressor effects over time. Diatoms play an immense role in all kinds of aquatic systems of the world from sea-ice communities at the poles to the Mangrove ecosystems of the tropics, from the freshest upland stream to the salty ocean plankton. Diatoms act as a favourite tool of modern ecological and evolutionary (Palaeo-climatic) researchers. Like many microorganisms, diatoms can bloom when conditions of nutrients, light and temperature are ideal. While diatom blooms are a major source of energy for the ocean, which ultimately supports the large predators such as fish and marine mammals and birds, severe blooms can be harmful. One diatom can kill farmed fish because its barbed spines can wound gills. Chemicals present in diatoms can build up in the bodies of animals that eat them and if concentrated enough, can affect people that eat the animals. In Australia, diatom blooms have been responsible for making shellfish unpleasant to taste and unsaleable for long periods of time. One diatom is responsible for Amnesic Shellfish Poisoning, a potentially fatal illness whose symptoms include memory loss. Worldwide outbreaks to date have been very rare; however there is concern that human activities are increasing the frequency of many sorts of micro-organism blooms.

Diatoms play an immense role in all kind of aquatic systems of the world. As the consumers of 20-25% of the world CO_2 they are thus of major significance for the study of our present climate. It would be fare to say ***"that every other breath of oxygen that we breathe comes from diatom"***. Diatoms have played a decisive role in the ecosystem for millions of years as one of the foremost set of oxygen synthesizers on earth and as one of the most important sources of biomass in oceans. Diatoms widely used soluble organic substance and so they can be sensitive "indicators" of a degree of the organic environment pollution including that due to men. Diatoms form the bulk of the periphyton communities in most of the streams. Diatoms are stationary and are therefore less able to avoid harmful conditions. Indices based on diatom composition give more accurate and valid predictions than benthic macroinvertebrates, as they react directly to pollutants. Invertebrates are more affected by substrate or current. Diatoms have much to offer as bioindicators of aquatic ecosystem health. Environmental monitoring, which requires both, short-term (acute) and long-term (chronic) effects, can be identified with this tool. The silica cell walls of diatoms furthermore make both sampling and preservation easy. River health can be assessed by using diatoms. Diatom assemblages in rivers and streams can be analysed by rigorous statistical techniques [multivariate analyses such as cluster, discriminant and canonical correspondence analysis] to establish their relationship to the environmental factors. The relative abundance of diatom species is used as the most valuable characteristics of diatom assemblages for bioassessment of river health.

ACKNOWLEDGEMENTS

The author acknowledges the academic support provided by the Higher Education Directorate, State of Uttarakhand during her research work.

REFERENCES

Science First Hand, Popular Science Journal, 5 (10) August 2006, pp. 1-96.
http://www.geog.ucl.ac.uk/ecrc

Diversity of Lower Plants (2014): pp. 325-330
Editors: Rajan Kumar Gupta & Mukesh Kumar
IK International Publishing House Pvt. Ltd.

25

Biodiversity and Biotechnology

RAVI RAJHANS*, RAJAN KUMAR GUPTA** AND MUKESH KUMAR***

*University of California, USA 95120
**Department of Botany, Govt. P.G. College, Rishikesh–249 201 (Dehradun) Uttarakhand, India
***Department of Botany, Sahu Jain P.G. College, Nazibabad–249 201 Uttar Pradesh, India

ABSTRACT

"Biodiversity and genomics are two key words governing bioscience in the 21st century. The Convention on Biological Diversity will be increasingly important". Modern biotechnology tools and techniques including tissue culture, micro-propagation, transgenic plants and animals, and genomics can be used to defend, conserve and propagate all forms of biodiversity. In short, biotechnologies should be used with other possible means to maintain biodiversity.

Keywords: Biodiversity, Biotechnology, Transgenic plants GM technology.

1. INTRODUCTION

The concept of biological biodiversity includes all organisms, species, and populations; the genetic variation among these; and all their complex assemblages of communities and ecosystems. It can also be defined as the interrelatedness of genes, species, and ecosystems and their interactions with the environment. Broadly biodiversity is discussed under genetic, species, and ecosystem diversity.

Genetic diversity includes different genes contained in all individual plants, animals, fungi, and microorganisms within a species as well as between species.

Species diversity includes all the differences within and between populations of species, as well as between different species. Ecosystem diversity is all the different habitats, biological communities, and ecological processes, as well as variation individual ecosystems.

Modern biotechnology applications have opened a new door to conserve the agricultural biodiversity. The best example is the transgenic crops that are more likely to increase agricultural biodiversity and help maintain native biodiversity rather than to endanger it.

**Communicating Author

2. CURRENT CONCERNS REGARDING BIODIVERSITY

The loss of biodiversity is a significant concern for all human beings since it is directly or indirectly related to us and our lives. Species are becoming extinct at the fastest rate and most of these extinctions have been done directly by human, for ex-rapid population growth and unsustainable patterns of resources consumption are major cause in the loss of species, populations, and ecosystems. The overexploitation and human generated pollution and contamination are other major concerns which are also directly related to global climate change. And in current scenario to world "global warming" is a real threat to the life is also a result of loss of biodiversity as well.

"Since the 1600s, over 250 species of birds, mammals, reptiles, and amphibians worldwide have become extinct as a result of human activities. In the past four years alone, 121 species have been added to the 11,167 already known to be threatened with extinction (i.e., those currently on the Red List of Threatened Species issued by the Species Survival Commission of the World Conservation Union [IUCN]) (Fonseca and Benson)."

Loss of biodiversity can be measured by a loss of individual species, groups of species or by decreases in numbers of organisms (Crane & Bateson, 2003). The major threats to global biodiversity are habitat loss (result of increasing industrialization and globalization). Habitats can also be damaged by flooding, lack of water and climate changes. Tropical humid forests are particularly rich in biodiversity and it is estimated that only half of the original 16 million km^2 of these forests a century ago are left, with about one million km^2 being destroyed every 5 to 10 years.

3. IMPORTANCE OF BIODIVERSITY

In one word it's our 'life' and we can have a very simple example justifying this statement that humans have always depended on the earth's biodiversity for food, shelter, and health. Biological resources that are the basic needs of the day-to-day life for human beings include food, shelter and medicines all came from biodiversity.

Food—species that are hunted, fished, and gathered, as well as those cultivated for agriculture, forestry, and aquaculture.

Shelter—timber and other forest products and fibres such as wool and cotton.

Medicines—both traditional medicines and those synthesized from biological resources and processes.

All food is directly or indirectly obtained from plants and other photosynthetic organisms. Apart from direct benefits of biodiversity from the harvest of domesticated or wild species for food, fibres, fuel, pharmaceuticals and many other purposes, humans also derive benefit from its influence on climate regulation, water purification, soil formation, flood prevention and nutrient cycling (i.e., ecological services); and the aesthetic and cultural impact is obvious (Daily 1997; Balmford *et al.,* 2002). The loss of populations, species, or groups of species from an ecosystem can unbalance the normal function and disrupt these ecological services. The earth's biodiversity contributes to the productivity of natural and agricultural systems. Insects, bats, birds, and other animals serve as pollinators. Parasites and predators can act as natural pest controls. Various biodegradable organisms are responsible for recycling organic materials and maintaining the productivity of soil. Genetic diversity is also important in terms of evolution. And being a biotechnologist, I am proud to say the fact that sometimes and for some diseases biodiversity provides medical models for research into solving human health problems. Biodiversity is thus fundamental for current and future social and economic livelihood.

4. BIOTECHNOLOGY AND BIODIVERSITY

The tools of biotechnology can be applied to maintain biodiversity. Biotechnology can be used as a great resource for knowledge by understanding its molecular level concepts by applying molecular biology advancements as well as using this knowledge practically and in applied way for the plant and animal breeding that is to transfer genetic information from one sort of organism to a particular crop, or to a farm animal to make it transgenic known as transgenic plants, and transgenic animals. And these became possible because of the excellent discovery in the field of biotechnology that is "Recombinant DNA Technology" (also known as Genetic Engineering) which in fact revolutionized the whole concepts and led to the discovery of new era to modern biotechnology. Biotechnology has also proven useful for following genetic markers in plant and animal breeding. Animal or plant varieties are crossed conventionally by sexual means then analyzing a few cells of the newly born calf or of the newly sprouted crop, one can look at the desired forms of gene. This gives scope to predict a phenotypic property, which will only show up later in life, for example, characteristics of a cow's milk or the crop's expected resistance to an infectious plant disease. *In vitro* fertilization is another revolutionary tool of biotechnology of animals through which the laboratory test can be done even before the embryo is implanted, which is referred as pre-implantation diagnostics. Genetic conservation of crops is another good example to prove the power of biotechnology for biodiversity conservation.

According to Dr. Channaputra S. Prakash, "Biotechnology improvements are in development that would allow hybrid rice to be colonized by bacteria that fix nitrogen from the atmosphere. Plants that are able to fix nitrogen improve productivity in the absence of synthetic fertilizers (which are typically unavailable to poor farmers). Further, improved tools such as cryopreservation, developed by biotechnologists, will help in the *ex situ* preservation of biodiversity, while creative techniques, such as gene shuffling, will help create more biodiversity and, perhaps, will even re-create extinct crop traits" (Prakash, C.S., 2000).

5. BIOTECHNOLOGICAL APPROACHES TO CONSERVATION

5.1 *Ex Situ* Conservation

It is the biological process of protecting an endangered species of plant or animal by removing part of the population from a threatened habitat and placing it out of the context of their natural habitats is referred to as *ex situ* conservation. Examples include zoos, botanical gardens, seed banks, etc.

Ex situ conservation in genebanks is a widely applicable method of preserving the genetic diversity of crops and wild species of plants. Cryogenic facilities are used for living sperms, eggs or embryos. The samples are frozen in liquid nitrogen. With *ex situ* conservation, the long-term safety and integrity of genetic resources can be maintained by collecting and preserving desired endangered seeds, living plants, cuttings, and tissue cultures. To save an endangered plant, it can be moved from *in situ* to a gene bank for *ex situ* conservation.

The functions of a gene bank include:

- Maintenance and expansion of germplasm collections
- Long-term conservation
- Characterization and evaluation of samples
- Data management
- Exchange of germplasm among researchers
- Promotion of germplasm to enhance crop productivity.

5.2 Transgenic Plants and Animals/GM Technology

We can improve food production and distribution in order to feed and fulfil the demand of growing world populations through GM technology which is also known as "transgenic plants" and the food and crop produced is known as transgenic crops and foods. GM technology can be applied to plants and microorganisms to address the current challenges. Foods can be produced through the use of GM technology with the following advantages:

- More nutritious,
- Stable in storage
- Health promoting
- Increased yield without increasing land under cultivation
- Virus resistant crops
- Pest resistant crops.

Since all genes consist of DNA, and the information in this DNA molecule is read in the same way in all organisms in order to make proteins, it is in principle possible to take any (single) gene from any organism and transfer of genetic materials that are DNA (gene of interest) from one organism to any other organism (so that the recipient produces a protein normally only made in the donor) is called transgenic. The first transgenic plants were made in 1985, and the first transgenic crop appeared in supermarkets in the USA, the "FlavrSavr" tomato.

6. OTHER BIOTECHNOLOGY TOOLS

The effective biotechnological tools available for conservation of endangered species are cryopreservation of gametes, artificial insemination, embryo transfer, *in vitro* fertilization, and cloning. These technologies have great feasible potential for the preservation of biodiversity. Embryo transfer technology is used to introduce fertilized embryos into surrogate mothers. Artificial insemination is a successful tool used in cattle farming. Frozen samples of sperm are thawed and introduced into females that are ovulating. The quality of cattle has been improved using selected males to breed entire generations. Cloning procedures have been widely used in animal farming and has proven to be very feasible.

7. BIOTECHNOLOGY, BIODIVERSITY AND DRUG DISCOVERY AND DEVELOPMENT

Natural products have been an important part of drug development from past to present and will continue in the future. The DNA sequencing of the human genome has laid new milestones to understand and identify the defective proteins expressed by defective genes. These proteins can be used as molecular targets/drug targets for testing and identifying the natural compounds through the applications of new technology for ex-high-throughput technology. "Natural product medicines have come from various source materials including terrestrial plants, terrestrial microorganisms, marine organisms, and terrestrial, vertebrates and invertebrates".

"An analysis of the origin of the drugs developed between 1981 and 2002 showed that natural products or natural product derived drugs comprised 28% of all new chemical entities (NCEs) launched onto the market" (Newman *et al*., 2003). In addition, "24% of these NCEs were synthetic or natural mimic compounds, based on the study of pharmacophores related to natural products" (Newman *et al*., 2000). This combined percentage (52% of all NCEs) suggests that "natural products are important sources for new drugs and are also good lead compounds suitable for further modification during drug development". (Young-Won Chin *et al*., 2006).

We have several successful histories of plants products which have produced a lot of important and effective drugs. Some examples are:

- Quinine – the antimalarial drug obtained from the bark of *Cinchona* species.
- Morphine – the analgesic obtained from the opium poppy.
- Digoxin – for heart disorders, from *Digitalis purpurea*.
- Reserpine – the antihypertensive agent, from *Rauwolfia serpentine*.
- Ephedrine – an anti-asthma agent, from *Ephredra sinica*.
- Tubocurarine – the muscle relaxant, from *Chondrodendron* and *Curarea* species.
- Taxol – the anti-cancer agents from the bark of *Taxus brevifolia*.
- Vinblastin and Vincristine – the anti-cancer agent from *Catharanthus rosesus*.

Microorganisms have also been extremely important in drug applications, ushering in the "golden age of antibiotics":

- Anti-bacterial agents from *Penicillium* species.
- Other antibacterial agents includes cephalosporins, aminoglycosides, tetracyclines, and polyketides.
- Immunosuppressants include cyclosporins and rapamycin, from *Streptomyces* species.
- Cholesterol lowering agents includes mevastatin and lovastatin, from *Penicillium* species.
- Anthelmintics and antiparasitic drugs include ivermectins, from *Streptomyces* species.

ACKNOWLEDGEMENTS

The authors are grateful to Uttarakhand State Council for Science and Technology (UCOST) and UGC for necessary lab facilities and financial assistance.

REFERENCES

Achard, F., Eva, H.D., Stibig, H.J., Mayaux, P., Gallego, J., Richards, T. and Malingreau, J.P. (2002). Determination of Deforestation Rates of the World's Humid Tropical Forests. In: *Science* **297(5583):** 999-1002.

Braun, R. and Ammann, K. (2002). Biodiversity: The Impact of Biotechnology. Encyclopedia of Life Support Systems. EOLSS Publishers, Oxford. http://www.botanischergarten.ch/EFB/UNESCO-Biodiv-Biotech-Final.pdf

Bryant, P.J. Biodiversity and Conservation: A Hypertext Book. Chapter 15: Temperate Forests and Deforestation.http://darwin.bio.uci.edu/~sustain/bio65/lec15/b65lec15.htm

Crane, Peter and Bateson, Patrick. (2003). Measuring biodiversity for conservation, The Royal Society. http://www.marinebiodiversity.ca/en/pdfs/royalsoc.pdf

Fonseca, G., and Benson, P.J., Biodiversity Conservation Demands Open Access, Center for applied biodiversity science. http://www.ncbi.nlm.nih.gov/pmc/articles/PMC261885/

Newman, D.J., Cragg, G.M. and Snader, K.M. (2000). The influence of natural products upon drug discovery. *Nat Prod. Rep.* **17:** 215-234.

Newman, D.J., Cragg, G.M. and Snader, K.M. (2003). Natural products as sources of new drugs over the period 1981-2002. *J. Nat. Prod.* **66:** 1022-1037.

Okuda, T. Trends in natural product-based drug discovery, and roles of biotech ventures and biological resource centers. Tamagawa University Research Institute, Japan. http://www.wfcc.nig.ac.jp/NEWSLETTER/newsletter35/a4.pdf

Prakash, C.S. Genetically Engineered Crops can Feed the World!. 21st Century Science and Technology Magazine. (2000). http://www.21stcenturysciencetech.com/articles/biotech.html

Transgenic Plants and World Agriculture, Royal Society, July 2000. http://www.biotech-info.net/GE_world_AG.pdf

Wildt, D.E. (1992). Genetic Resource Banks for conserving wildlife species: justification, examples, and becoming organized on a global basis, *Animal Reproduction Science* **28:** 247-257.

Young-Won Chin, Marcy, J. Balunas, Hee Byung Chai, and A. Douglas Kinghorn Drug Discovery from Natural Sources Submitted: January 24, 2006. Accepted: February 19 , 2006; Published: April 14, 2006. The AAPS Journal 2006, 8(2) Article 28. (http://www.aapsj.org). http://www.aapsj.org/articles/aapsj0802/aapsj080228/aapsj080228.pdf

Diversity of Lower Plants (2014): pp. 331-335
Editors: Rajan Kumar Gupta & Mukesh Kumar
IK International Publishing House Pvt. Ltd.

26

Mycorrhiza: A Symbiotic Fungi

DEVENDRA SINGH[1&2], RAJAN KUMAR GUPTA[2] AND MOHD IRFAN[2]

[1]*Uttarakhand Organic Commodity Board, State Training Centre for Organic Farming Majkhali, Almora, Uttarakhand.*
[2]*Govt. P.G. College Rishikesh, Dehradun, Uttarakhand*

ABSTRACT

The mycorrhizal fungi absorb nutrients from soil and translocate them to the host plants. The host provides photosynthates to the mycorrhizal fungi. The mycorrhizal symbiosis is the natural means by which most plants get the phosphorus they need from the soil. Their importance has been demonstrated in numerous experiments in which normally mycorrhizal plants have been compared to plants without symbionts. These experiments repeatedly show as much as sixtyfold differences in growth rate, with the enhancement usually attributed to improved phosphorus nutrition. Chemical fertilization of non-mycorrhizal plants can partially replace mycorrhizae, but cannot replace the soil-building and other ancillary effects of the symbionts. Because heavy fertilization prevents establishment of the symbiosis, it creates independence on the chemicals.

Since the earliest agricultural times, mycorrhizal fungi have been taking care of themselves. In the last few decades, the widespread use of fertilizers and pesticides has created conditions unfavourable for mycorrhizae. The situation is especially acute in nurseries producing container stock. Grown in sterilized potting soil, heavily fertilized, and freely treated with pesticides, these non-mycorrhizal plants require continued intensive care after outplanting.

There has been in recent years a worldwide movement toward controlled and sustainable agriculture. In the tropics the movement is fuelled primarily by the economics of importing fertilizer. In developed countries, it arises from a desire to protect and build soil resources so that high yields may be maintained far into the future. There will be an important place in this kind of agriculture for tree crops, with their ability to use soil resources conservatively, and for the mycorrhizal fungi that help them do it.

Keywords: Mycorrhiza, Fertilizer, Symbiotic fungi, VAM.

*Communicating Author

1. INTRODUCTION

We tend to think of fungi as harmful, but in fact the fungi that infect more plant tissue than any other kind are the beneficial mycorrhizal fungi. The relationship is a symbiosis, meaning that both partners benefit. The fungus receives sugars from the plant, and the plant receives phosphorus from the fungus. The word mycorrhiza means 'fungus-root"; its plural is mycorrhizae or mycorrhizas. "Mycorrhiza" denote an association of fungi with the roots of plants. Through the association the plants and fungi establish a symbiotic relationship with each other. Such association may also be defined as a naturally occurring non-pathological state in which a prolonged association of a symbiotic nature develops.

2. MYCORRHIZAE ARE SYMBIOTIC FUNGI

These little-appreciated and inconspicuous microorganisms live partly in the soil and partly inside the roots. Mycorrhizal fungi are allies of both wild and domesticated plants, aiding them in their struggle to extract nutrients from the competitive and sometimes hostile environment of the soil. Nutrient deficiencies are almost never found in wild plants, in large part because mycorrhizal fungi are abundant in nature. Mycorrhizae are also present, or ought to be present, in a great variety of cultivated plants. They can be found in almost every soil in the world, except where human activities have suppressed the symbiosis.

Mycorrhizae lie hidden below ground, and usually are invisible unless specially stained. They were known to science by the turn of the century, but have not been studied in detail until the last two decades. We classify them into several distinct types, each of which has a different kind of fungus and host plant. The most important types of mycorrhizal symbiosis are given below.

1. **Vesicular-arbuscular mycorrhizae (VAM):** Sometimes called "endomycorrhizae," these are by far the most widespread. VAM are the mycorrhizae of crop plants and most annual and woody natives. They do not produce visible mushroom-type reproductive structures, but form spores that are the largest of any fungi. They cannot be grown in laboratory conditions, but will grow with a wide variety of host plants. The same species of fungus may be associated with liverworts, ferns, Redwood trees, and garden vegetables.
2. **Ectomycorrhizae (ECM):** The mycorrhizae of forest trees: the Pines, Firs, Spruces, Oaks, and Willows. The Eucalypts are primarily ectomycorrhizal, as are some tropical forest trees. Among crop plants, Chestnut, Hazelnut, Pecan and possibly some tropical tree crops have ECM. The fungi of ECM form some of the mushrooms that we see in coniferous and Oak forests; truffles and the deadly *Aminitas* are examples. The fungi form a layer around the short side roots, and can often be seen with the unaided eye. ECM fungi are much easier to grow in laboratory conditions than VAM fungi.
3. **Other types:** The family Ericaceae has mycorrhizae which are similar in many ways to ECM. Blueberries and their relatives are examples of crop plants with the emicoid kind of mycorrhiza. The Orchidaceae and some families of small tropical plants also have distinct kinds of mycorrhizae.

3. FUNCTION OF MYCORRHIZA

Emphasis in mycorrhizal research has for long time been concentrated on the uptake and transfer of phosphate ion and mycorrhiza have been considered to function simply as an extension of the root system, which increases the efficiency of foragating for mineral nutrients.

4. NUTRITIONAL EFFECT OF MYCORRHIZA

4.1 Mycorrhizae: Phosphorus Uptake

The source of phosphate for mycorrhizae is inorganic orthophosphate present in the soil solution or readily realeased from the absorption. Both VAM and ECM can greatly improve uptake of phosphorus and micronutrients, especially zinc and copper. We also know that mycorrhizal plants are better able to withstand drought than non-mycorrhizal plants. Mycorrhizal container stock has a higher rate of survival and better performance after transplanting. Mycorrhizal plants are more resistant to some pathogens, have altered production of plant hormones, and have more highly branched root systems than non-mycorrhizal plants. Cuttings of some species have an improved rooting ability when the medium contains mycorrhizal fungi. Many of the benefits, even those which seem unrelated to phosphorus, appear to actually be side benefits of improved phosphorus nutrition.

4.2 Carbon Metabolisms in Mycorrhizal Fungi and Roots

The synthesis of specifically fungal carbohydrates (trehalose, manniton and glycogen) is one way by which ascomycetous and basidiomycetous fungi of mycorrhizal associations can act as "sinks" for photosynthate derived from their host.

4.3 Nitrogen Assimilation by Mycorrhiza

Mycorrhizal symbiosis results in growth stimulation of mycorrhizal plants, particularly when the soils are poor in nutritious elements. This positive effect is due to a mineral nutritious improvement. Particularly the phosphate nutrition. Mycorrhizal fungi can be involved in different stages of nitrogen metabolisms; soil organic nitrogen breakdown, uptake and translocation to the host of soluble organic nitrogen forms (amino acid) or mineral forms (NH_4, NO_3) and in the assimilation of these last nitrogen sources.

- (i) Breakdown, uptake and translocation of organic forms of nitrogen.
- (ii) Uptake and translocation of mineral nitrogen.
- (iii) Assimilation of mineral nitrogen: Nitrate assimilation.
- (iv) Ammonium assimilation.

5. NON-NUTRITIONAL EFFECT OF MYCORRHIZA

5.1 Mycorrhiza and Plant Water Relations

A major advance towards obtaining a distinction between the effects of enhancement of phosphorus (P) and of H_2O uptake can be expected from further studies of *Agropyron smithii* because this plant shows large increase of transpiration, photosynthesis and stomatal conductance following infection, with no increase of tissue concentration.

5.2 Mycorrhiza and Heavy Metal Resistance

While the production of a prolific external mycelium is advantageous in terms of providing access of growth limiting resources such as water it could be disadvantageous in situation where the soil solution contains elevated levels of potentially toxic metal elements.

5.3 Mycorrhiza and Resistance to Organic Acid Toxicity

With gradient of decreasing pH, the turnover of organic matter declines and the concentration of aromatic and aliphatic residues increase in the accumulating parent material.

Mycorrhizal infection provides protection against most of the phenolic acids and the root growth can actually be stimulated by addition of these compounds relative to that seen in controls.

5.4 Mycorrhiza and Disease Resistance

Mycorrhiza might decrease the disease resistance simply by additionally improving the vigour of the host plant and dominating in the environment around the root. Mycorrhizal fungi may exclude pathogens simply by their physical presence, by the production of antibodies, or by competition for resources.

6. MYCORRHIZAE MAY BE LOST FROM THE SOIL

Mycorrhizae may be lost through the use of pesticides. Pesticides are labelled as specific to a particular group of organisms; i.e., nematodes, insects, herbs, etc. However, all have a certain amount of activity against non-target organisms. Human beings and mycorrhizal fungi are examples of non-target organisms that can be seriously harmed by pesticides. Fortunately, not all pesticides are inhibitory, and appropriate compounds can be found to treat most pest problems without destroying the mycorrhizal fungi. Heavy and indiscriminate use of pesticides is most likely to suppress the symbiosis, as well as other beneficial soil organisms.

Heavy fertilization may lead to faster growth or a larger yield over the short term, and can compensate for a lack of mycorrhizae, but creates an "addiction" from which there is no easy return. Mycorrhizae are a phenomenon of natural soils; they can only operate in the range of mineral availability found under natural conditions. A high concentration of phosphorus or nitrogen triggers a defence mechanism in the plant, and the symbionts are excluded. The plant's phosphorus needs are met by the fertilizer, but mycorrhizae are excluded and can no longer aid in the uptake of zinc. The non-nutritional benefits of mycorrhizae, such as improved soil structure, are also lost. Until mycorrhizae, and conditions allowing their survival, are reintroduced, the plant will grow only with heavy chemical fertilization.

Mycorrhizae may be absent because at some time in the past the native fungi in the soil were all killed. The fungal partners of mycorrhizae are widespread in native soils, being absent only in certain desert areas and in places devoid of plants. However, soils disturbed by human activities often lack mycorrhizal fungi. Removal of the plant cover, handling and treatment of soil, and inversion of soil profiles all reduce or eliminate natural inoculum, the spores and other fungal parts that can propagate the symbiosis.

7. MYCORRHIZAE MAY BE RESTORED TO CULTIVATED PLANTS

Fortunately, once the lack of fungus has been identified as a problem, it is often possible to put it back into the soil. In the past few years, a great deal of research has been directed to the problem of enriching agricultural soil with VAM fungi.

At present, commercial use of VAM is limited to certain kinds of container plants, especially those destined for use in revegetation of disturbed sites. The enhancement of transplant survival and the subsequent ability of the plants to thrive with minimal care make the introduction of VAM at the nursery worthwhile. There is also much interest in inoculating new plantings of *Citrus,* because it is planted into fumigated soil that lacks native fungi. In tropical and low-input agriculture, a combination of natural symbionts and rock phosphate may be preferred to expensive chemical fertilizer. It appears that rock phosphate has no fertilizer value except to mycorrhizal plants. Unfortunately, inoculum is only now becoming available. The best strategy is to encourage the spread of native fungi that may already be present.

We can encourage the natural spread of mycorrhizae by weaning our plants away from heavy fertilization. Phosphorus, in particular, is harmful. Mycorrhizal plants in most California soils can obtain all the phosphorus they need from native soil phosphorus, bone meal or rock phosphate. In moderate quantities, these substances seem to encourage rather than inhibit the mycorrhizal symbiosis. Nitrogen should be applied in organic forms where possible. Chemical nitrogen should be in the form of nitrate rather than ammonium. Organic soil amendments, such as manure, should be well-aged. Fresh organic materials may encourage large microbial populations that can be inhibitory to mycorrhizal fungi.

In container plants growing in soil-free or sterilized potting mix, mycorrhizal inoculum is entirely absent. Mycorrhizae may be introduced to containers by taking advantage of the wide host range and ubiquitous presence of the fungi. Mycorrhizal roots from field-grown plants constitute a very effective source of inoculum. Even dead roots from last year's crop should quickly infect potted plants if placed correctly in the soil and used in mycorrhiza-compatible conditions.

By inoculating container plants with these methods, one accepts a few risks. Roots and soil commonly carry with them nematodes, pathogenic fungi, and other unwanted hitchhikers. The advantages of pine-sterilized potting mix will be lost. Further, not all species of mycorrhizal fungi are equally effective in aiding the host plants. You will probably introduce a grab-bag of several species, and there is no guarantee that particularly effective ones will be among them. Most experimental and nursery inoculation is done with fungi that have been proven under the local conditions. A final class of complications involves unexpected and sometimes subtle things that can go wrong. The inoculum must be placed properly in the soil, so that new roots will pass through it. Soil temperature must be neither too high or too low. Colonization of roots usually will not take place above about 35°C or below 10°C. The plant should be well illuminated to allow optimum photosynthesis. If it is in a dark corner it may not have any extra sugars available to give to the fungus. Inoculation attempts often fail during the winter, even in greenhouses, because the day length is too short to allow optimum photosynthesis. All of these factors vary widely with the species of fungus, of host plant, and with environmental conditions, making the use of mycorrhizae as much art as it is science.

ACKNOWLEDGEMENTS

The authors are grateful to Uttarakhand State Council for Science and Technology (UCOST) and UGC for necessary lab facilities and financial assistance.

Index